Frontiers in
Nuclear Dynamics

ETTORE MAJORANA INTERNATIONAL SCIENCE SERIES

Series Editor:
Antonino Zichichi
European Physical Society
Geneva, Switzerland

(PHYSICAL SCIENCES)

Recent volumes in the series:

A Continuation Order Plan is available for this series. A continuation order will bring delivery of each new volume immediately upon publication. Volumes are billed only upon actual shipment. For further information please contact the publisher.

Frontiers in Nuclear Dynamics

Edited by

R. A. Broglia

and

C. H. Dasso

The Niels Bohr Institute
Copenhagen, Denmark

Springer Science+Business Media, LLC

Library of Congress Cataloging in Publication Data

Main entry under title:

Frontiers in nuclear dynamics.

(Ettore Majorana international science series. Physical sciences; 25)
"Proceedings of the first course of the International School of Heavy Ion Physics on Frontiers in Nuclear Dynamics, held July 16–27, 1984, in Erice, Sicily, Italy"—T.p. verso.
Bibliography: p
Includes index.
1. Heavy ion collisions—Congresses. 2. Nuclear structure—Congresses. I. Broglia, R. A. II. Dasso, C. H. (Carlo H.) III. Title: Nuclear dynamics. IV. Series: Ettore Majorana international science series. Physical sciences; v. 25.
QC794.6.C6F76   1985   539.7   85-12082
ISBN 978-1-4899-5360-5

ISBN 978-1-4899-5360-5   ISBN 978-1-4899-5358-2 (eBook)
DOI 10.1007/978-1-4899-5358-2

Proceedings of the first course of the International School of Heavy Ion Physics on Frontiers in Nuclear Dynamics, held July 16–27, 1984, in Erice, Sicily, Italy

© 1985 Springer Science+Business Media New York
Originally published by Plenum Press, New York in 1985
Softcover reprint of the hardcover 1st edition 1985

DEDICATION

Francesco Resmini had agreed to lecture at the first Course of the International School of Heavy Ion Physics. Most unfortunately he died in February 1984.

He was a highly regarded member of the nuclear physics community, possessing unique drive, imagination, and taste for what is relevant in our field of research. Throughout his career he was engaged in the study of nuclear structure as probed by light and heavy ion reactions. At the same time he made major contributions to the development of the accelerators used in this quest, and he was a recognized leader in the field. He was one of the first to realize the feasibility of superconducting cyclotrons and the new perspectives these machines opened up in the field of heavy ion physics. Having played a central role in the MSU project he became the Director of the CS project in Milano, a position which he occupied until his untimely death.

Francesco Resmini was a towering figure within Italian and international physics. It seems very natural to dedicate this School to him.

Ricardo A. Broglia

PREFACE

Nuclear physics plays a central role in the natural
sciences, as it deals with the massive constituents of
all matter. The picture of the atomic nucleus that has
emerged after almost fifty years of research is a rich
one, displaying a wealth of static and dynamical facets.
The large diversity of structural features results from
the complexity of the interactions between the nucleons,
themselves composite systems of quarks bound together by
strong forces.

The progress in the understanding of nuclear
structure has come about from the concerted efforts of
many groups, attacking the problem with a variety of
techniques. One has thereby gained insight into a world
where nucleons moving rather independently in a common
central potential reflect, nonetheless, the presence of
residual interactions as they are capable of arranging
themselves into organized collective motion.

The study of many-body systems is a theme that
unifies the work in almost all branches of contemporary
quantal physics. They range from macroscopic phenomena
in condensed matter through molecular and atomic structures to nuclei and elementary particles. Nuclear physics
has a special position in this common enterprise because
it is concerned with systems of sufficient complexity to
exhibit a variety of collective phenomena and symmetries,
but yet simple enough for their spectra to contain sharp,
specific states which can be studied in detail. The
fundamental tool which is at the basis of the study of
the nuclear structure is the shell model and its dynamical
extensions; this subject was covered in the lectures of
C. Mahaux.

During the last decade, research in nuclear physics
has, to a large extent, been influenced by the possibilities that have been created by the development of new

experimental tools. Here especially the heavy ion
facilities have opened new frontiers. With the new
generation of accelerators one is able to give relative
velocities to heavy nuclei which are high enough to
overcome their mutal Coulomb repulsion and reach distances where nuclear reactions take place. In grazing
collisions, where few degrees of freedom are excited,
the outgoing particle carries detailed nuclear structure
information. How to extract this information from the
detection of the resulting fragments was the subject of
the lectures of A. Winther, W. Von Oertzen, and to some
extent, those of D. Schwalm. The task of reviewing
topics of heavy ion collisions at low and intermediate
bombarding energies was accomplished by G. Bertsch and
C. Ngô in their contributions.

In many cases nuclei will fuse, making use of their
collective and single-particle degrees of freedom, to
penetrate the barrier even at sub-Coulomb energies. This
topic was covered by D. Brink. At higher energies fusion
reactions are studied to explore nuclear composite systems
as a function of total mass, charge to mass ratio, and
total angular momentum. Part of C. Ngô's lectures
focused on these aspects of the process.

Systems that avoid fusion relax towards stable
conditions through the emission of a sequence of neutrons,
followed by long cascades of gamma transitions. Due to
the large angular momentum content of the fused systems
these cascades have been observed to be made of up to
20 or 30 consecutive transitions. Their spectroscopy
provides a wealth of information on the response of
nuclei under intense Coriolis and centrifugal forces.
This subject was the main theme of the lectures of
J.J. Gaardhøje, D. Schwalm, and F. Stephens.

There has been in recent years increasing interest
in the study of heavy ion collisions at higher energies.
These collisions, which are often followed by a complete
break-up of the two nuclei, lead to the fascinating
possibility of studying nuclear matter at very high
temperatures and pressures. At energies of several GeV
per particle, nucleons participating in central collisions
may lose their identity and give way to a plasma made of
their constituent quarks and gluons. The established
results and the future possibilities of learning about
nuclear matter at these extreme conditions were dealt
with in lectures by D. Scott and H. Specht.

In addition to the invited lectures, which are contained in this volume, a series of complementary seminars were given by participants to the School. We would like to acknowledge the presentations made by F. Barranco, P.F. Bortignon, P. Fröbrich, M. Gallardo, N. Lo Iudice, P. Jacobs, S. Leray, G. Löbner, D. Love, A. Macchiavelli, J.M. Quesada, S. Stringari, and M. di Toro.

Heavy ion physics is characterized by its outstanding breadth. It allows the study of properties of individual nuclear levels as well as makes it possible to recreate conditions under which nuclear matter existed in the early universe. If this School can be used as a gauge, the standard of the students, the quality of the lectures, and the lively discussions that took place during the Course testify to the vitality of the field of research which goes under the name of heavy ion physics.

We take pleasure in thanking both students and lecturers for the exciting atmosphere they were able to create, which made this first Course of the International School of Heavy Ion Physics a most rewarding intellectual experience. The efficiency and kindness of Dr. A. Gabrielle and his staff contributed in an important way to this success.

We wish especially to thank Prof. A. Zichichi, Director of the Centre for Scientific Culture "Ettore Majorana," who made it possible for this School to take place.

        C.H. Dasso   Ricardo A. Broglia

CONTENTS

THE DYNAMICAL SHELL MODEL

C. Mahaux

Institut de Physique B5, Université de Liège au

Sart Tilman, B-4000 Liège 1, Belgium

1. INTRODUCTION

Many nuclear properties can be reproduced in the framework of the shell model in which each nucleon is assumed to move in a mean field. In its simplest version, the shell model neglects the nucleon-nucleon collisions, i.e. considers that the mean free path of each nucleon is infinite. The corresponding mean field is given by the Hartree-Fock approximation.

In the present lectures, we exhibit and discuss some empirical properties which point to the possibility of including into the mean field some effects of the nucleon-nucleon collisions. This leads to the dynamical shell model.

The energy of the nucleons can be positive or negative. At positive energy the dynamical shell model can be identified with the optical model. In effect, these lectures are thus devoted to the extension of the optical model towards negative energies. We make no attempt at completeness and only give a few illustrative examples. A fairly exhaustive discussion and list of references can be found in a recent review article.[1]

Section 2 contains a description of some properties of the empirical mean field and of their phenomenological interpretation. Section 3 gives a few basic definitions and equations. The Hartree-Fock approximation to the mean field is briefly described in Sect. 4; it corresponds to the independent particle model. Section 5 is devoted to the dynamical corrections to the Hartree-Fock approximation and Sect. 6 to the ground state correlations.

2. EMPIRICAL PROPERTIES OF THE MEAN FIELD

2.1. The Valence Shells

As a first approximation, one can assume that the central part of the single-particle mean field is a local Woods-Saxon potential as depicted in Fig. 1 :

$$V(r) = \frac{V_o}{1 + \exp[(r-R_V)/a_V]} \quad . \tag{2.1}$$

In the independent particle model for the ground state, all bound single-particle states are filled up to the Fermi level; this constitutes the Fermi sea.

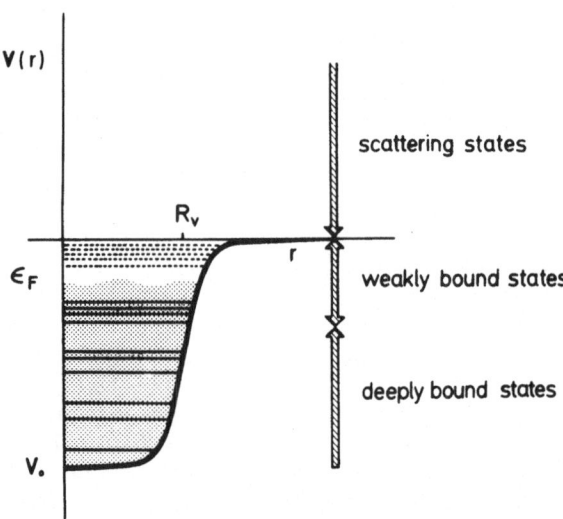

Fig. 1. *Taken from ref.* [1]. *Schematic representation of the mean field and of single-particle energies in the independent particle approximation. The single-particle orbits which are occupied in the ground state are represented by horizontal full lines; they form the Fermi sea (shaded area). The unoccupied orbits can be bound (horizontal dashed lines) or unbound. In the latter case they correspond to scattering states.*

Let us consider the example of ^{208}Pb . In that case, the energy ε_F of the Fermi level is $\varepsilon_F \approx -6$ MeV for both protons and neutrons. The energies of the single-particle states which belong to the neutron valence shell located right above ε_F can be determined from stripping reactions, e.g. ^{208}Pb(d,p)^{209}Pb . They correspond to the ground state and to excited states of ^{209}Pb . The energies of the levels which belong to the neutron valence shell located right below ε_F can be determined from pick-up reactions, e.g. ^{208}Pb(p,d)^{207}Pb . They correspond to the ground state and to excited states of ^{207}Pb .

The observed single-particle energies of the valence shells are represented on the right-hand side of Fig. 2. They are in fair agreement with the energies shown on the left-hand side, which have been calculated from the Woods-Saxon potential (2.1) with the following parameter values

$$V_o = -44 \text{ MeV} , \quad R_V = 7.51 \text{ fm} , \quad a_V = 0.67 \text{ fm} . \quad (2.2)$$

The empirical potential depth V_o depends upon the values assumed for the potential radius R_V and for the diffuseness a_V . The experimental data appear to mainly determine the volume integral per nucleon J_V , which is defined as follows

$$J_V = A^{-1} \int d^3r \, V(r;E) = G_V \, V_o . \quad (2.3)$$

Here, G_V is a geometrical coefficient which only depends upon R_V and a_V . In the case of the geometry (2.2), one has

$$G_V = 9.25 \text{ fm}^3 . \quad (2.4)$$

From this brief description, one should keep in mind the following two main features.

(a) The mean field carries information on excited states of both ^{207}Pb and ^{209}Pb . Single-particle states which correspond to ^{209}Pb are referred to as "particle" states; their energies are larger than the Fermi energy ε_F . Single-particle states which correspond to levels in ^{207}Pb are referred to as "hole" states; their energies are smaller than the Fermi energy ε_F .

(b) The single-particle energies of the two valence shells can be reproduced with a potential well whose depth is constant i.e. is independent of the single-particle energy.

2.2. Scattering States

For single-particle states with positive energy, the mean field can be investigated by means of elastic scattering experiments. The full curves in Fig. 3 represent measured neutron total cross sections.

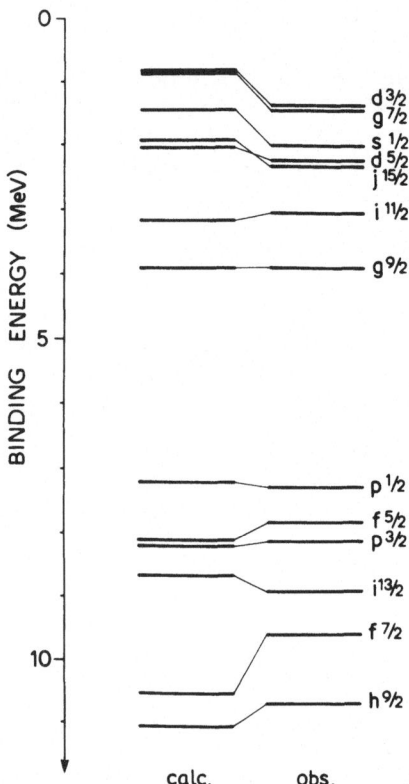

Fig. 2. *Adapted from ref.* [2]. *The column at the right-hand side shows the neutron single-particle of the valence shells of* ^{208}Pb *as observed from pick-up and stripping reactions. The energies contained in the column on the left-hand side have been calculated from the Woods-Saxon potential well (2.1), (2.2).*

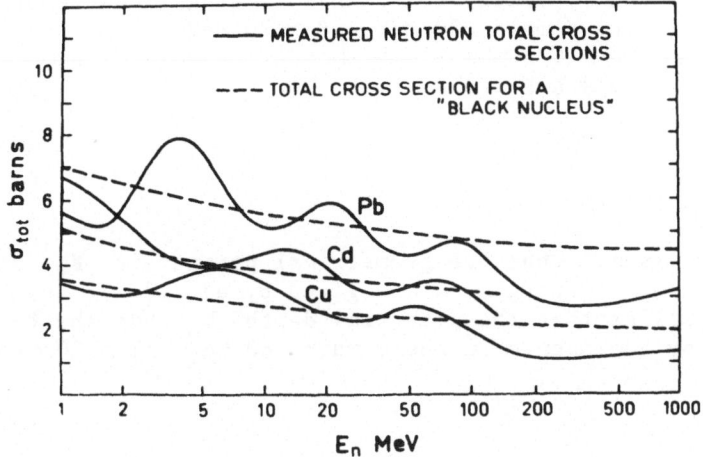

Fig. 3. Taken from ref. [2]. The full curves show the energy dependence of the total neutron cross section on Pb , Cd and
Cu . The dashed lines represent cross sections which are
calculated from the assumption that the nucleus is black
i.e. is a perfectly absorbing medium for the incoming wave.

They exhibit maxima and minima which are interpreted as arising from the interference between the part of the beam which goes through the nucleus and the part of the beam which passes by the target, see Fig. 4. The phase difference between parts (1) and (2) of the beam depends upon the depth of the single-particle potential. This yields a way of determining the potential depth as a function of energy.

As in the case of bound single-particle states, the scattering data mainly determine the volume integral per nucleon of the mean field, namely

$$J_V(E) = A^{-1} \int d^3r \, V(r;E) \quad , \tag{2.5}$$

where we have now explicitly indicated that J_V may depend upon energy. One finds typically that the following linear energy dependence holds in the domain $80 \text{ MeV} > E > 20 \text{ MeV}$:

$$J_V(E) = J^0 + \alpha E \quad , \tag{2.6}$$

with

$$\alpha \approx 2.7 \text{ fm}^3 \quad . \tag{2.7}$$

If one assumes that the geometrical parameters R_V and a_V are independent of energy and are given by eq. (2.2), the linear law (2.6) implies that the potential depth V_o has the following linear energy dependence in the domain $80 \text{ MeV} > E > 20 \text{ MeV}$:

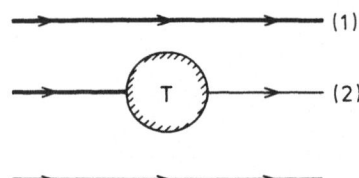

Fig. 4. *Taken from ref.* [3]. *The optical theorem relates the total cross section to the imaginary part of the forward scattering amplitude. Maxima occur in the total cross section when the phase difference between parts (1) and (2) of the incident beam is an odd multiple of* π . *In optics, this is known as the Ramsauer effect.*

$$V_o(E) \approx V^0 + 0.3 E \quad , \tag{2.8}$$

where V is a constant. This property is to be contrasted with the results shown in Fig. 2 which indicate that V_o and J_V are approximately independent of energy in the domain $0 > E > - 10$ MeV which corresponds to the valence shells.

This dependence upon energy of the volume integral of the empirical mean field is illustrated in Fig. 5 in the case of protons in ^{208}Pb . In the domain $- 10$ MeV $< E-\varepsilon_F < 20$ MeV (we recall that $\varepsilon_F \approx 6$ MeV) the empirical values of $J_V(E)$ systematically deviate from the extrapolation of the linear energy dependence (2.6). This is exhibited in the lower part of Fig. 5 by the difference $\Delta J_V(E)$ between the values represented by the full and the dashed curves in the upper part.

2.3. Average Effective Masses

The energy dependence of $J_V(E)$ can be characterized by the average effective mass $<m^*(E)>$ which is defined as follows

$$<m^*(E)/m> = 1 - G_V^{-1} \frac{d}{dE} [J_V(E)] \quad , \tag{2.9}$$

where m is the nucleon mass. The symbol $< >$ refers to the fact that $<m^*/m>$ is defined in terms of the volume integral of the potential, i.e. is a radial average.

The empirical value of the quantity $<m^*(E)/m>$ is represented by the full curve in Fig. 6. In the domain 70 MeV $> E-\varepsilon_F > 20$ MeV , it is approximately independent of energy and equal to

$$<m_k/m> = 1 - G_V^{-1} \alpha \approx 0.7 \quad . \tag{2.10}$$

In the domain 10 MeV $> E-\varepsilon_F > - 10$ MeV , the average effective mass displays an enhancement. This enhancement can be characterized by the average ω-mass which is defined as follows

$$<m_\omega(E)/m> = 1 - G_V^{-1} \frac{d}{dE} [\Delta J_V(E)] \quad , \tag{2.11}$$

where $\Delta J_V(E)$ is represented at the bottom of Fig. 5. The interest of the average ω-mass in the present context is that, we shall see in Sect. 4, $\Delta J_V(E)$ is directly related to the dynamical corrections to the Hartree-Fock approximation. Note that one has

$$<m^*(E)/m> = <\tilde{m}/m> + <m_\omega(E)/m> - 1 \quad . \tag{2.12}$$

In order to exhibit the origin of the expression "effective mass", let us consider a uniform one-dimensional medium. Because of translational invariance, the single-particle wavefunction is a plane wave

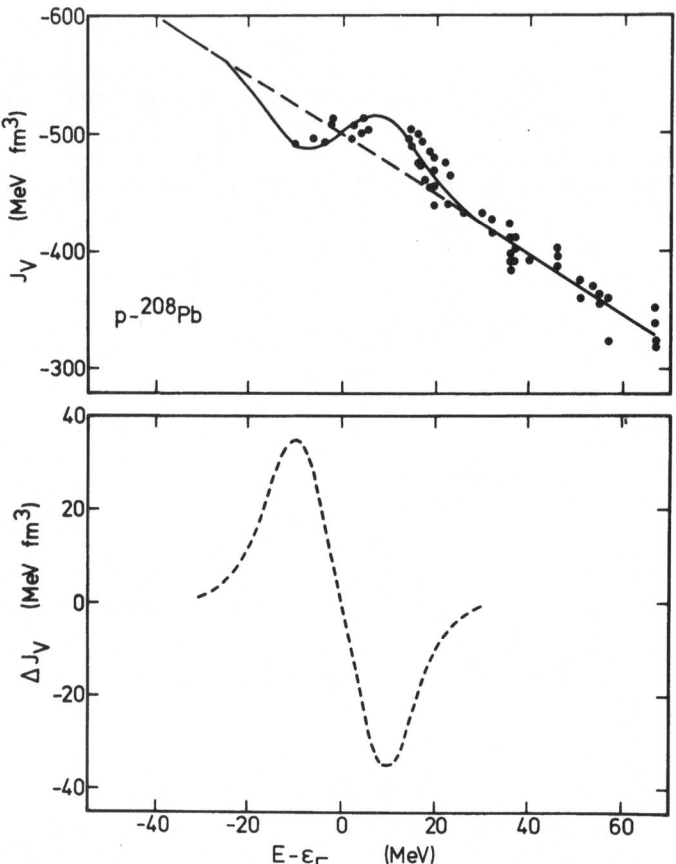

Fig. 5. In the upper part, the full dots represent the volume inte-
       gral per nucleon of the potential well required to fit
       p-^{208}Pb elastic scattering data ($E-\varepsilon_F > 15$ MeV) or the
       observed proton single-particle energies in the valence
       shells (10 MeV > $E-\varepsilon_F$ > - 10 MeV) , see ref. [4]. The full
       curve has been drawn by eye through these empirical dots.
       The dashed line is a linear extrapolation towards negative
       energies. The difference $\Delta J_V(E)$ between the values repre-
       sented by the full curve and by the dashed straight line is
       plotted in the lower part.

$$\psi_k(z) = \exp(ikz) \quad .$$

(2.13)

The wave equation reduces to the energy-momentum relation (we set $\hbar = 1$)

$$E = k^2/2m + V \quad .$$

(2.14)

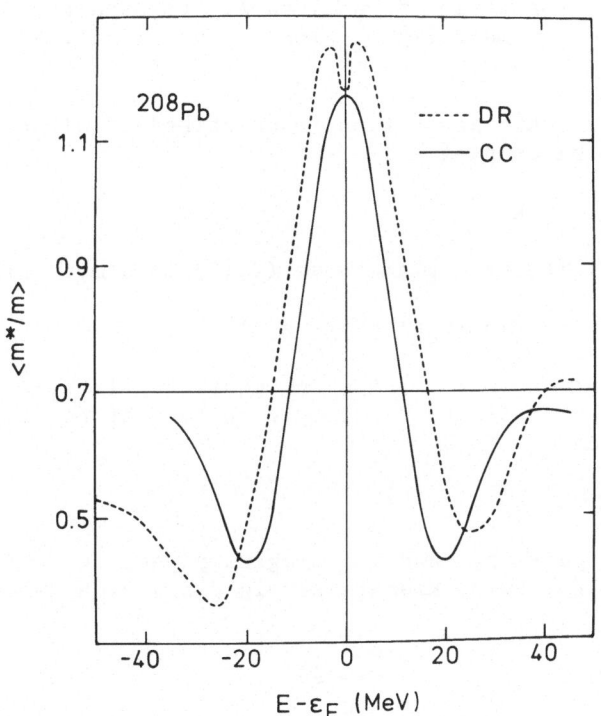

Fig. 6. Taken from ref. [4]. The full curve represents the dependence upon $E-\varepsilon_F$ of the average effective mass as defined by eq. (2.9), where $J_V(E)$ is given by the full curve in the upper part of Fig. 5. The dashed curve is the result of a theoretical calculation based on the dispersion relation approach, see Sect. 5.2.

If V depends upon the nucleon energy E , the velocity of a wave packet is given by the group velocity which is equal to

$$\frac{dE}{dk} = \frac{k}{m^{\ast}} , \tag{2.15}$$

where

$$m^{\ast}(E)/m = 1 - \frac{d}{dE}[V(E)] . \tag{2.16}$$

The expression (2.15) of the group velocity has the same form as the velocity $v_f = k/m$ of a nucleon in free space, except that the mass m is replaced by the effective mass $m^{\ast}$.

2.4. Imaginary Part of the Mean Field

From the analysis of the amplitude of the oscillations exhibited by the total cross sections (Fig. 3) one can show that only a fraction of the beam which enters the target emerges with the same energy. Hence, the target is only partly transparent to the incoming wave and acts very much like a cloudy crystal ball does when it scatters a light beam.

In optics, this absorption is described by allowing the wave-number k to be complex

$$k = k_R + i k_I , \tag{2.17}$$

with $k_I > 0$. Then the plane wave (2.15) is damped since

$$\exp(ikz) = \exp(ik_R z) \exp(- k_I z) . \tag{2.18}$$

The penetration depth is finite and equal to $(2 k_I)^{-1}$. The complex character of k reflects the complex nature of the dielectric constant ε , since

$$k = E c/\sqrt{\varepsilon} . \tag{2.19}$$

The real part ε_R and the imaginary part ε_I of ε are connected by the following Kramers-Kronig dispersion relation (see, e.g., [5])

$$\varepsilon_R(E) = 1 + \frac{P}{\pi} \int_{-\infty}^{\infty} \frac{\varepsilon_I(E')}{E' - E} dE' , \tag{2.20}$$

where P denotes a principal value integral. This dispersion relation directly derives from the property that $\varepsilon(E)$ is analytic in the upper half of the complex E-plane. The physical origin of this analyticity lies in the causality principle which states that the polarization due to an incoming wave cannot occur before the arrival of the wave.

The dispersion relation (2.20) shows that the frequency (or energy) dependence of $\varepsilon_R(E)$ necessarily implies that $\varepsilon(E)$ is complex, and conversely. This energy dependence of ε implies that the relation between the displacement $\vec{D}(\vec{x};t)$ and the electric field $\vec{E}(\vec{x};t)$ is nonlocal in time. From $\vec{D}(x;\omega) = \varepsilon(\omega) \vec{E}(x;\omega)$, one indeed derives that [5]

$$\vec{D}(\vec{x};t) = E(\vec{x};t) + \int_0^\infty G(t') \vec{E}(\vec{x};t-t') \, dt' \quad , \qquad (2.21)$$

where

$$G(t') = (2\pi)^{-1} \int_{-\infty}^\infty [\varepsilon(E) - 1] \, e^{-iEt'} \, dE \quad , \qquad (2.22a)$$

$$\vec{D}(\vec{x};t) = (2\pi)^{-1/2} \int_{-\infty}^\infty \vec{D}(\vec{x};E) \, e^{-iEt} \, dE \quad . \qquad (2.22b)$$

The causality principle entails that $G(t') = 0$ for $t' < 0$.

In the nuclear optical model, one also introduces complex wave-numbers by allowing the mean field M to be complex $(W > 0)$

$$M = V - iW \quad . \qquad (2.23)$$

The imaginary part W reflects the existence of collisions inside the target. In the uniform one-dimensional model considered above the energy-momentum relation (2.14) then becomes

$$E = k^2/2m + V - iW \quad . \qquad (2.24)$$

As in the case of the dielectric constant the fact that M is complex necessarily implies that V and W depend upon the energy. This is so because one can demonstrate that the following dispersion relation holds

$$V(E) = V^0 - \frac{P}{\pi} \int_{-\infty}^\infty \frac{W(E')}{E' - E} \, dE' \quad , \qquad (2.25)$$

where V^0 is independent of E.

As in the case of optics, this dispersion relation corresponds to the fact that M is analytic in the upper half of the complex E-plane and expresses a causality property : the mean field that a nucleon feels at time t only depends upon the earlier history of the system. The energy (or frequency) dependence of $M(E)$ is indeed equivalent to a nonlocality in time, since the energy momentum relation

$$E = k^2/2m + M(E) \qquad (2.26)$$

derives from the single-particle wave equation

$$- (2m)^{-1} \frac{d^2}{dz^2} \psi(z,t) + \int M(z,t-t') \, \psi(z,t-t') \, dt' \; =$$

$$i \frac{\partial}{\partial t} \psi(z,t) \quad , \tag{2.27}$$

where (see eq. (2.13))

$$\psi(z,t) \; = \; \psi_k(z) \, e^{-iEt} \quad . \tag{2.28}$$

2.5. Single-Particle Spreading Width

Instead of interpreting eq. (2.24) as defining a complex wave-number k for a given frequency E , one may consider that it defines a complex E for k real. Then, the frequency is complex, which means that the amplitude of the wave decays exponentially in time

$$\exp(- iEt) \; = \; \exp(- iE_R t) \, \exp(- Wt) \quad . \tag{2.29}$$

The decay time is equal to $\tau = (2W)^{-1}$. It is related to an energy spread $\Gamma^\downarrow$ by the uncertainty relation

$$\tau . \Gamma^\downarrow \; = \; 1 \quad , \tag{2.30}$$

which yields

$$\Gamma^\downarrow \; = \; 2W \quad . \tag{2.31}$$

The quantity $\Gamma^\downarrow$ is called the single-particle spreading width. Its interpretation is the following. In the independent particle model, i.e. in the absence of collisions, a nucleon with momentum k has a well-defined energy $E_R(k)$ given by eq. (2.14). When collisions are taken into account, the probability of finding inside the nucleus a nucleon with momentum k appreciably differs from zero in an energy band of width $\Gamma^\downarrow$ centered on $E_R(k)$. This is illustrated in Fig. 7, which shows that the probability of finding in ^{208}Pb a neutron with quantum numbers $1h11/2$ differs from zero in an energy domain which ranges from 6 to 11 MeV excitation energy in ^{207}Pb .

2.6. Particles and Holes

In Sect. 2.1 we emphasized that the real part of the mean field felt by particle states (i.e. by single-particle states in ^{209}Pb in the example of neutrons in ^{208}Pb) is a continuous extrapolation of the real part of the mean field felt by hole states (i.e. by single-particle states in ^{207}Pb) . Figure 8 shows that a similar treatment holds for the imaginary part of the mean field. Indeed, this imaginary part is approximately symmetric with respect to the Fermi energy. Within that approximation, one can write the dispersion

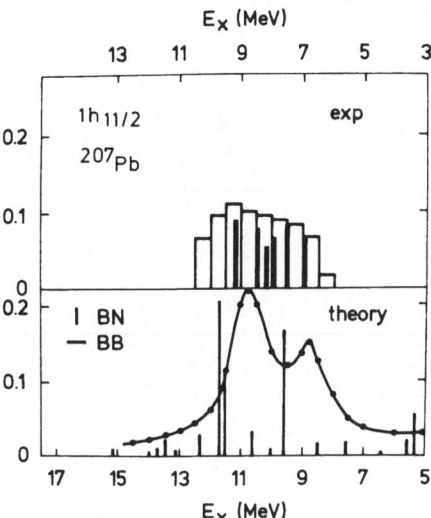

Fig. 7. Adapted from ref. [6]. *The histogram in the upper part represents the distribution of the 1h11/2 neutron single-particle strength in* ^{207}Pb *as measured from the pick-up
reaction* $^{208}Pb(^3He,\alpha)$ *. The lower part shows the result of
theoretical calculations performed by V. Bernard and Nguyen
Van Giai (bars) and by P.F. Bortignon and R.A. Broglia
(curve).*

relation (2.25) in the form

$$V(E) = V^0 + \frac{2}{\pi} (E-\varepsilon_F) \, P \int_{\varepsilon_F}^{\infty} \frac{W(E')}{(E-\varepsilon_F)^2 - (E'-\varepsilon_F)^2} \, dE' \quad . \quad (2.32)$$

This can be compared with the standard way of writing the Kramers-
Kronig relation (2.20), namely [5]

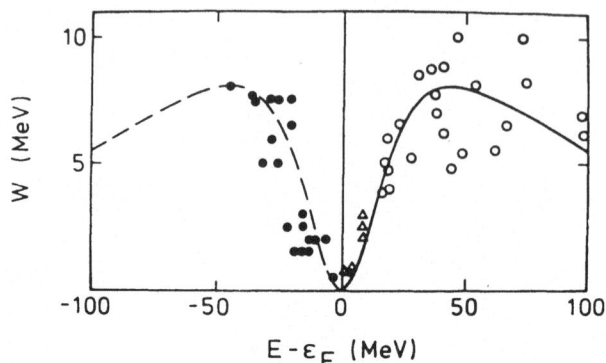

Fig. 8. Taken from ref. [7]. Dependence upon $E-\varepsilon_F$ of the strength of the imaginary part of the mean field for various nuclei with mass number $11 < A < 61$. The full dots are obtained from single-particle spreading widths. The open dots and triangles are derived from the analysis of scattering experiments.

$$\varepsilon_R(E) = 1 + \frac{2}{\pi} P \int_0^\infty \frac{E' \, \varepsilon_I(E')}{E'^2 - E^2} \, dE' \quad , \qquad (2.33)$$

which is a consequence of (2.20) and of the property that $\varepsilon_I(E)$ is an odd function of the frequency E.

Returning to the example of ^{208}Pb, the approximate symmetry of $W(E)$ with respect to $E = \varepsilon_F$ reflects the physical property that the spreading width of a single-particle excitation is approximately the same in ^{207}Pb and in ^{209}Pb, provided that the excitation energy be the same in both cases.

3. THEORETICAL DEFINITION OF THE MEAN FIELD

3.1. Introduction

The theoretical definition of the mean field is not a trivial matter, and we shall not attempt to discuss it here. We only make a few statements.

One can define a single-particle operator $M(\vec{r},\vec{r}';E)$ called the self-energy which has the following property. The asymptotic behaviour of the solution $\psi(\vec{r};E)$ of the wave equation

$$- (2m)^{-1} \nabla^2 \psi(\vec{r};E) + \int d^3r' \, M(\vec{r},\vec{r}';E) \, \psi(\vec{r}';E) \; = \; E \, \psi(\vec{r};E) \qquad (3.1)$$

yields the exact value of the diagonal (elastic) element of the
scattering matrix which describes the reactions induced by an incoming nucleon.[8] The same operator is intimately related to the description of knockout reactions which detect hole states.[9] It is
therefore a good candidate for being identified with the mean field.

The dependence of M upon $\vec{r}$ and $\vec{r}'$ implies that the self-
energy is nonlocal in the spatial coordinates. In finite systems
the dependence of M upon E is very complicated. However, one
can show that the quantity

$$M(\vec{r},\vec{r}';E+i\Delta) \qquad , \qquad\qquad\qquad\qquad\qquad (3.2)$$

where Δ is an energy averaging interval, describes the average
properties of the single-particle excitations of the $(A-1)$ and of
the $(A+1)$ particle systems, i.e. of the holes and particle states.[10]
This quantity is a smooth function of E and its dependence upon
$\vec{r}$ and $\vec{r}'$ also appears to be rather simple.[11] Henceforth we shall
for simplicity drop any explicit reference to the averaging interval Δ .

We now exhibit that the dependence of M upon the frequency E
implies that the mean field is nonlocal in time with the example of
a spherically symmetric field which is local in the spatial coordinates. We write

$$\varphi(\vec{r};t) \; = \; \psi(\vec{r};E) \, e^{-iEt} \qquad . \qquad\qquad\qquad\qquad (3.3)$$

The equations

$$- (2m)^{-1} \nabla^2 \psi(\vec{r};E) + M(r;E) \, \psi(\vec{r};E) \; = \; E \, \psi(\vec{r};E) \qquad (3.4a)$$

$$- (2m)^{-1} \nabla^2 \varphi(\vec{r};t) + \int dt' \, M(r;t-t') \, \varphi(\vec{r};t')$$
$$= \; i \, \frac{\partial}{\partial t} \, \varphi(\vec{r};t) \qquad\qquad (3.4b)$$

are equivalent, with

$$M(r;E) \; = \; \int M(r;t-t') \, e^{iE(t-t')} \, d(t-t') \qquad . \qquad (3.5)$$

Hence, the self-energy is nonlocal with respect to both the spatial
and the time coordinates.

Let us decompose M into its real and imaginary parts

$$M(\vec{r},\vec{r}';E) \; = \; V(\vec{r},\vec{r}';E) - i \, W(\vec{r},\vec{r}';E) \qquad . \qquad (3.6)$$

The following dispersion relation holds

$$V(\vec{r},\vec{r}';E) = V_{HF}(\vec{r},\vec{r}') - \frac{P}{\pi} \int_{-\infty}^{\infty} \frac{W(\vec{r},\vec{r}';E')}{E' - E} dE' \quad . \tag{3.7}$$

It appears most likely that this dispersion relation is a consequence of the causality principle, but this physical origin does not seem to have been explicitly exhibited yet.

3.2. Nuclear Matter

In a finite system the self-energy $M(\vec{r},\vec{r}';E)$ depends upon many variables; this dependence can moreover be quite complicated. This is why it is of interest to first study nuclear matter. The latter is a uniform medium which is made up of protons and neutrons, and where the Coulomb interaction has been turned off. Because of translational invariance, any single-particle wavefunction is then a plane wave $\exp(i\vec{k}.\vec{r})$ and any single-particle operator can only depend upon $|\vec{r}-\vec{r}'|$ rather than upon $\vec{r}$ and $\vec{r}'$ separately. The single-particle wave equation (3.1) then reduces to the energy-momentum relation

$$E = k^2/2m + M(k;E) \quad , \tag{3.8}$$

where

$$M(k;E) = \int d^3r' \, M(|\vec{r}-\vec{r}'|;E) \, e^{i\vec{k}.(\vec{r}'-\vec{r})} \quad . \tag{3.9}$$

Note that the dependence upon k of the mean field $M(k;E)$ reflects its nonlocality in space while its dependence upon E reflects its nonlocality in time.

In the case of nuclear matter, the dispersion relation (3.7) reduces to

$$V(k;E) = V_{HF}(k) - \frac{P}{\pi} \int_{-\infty}^{\infty} \frac{W(k;E')}{E' - E} dE' \quad , \tag{3.10}$$

with $W(k;E') \geqslant 0$. One has

$$W(k;\varepsilon_F) = 0 \quad . \tag{3.11}$$

4. THE HARTREE-FOCK APPROXIMATION

The Hartree-Fock approximation amounts to the independent particle model. In this approximation the ground state of nuclear matter is a Slater determinant constructed with plane waves whose momentum $|\vec{j}| = j$ ranges from zero up to a maximum value k_F which is called the Fermi momentum and is related to the density ρ by

$$\rho = \frac{2}{3\pi^2} k_F^3 \quad . \tag{4.1}$$

Let v denote some effective nucleon-nucleon interaction. The Hartree-Fock approximation to the mean field is given by

$$V_{HF}(k) = \sum_{j<k_F} \langle \vec{k},\vec{j}|v|\vec{k},\vec{j}-\vec{j},\vec{k}\rangle \quad , \tag{4.2}$$

where

$$|\vec{k},\vec{j}\rangle = \exp i(\vec{k}.\vec{r}_1 + \vec{j}.\vec{r}_2) \quad .$$

We note that V_{HF} is real and that it is independent of E . Equation (3.10) shows that one of these two properties entails the other.

The Hartree-Fock potential depends upon the nucleon wave number k , i.e. is nonlocal in the spatial coordinates. The energy-momentum relation (3.8) reads

$$E = k^2/2m + V_{HF}(k) \quad , \tag{4.3}$$

which defines a function k(E) . The corresponding effective mass is defined by eq. (2.15) :

$$\frac{dE}{dk} = \frac{k}{m^{**}_{HF}} \quad , \tag{4.4}$$

where

$$m^{**}_{HF}(E)/m = [1 + \frac{m}{k}\frac{dV_{HF}}{dk}]^{-1}_{k=k(E)} \quad . \tag{4.5}$$

The function

$$V_{HF}(E) = V(k(E)) \tag{4.6}$$

is equivalent to V(k) in the sense that the energy-momentum relation (4.3) has the same root k(E) if one replaces $V_{HF}(k)$ by $V_{HF}(E)$. Since $V_{HF}(E)$ does not explicity depend upon k , it is local in the spatial coordinates. It is called the "local equivalent" to the Hartree-Fock potential. In finite nuclei one can construct local potentials which are approximately equivalent to a nonlocal field by using a transformation similar to eq. (4.6).[12] In this case, the word "equivalent" refers to the properties that the local and the nonlocal potentials yield approximately the same scattering phase shifts. However, their wavefunctions differ from one another at finite distance. A remnant of the latter feature also exists in nuclear matter in the sense that wavepackets moving in the uniform fields $V_{HF}(E)$ or $V_{HF}(k)$ have different natural spreading.[13]

Equation (4.5) can be written in the form

$$m^{**}_{HF}(E) = 1 - \frac{d}{dE} V_{HF}(E) \quad . \tag{4.7}$$

For most effective interactions, $m_{HF}^{\ddot{}}$ is approximately independent
of E . In particular, $m_{HF}^{\ddot{}}$ is exactly independent of E for
Skyrme-type interactions. This justifies the identification of the
straigth line in the upper half of Fig. 5 with the result of a
Hartree-Fock calculation. Hence, the difference between this linear
law and the empirical values, which is plotted at the bottom of Fig.
6, should be ascribed to dynamical corrections i.e. to effects
which are omitted in the independent particle model.

Since $m_{HF}^{\ddot{}}$ is approximately independent of E , one can identify it with the quantity m_k defined by eq. (2.10) :

$$m_k/m \approx m_{HF}^{\ddot{}}/m \quad ; \tag{4.8}$$

no radial average < > is needed in the case of nuclear matter.

Figure 5 shows that the Hartree-Fock potential is not sufficiently attractive for the valence shell located above the Fermi
energy, while it is too attractive for the valence shell below the
Fermi energy. Hence, the average energy separation between the two
valence shells as calculated in the Hartree-Fock approximation is
expected to be larger than the empirical value. This is confirmed
in Fig. 9. Note that the density of single-particle energies increases with increasing $m_{HF}^{\ddot{}}$. This can be understood from eq. (4.4)
which shows that the energy variation dE associated with a wavelength variation dk is inversely proportional to $m_{HF}^{\ddot{}}$.

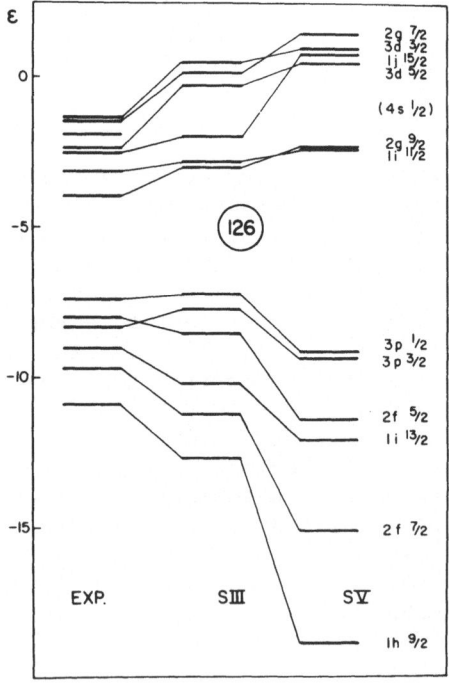

Fig. 9. Taken from ref. [14]. Neutron
single-particle energies in
the valence shells of ^{208}Pb.
The left-hand column shows
the observed values (see
Fig. 2). The middle and the
right-hand columns have been
computed in the Hartree-Fock
approximation with two Skyrme-
type interactions which correspond to the effective mass
$m_{HF}^{\ddot{}}/m = 0.76$ (SIII) and
$m_{HF}^{\ddot{}}/m = 0.38$ (SV) respectively.

5. DYNAMICAL CORRECTIONS

5.1. Introduction

In order to perform a microscopic calculation of the corrections to the Hartree-Fock approximation for the nuclear mean field at positive energy, one should evaluate the effect of the collisions between the incoming nucleon and the target nucleons. These collisions may give rise to two particle-one hole states, or excite surface vibrations, or giant resonances, etc... Calculations of this type exist.[1] They are unavoidably fairly involved and yield rather complicated results for the self-energy. This renders somewhat difficult the comparison of their results with empirical mean fields or with empirical single-particle properties.

The observation that the empirical mean field can be approximated by a local potential which has rather simple properties suggests the use of less microscopic approaches in which the simple nature of the mean field would be postulated from the outset. This is the spirit of the dispersion relation approach.

5.2. The Dispersion Relation Approach

The second term on the right-hand side of eq. (3.7) gives the dynamical correction to the Hartree-Fock potential. At low energy, the nonlocality of the imaginary part $W(\vec{r},\vec{r}';E)$ is likely to be complicated and to fluctuate rapidly with energy. Hence, it appears plausible that the effect of this nonlocality disappears when one only considers average trends. Besides, the nonlocality of $W(\vec{r},\vec{r}';E)$ probably becomes simple at high energy. One can then introduce local equivalent potentials, very much in the same way as we have discussed in connection with eq. (4.6). In view of this, it appears reasonable to assume that one can evaluate a local correction to the Hartree-Fock field by substituting in the integral on the right-hand side of eq. (3.7) the modulus of the empirical value $W(r;E)$ of the imaginary part of the local optical-model potential. In view of the approximate symmetry of $W(E)$ exhibited in Fig. 9, one can furthermore assume that $W(r;E)$ is symmetric with respect to ε_F . Then, the correction to the Hartree-Fock potential can be approximated by the expression (see eq. (2.32))

$$\Delta V(r;E) \;=\; \frac{2}{\pi}\,(E-\varepsilon_F)\; P \int_{\varepsilon_F}^{\infty} \frac{W(r;E')}{(E-\varepsilon_F)^2 - (E'-\varepsilon_F)^2}\; dE' \qquad , \qquad (5.1)$$

where $W(r;E)$ is the modulus of the imaginary part of the empirical local optical-model potential.

The volume integral per nucleon of eq. (5.1) yields

$$\Delta J_V(E) \;=\; \frac{2}{\pi}\,(E-\varepsilon_F)\; P \int_{\varepsilon_F}^{\infty} \frac{J_W(E')}{(E-\varepsilon_F)^2 - (E'-\varepsilon_F)^2}\; dE' \qquad , \qquad (5.2)$$

where $J_W(E)$ is the volume integral per nucleon of the absolute value of the imaginary part of the optical-model potential. Since $J_W(E) \geqslant 0$ and $J_W(\varepsilon_F) = 0$, $\Delta J_V(E)$ is positive (repulsive) for E somewhat smaller than the Fermi energy and is negative (attractive) for E somewhat larger than ε_F. This is in qualitative agreement with the empirical features shown at the bottom of Fig. 5. This agreement also holds at the semi-quantitative level. This is shown by Fig. 6. There, the dashed curve represents the radial average ω-mass $\langle m_\omega/m \rangle$ as obtained from eq. (2.11) where $\Delta J_V(E)$ has been computed from eq. (5.2) by replacing in the dispersion integral $J_W(E)$ by its empirical value.

The close agreement between the dashed and the full curves in Fig. 6 gives confidence in the reliability of the dispersion relation approach in the case of ^{208}Pb. Let us now turn to the medium-light nuclei. The curve drawn through the empirical values of the strength W in Fig. 8 has been reproduced at the top of Fig. 10. When inserted in the dispersion integral on the right-hand side of eq. (5.1), they yield the dynamical correction ΔV shown in the middle part and the ω-mass plotted at the bottom. These results are in fair agreement with the empirical values of ΔV.[7]

The energy dependence of the real part of the dynamical correction to the real part of the mean field near the Fermi energy is similar to that of the real part ε_R of the dielectric constant ε in the vicinity of the energy ω_o at which the imaginary part ε_I shows a peak (anomalous dispersion, see Fig. 11). This reflects the similitude between the dispersion relations (2.20) and (2.25). Note, however, that in the nuclear case the dispersion correction is due to a dip of $W(E)$ near the Fermi energy (see Fig. 8), while in the case of optics it is due to a peak of $\varepsilon_I(E)$ near the resonance frequency $E = \omega_o$. This is related to the difference between the sign in front of the integral which appears in the dispersion relations (2.20) and (2.25), respectively.

5.3. Radial Dependence of the Dynamical Corrections

The calculated values shown in Fig. 6 correspond to the volume integral per nucleon of the dynamical correction, from which the strength shown in Fig. 10 differs by a geometrical factor only. One can investigate the dynamical correction in more detail. Indeed the approximate dispersion relation (5.1) enables one to estimate the radial dependence of the dispersion correction to the mass operator.

Let us define the radial- and energy-dependent ω-mass by the relation

$$m_\omega(r;E)/m = 1 - \frac{d}{dE}[\Delta V(r;E)] \quad . \tag{5.3}$$

The energy dependence of m_ω in the case of ^{208}Pb is represented

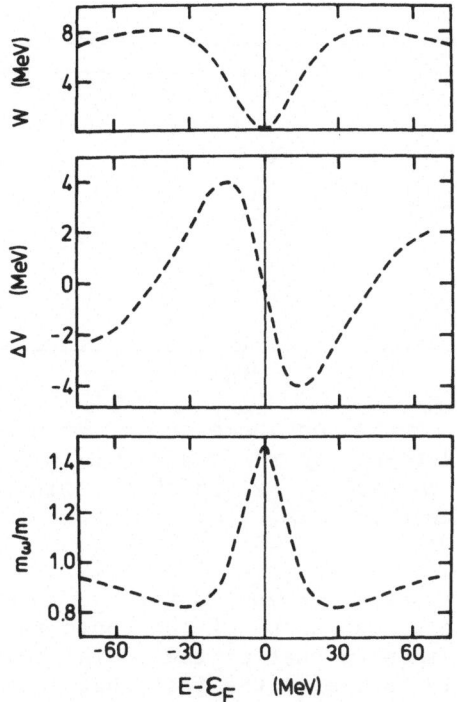

Fig. 10. Adapted from ref. [7]. The top part represents the energy
dependence of the strength W(E) of the imaginary part of
the optical-model potential for nuclei with mass number
11 < A < 61, see Fig. 8. The middle part shows the strength
of the dispersion correction to the real part as calculated
from eq. (5.1). The bottom part gives the corresponding
value of the ω-mass defined by eq. (2.11).

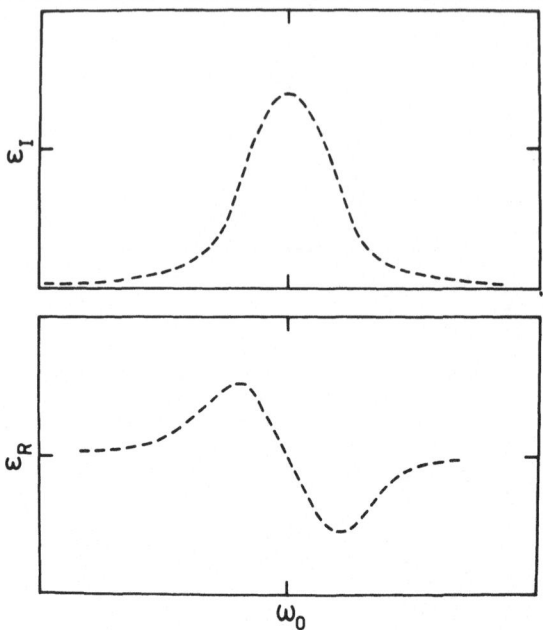

Fig. 11. *The bottom part gives a sketch of the energy dependence of the real part ε_R of the dielectric constant in the vicinity of an energy ω_O at which the imaginary part ε_I has an absorption peak.*

by the full curves in Figs. 12 and 13 for $r = 0$ and for $r = R_W = 7.54$ fm , respectively. The shape of the enhancement peak of $m_\omega(R_W;E)$ is quite close to that of the radial average $\langle m_\omega/m \rangle$ shown in Fig. 6; this is due to the fact that a weighting factor r^2 enters in the definition of the radial average. The dashed curves in Figs. 12 and 13 represent the quantities

$$m_\omega^{(a)}(r;E)/m = 1 - \frac{d}{dE}[\Delta V_a(r;E)] \quad , \tag{5.4a}$$

$$m_\omega^{(b)}(r;E)/m = 1 - \frac{d}{dE}[\Delta V_b(r;E)] \quad , \tag{5.4b}$$

where

$$\Delta V_a(r;E) = -\frac{P}{\pi}\int_{\varepsilon_F}^{\infty} \frac{W(r;E')}{E'-E}dE' \quad , \tag{5.5a}$$

$$\Delta V_b(r;E) = -\frac{P}{\pi}\int_{-\infty}^{\varepsilon_F} \frac{W(r;E')}{E'-E}dE' \quad . \tag{5.5b}$$

The contributions ΔV_a and ΔV_b are associated with two different

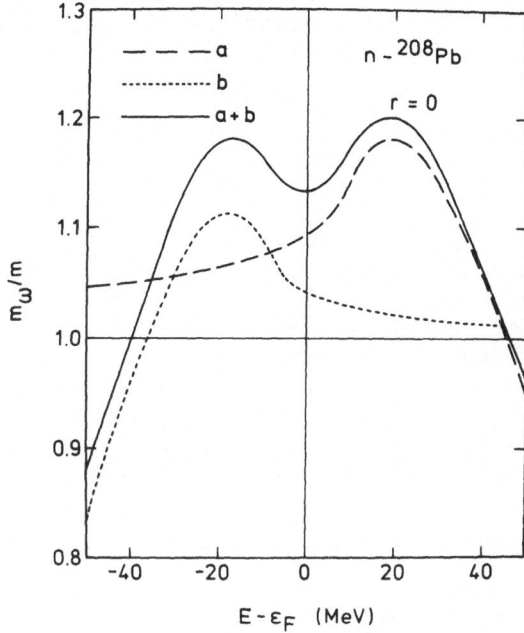

*Fig. 12. Taken from ref.
[15]. The full curve represents the energy dependence
of the ω-mass at the centre
of ^{208}Pb . The dashes give
the values of $m_\omega^{(a)}/m$ and
of $m_\omega^{(b)}/m$ defined by eqs.
(5.4a,b) . In the present
case it has been assumed
that $W(r;E)$ is slightly
asymmetric with respect
to $E = \epsilon_F$.*

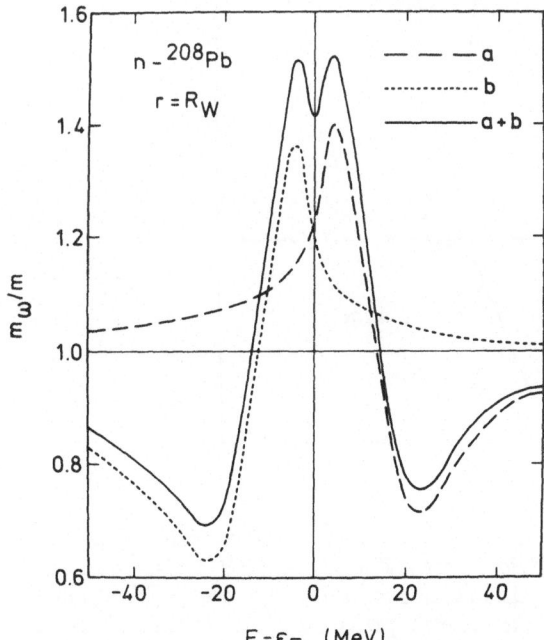

*Fig. 13. Taken from ref.
[15]. Energy dependence of
$m_\omega(R_W;E)/m$ (full curve),
$m_\omega^{(a)}(R_W;E)/m$ (long dashes)
and $m_\omega^{(b)}(R_W;E)/m$ (short
dashes) for ^{208}Pb ($R_W =
7.54$ fm) .*

physical processes, often called polarization and correlation corrections. If one assumes that $W(r;E)$ is symmetric with respect to
$E = \varepsilon_F$, one has

$$\Delta V_a(r;E-\varepsilon_F) \;=\; -\,\Delta V_b(r;\varepsilon_F-E) \quad , \tag{5.6a}$$

$$m_\omega^{(a)}(r;E-\varepsilon_F) \;=\; m_\omega^{(b)}(r;\varepsilon_F-E) \quad . \tag{5.6b}$$

Figures 12 and 13 show that the enhancement peak of m_ω is
much broader and less pronounced at the nuclear centre than at the
nuclear surface. At the nuclear centre, the peak of $m_\omega(0;E)$ is
similar to the enhancement found in the case of nuclear matter.
This is a hint that the enhancement of the empirical value $\langle m_\omega/m \rangle$
near the Fermi energy is mainly due to the property that the additional particle (or hole) excites surface degrees of freedom of the
core. This is confirmed by Fig. 14, which shows that for fixed E
close to ε_F the ω-mass $m_\omega(r;E)$ has a peak at the nuclear surface.

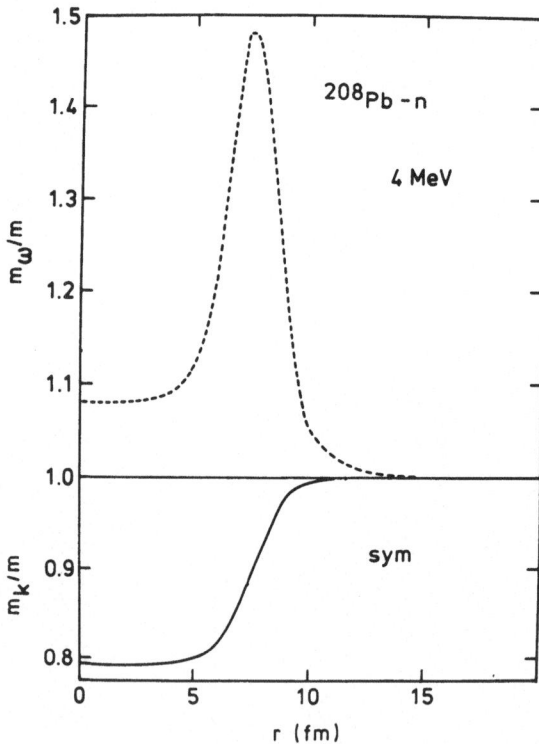

Fig. 14. Adapted from ref. [16]. The dashes show the radial dependence
of $m_\omega(r;E)/m$ in ^{208}Pb for $E-\varepsilon_F = 4\ MeV$. The full
curve represents the radial dependence of the effective
mass associated with the local potential equivalent to a
Skyrme-Hartree-Fock potential.

5.4. The Full Mean Field

The local potential $V(r;E)$ which is equivalent to the real part of the full mean field can be approximated by the expression

$$V(r;E) = V_{HF}(r;E) + \Delta V(r;E) \quad , \tag{5.7}$$

where $V_{HF}(r;E)$ is the local equivalent to the Hartree-Fock potential. One can define a radial dependent effective mass by the relation

$$\frac{m^{*}(r;E)}{m} = 1 - \frac{d}{dE} [V(r;E)] \quad . \tag{5.8}$$

Equations (5.7) and (5.8) yield

$$m^{*}(r;E)/m = [m_{k}(r;E)/m] + [m_{\omega}(r;E)/m] - 1 \quad , \tag{5.9}$$

where

$$m_{k}(r;E)/m = 1 - \frac{d}{dE} V_{HF}(r;E) \tag{5.10}$$

is the effective mass in the Hartree-Fock approximation. In the case of a Skyrme-type effective nucleon-nucleon interaction, $m_{k}(r;E)$ is independent of E. This remains approximately true for other interactions. We therefore drop the variable E in m_{k}. The radial dependence of $m_{k}(r)$ is represented by the full curve in Fig. 14.

The dependence of $\Delta V(r;E)$ upon r and E is fairly complicated for E close to ε_{F}. Hence, the shape of the full potential $V(r;E)$ depends upon E. This energy dependence is illustrated in Fig. 15. Note that $V(r;\varepsilon_{F})$ is equal to the local equivalent of the Hartree-Fock potential since $\Delta V(r;\varepsilon_{F}) = 0$.

5.5. Level-Dependent Quantities

Until now we discussed the dynamical correction to the Hartree-Fock approximation to the single-particle field. Many observables are related to expectation values of this correction in a single-particle state. For instance, the dynamical correction to the energy $\varepsilon_{n\ell j}$ of a Hartree-Fock single-particle state $\phi_{n\ell j}$ can be estimated by the expression

$$\Delta V_{n\ell j} = <\phi_{n\ell j}| \Delta V(r;\varepsilon_{n\ell j}) |\phi_{n\ell j}> \quad . \tag{5.11}$$

Figure 16 shows that on the average the energies $\varepsilon_{n\ell j} + \Delta V_{n\ell j}$ are in close agreement with the observed single-particle energies.

Other level-dependent quantities are of interest. For instance, one can define a level-dependent ω-mass by the expression

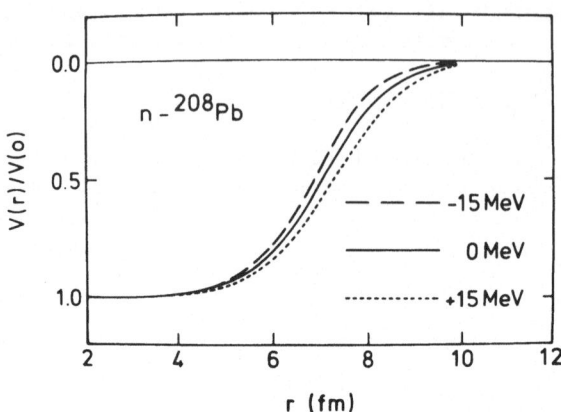

Fig. 15. *Taken from ref.* [15]. *Shape of the local equivalent to the full mean field in* ^{208}Pb *, for* $E-\varepsilon_F = 0$ *(full curve), - 15 MeV (long dashes) and + 15 MeV (short dashes).*

$$m_\omega^{n\ell j}/m \; = \; <\phi_{n\ell j}| \; m_\omega(r;\varepsilon_{n\ell j})/m \; |\phi_{n\ell j}> \quad . \qquad\qquad (5.12)$$

When plotted versus $\varepsilon_{n\ell j}$, the quantities (5.12) display an average trend which shows an enhancement peak near ε_F .[18] The shape of this peak is close to the one found in microscopic calculations.[1]

6. GROUND STATE CORRELATIONS

Section 5 was essentially devoted to the dynamical contributions to the mean field felt by a nucleon (or by a hole) added to a ^{208}Pb core. Because of the existence of deviations from the independent particle model, the nucleons in the ^{208}Pb core do not move independently of one another. One consequence of these correlations is that the Fermi sea is partly depleted, while the single-particle levels located above the Fermi surface are partly occupied.

The occupation probabilities of the Hartree-Fock single-particle levels in ^{208}Pb have been computed by Decharge, Gogny and Sips [19] in the framework of the microscopic random phase approximation, using Gogny's D1 effective interaction. Their results are represented by the full dots in Fig. 17.

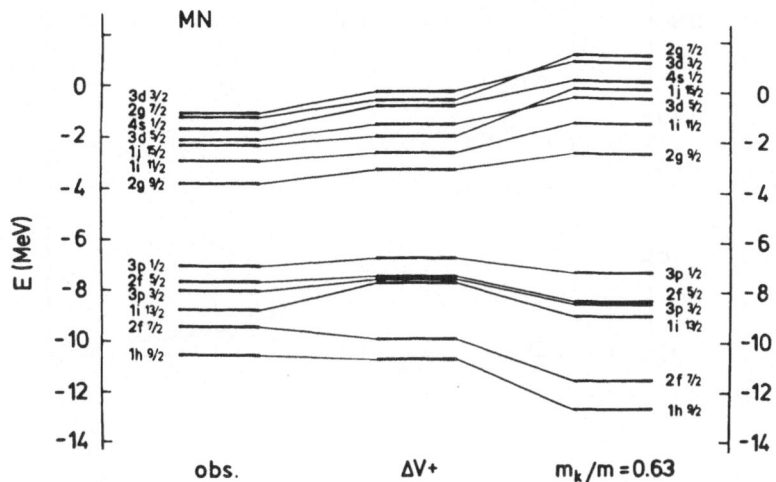

Fig. 16. *Adapted from ref. [17]. The first column gives the observed single-particle energies for neutrons in ^{208}Pb , see Figs. 2 and 9. The third column shows the Hartree-Fock values, for $m_{HF}^{\ddot{}}/m = 0.63$. The middle column includes the dynamical corrections as evaluated from eq. (5.11).*

The calculated depletion of the 3s1/2 orbit is approximately equal to ten per cent. This is significantly smaller than the thirty per cent depletion derived[20] from the interpretation of the difference between the experimental charge density distributions of ^{206}Pb and ^{205}Tl . It has been proposed[21] that the difference between the empirical and the theoretical values of the depletion is due to short range correlations, whose effect cannot be described with the help of effective nucleon-nucleon interactions which are weak by requirement.

In the dispersion relation approach, the occupation probability of a level $\phi_{n\ell j}$ is given by[18]

$$n_{n\ell j} = <\phi_{n\ell j}|\ 2 - [m_\omega^{(a)}(\varepsilon_{n\ell j})/m]\ |\phi_{n\ell j}> \qquad (6.1)$$

for $\varepsilon_{n\ell j} < \varepsilon_F$, and by

$$n_{n\ell j} = <\phi_{n\ell j}|\ [m_\omega^{(b)}(\varepsilon_{n\ell j})/m] - 1\ |\phi_{n\ell j}> \qquad (6.2)$$

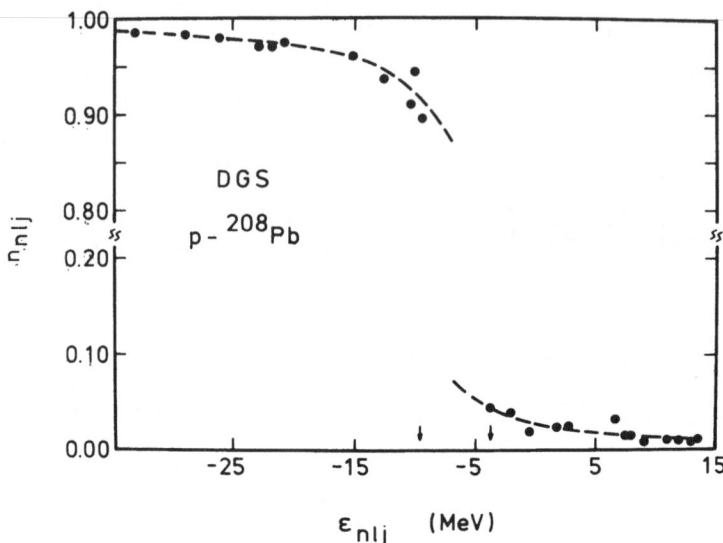

*Fig. 17. Derived from ref. [19]. The full dots represent the proton
occupation probabilities $n_{n\ell j}$ of Hartree-Fock single-
particle levels in the correlated ground state of ^{208}Pb,
as calculated in the framework of the random phase approximation. The curves have been drawn by eye through the dots.*

for $\varepsilon_{n\ell j} < \varepsilon_F$. The quantities $m_\omega^{(a)}$ and $m_\omega^{(b)}$ are defined by
eqs.(5.4a), (5.4b) . The value of $\hat{n}_{n\ell j}$ depends upon the behaviour
which is assumed for $W(r;E)$ for large $|E|$. This behaviour is
sensitive to the short range part of the nucleon-nucleon interaction.
This could be exploited to investigate the effect of short range
correlations on the occupation probabilities and on the density
distribution in the correlated ground state.

7. DISCUSSION

The most spectacular effect of the dynamical corrections to the
real part of the mean field consists in the existence of a peak in
the effective mass when the nucleon energy E is close to the Fermi
energy ε_F . This maximum can be ascribed to the fact that near the
Fermi energy the nucleon is heavier because it drags collective
excitations. The latter become decoupled from the single-particle
motion once the difference $|E-\varepsilon_F|$ becomes larger than the excita-
tion energy of the most collective excitations. This gives an estimate of the width of the enhancement peak of the effective mass.
This width is approximately equal to 15 MeV in heavy nuclei.

A similar enhancement is encountered in other physical systems.
For instance, Fig. 18 shows the calculated effective mass in normal

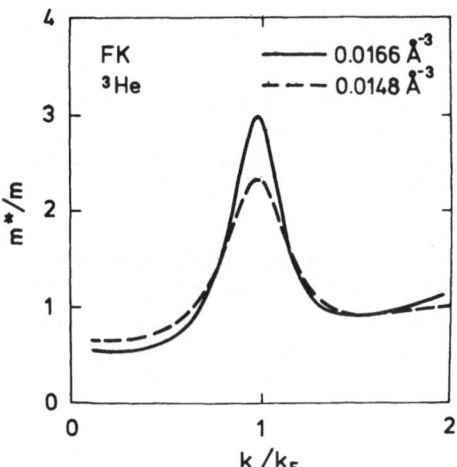

Fig. 18. Adapted from ref. [22]. Dependence upon k/k_F of the effective mass in normal liquid ^{3}He, for two different densities.

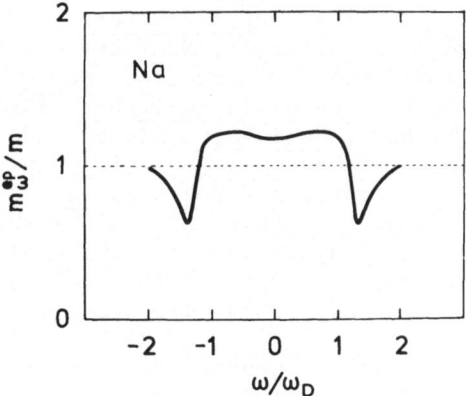

Fig. 19. Adapted from ref. [25]. Dependence upon $\omega/\omega_D = |E-\varepsilon_F|/\omega_D$ of the effective mass of electrons in the case of Na. The Debye energy $\omega_D = 0.013$ eV is a typical phonon energy.

liquid ^{3}He . In this case, the collective excitations which are responsible for the enhancement are spin fluctuations, whose energy is of the order of 10^{-5} eV . The enhancement of the effective mass in liquid ^{3}He has a strong influence on the value and on the temperature dependence of the specific heat at low energy.[23]

The electron-phonon coupling gives rise to a strong enhancement of the electron effective mass near the Fermi energy.[24] The width of the enhancement peak is of the order of a typical phonon energy, i.e. 10^{-2} eV , see Fig. 19.

I am grateful to Hélène Ngô with whom I enjoyed a pleasant and fruitful collaboration on the dispersion relation approach on which these lectures have focused. I apologize to the authors of the careful microscopic evaluations of the dynamical corrections to whom I did not give due credit here. A more balanced and much more complete discussion can be found in ref. [1].

REFERENCES

1. C. Mahaux, P.F. Bortignon, R.A. Broglia and C.H. Dasso, Physics Reports (in press).
2. A. Bohr and B.R. Mottelson,"Nuclear Structure", vol. 1, Benjamin, New York (1969).
3. C. Mahaux, in: "Fundamental Nuclear Physics", World Scientific Publ. Singapore (in press).
4. C. Mahaux and H. Ngô, Phys.Lett. 126B:1 (1983).
5. J.D. Jackson, "Classical Electrodynamics", John Wiley & Sons, New York (1975).
6. S. Galès, in: "Proceedings of the 1980 Masurian Summer School," Mikolajki, Poland.
7. C. Mahaux and H. Ngô, Phys.Lett. 100B:285 (1981).
8. J.S. Bell and E.J. Squires, Phys.Rev.Lett. 3:96 (1959).
9. D.H.E. Gross and R. Lipperheide, Nucl.Phys. A150:449 (1970).
10. C.A. Engelbrecht and H.A. Weidenmüller, Nucl.Phys. A184:385 (1972).
11. V. Bernard and Nguyen Van Giai, Nucl.Phys. A327:397 (1979).
12. F.G. Perey and B. Buck, Nucl.Phys. 32:353 (1962).
13. J.P. Jeukenne and C. Mahaux, Z.Phys. (in press).
14. P. Quentin and H. Flocard, Ann.Rev.Nucl.Sci. 28:523 (1978).
15. C. Mahaux and H. Ngô, Nucl.Phys. A378:205 (1982).
16. C. Mahaux and H. Ngô, Physica Scripta T5:74 (1983).
17. C. Mahaux and H. Ngô, Nucl.Phys. A410:271 (1983).
18. C. Mahaux and H. Ngô, Nucl.Phys. (in press).
19. J. Decharge and L. Sips, Nucl.Phys. A407:1 (1983).
20. J.M. Cavedon, B. Frois, D. Goutte, M. Huet, Ph. Leconte, C.N. Papanicolas, H.H. Phan, S.K. Platchkov, S. Williamson, W. Boeglin and I. Sick, Phys.Rev.Lett. 49:978 (1982).
21. C. Mahaux, in: "Perspectives in Nuclear Physics at Intermediate Energies", S. Boffi, C. Ciofi degli Atti and M.M. Giannini, eds.,

World Scientific Publ. Singapore, p. 1 (1984).
22. B.L. Friman and E. Krotscheck, Phys.Rev.Lett. 49:1705 (1982).
23. V.K. Mishra, G.E. Brown and C.J. Pethick, Journ.Low Temp.Phys.
    52:379 (1983).
24. M.J. Buckingham, Nature 168:281 (1951).
25. G. Grimvall, J.Phys.Chem.Solids 29:1221 (1968).

PHYSICS WITH MODULARIZED 4π NaI DETECTOR SYSTEMS[*]

D. Schwalm, Ch. Ender, H. Gröger, D. Habs, U.v. Helmolt,
W. Hennerici, H.J. Hennrich, H.W. Heyng, W. Korten,
R. Kroth, M. Mušič, J. Schirmer, B. Wchwartz, and W. Wahl

Physikalisches Institut der Universität Heidelberg
and Max-Planck-Institut für Kernphysik
D-69 Heidelberg, F.R. Germany

R.S. Simon
Gesellschaft für Schwerionenforschung
D-61 Darmstadt, F.R. Germany

W. Kühn and V. Metag
2. Physikalisches Institut der Universität Giessen
D-63 Giessen, F.R. Germany

C. Broude
Weizmann Institute of Science
76100 Rehovot, Israel

1. INTRODUCTION

Nuclear γ-ray spectroscopy has always been an important tool
in nuclear research. It constitutes not only one of our main sources
of information about the structure of nuclei, it also plays an
increasingly important part in heavy ion reaction studies. There
are several reasons for this exceptional role of γ-ray spectroscopy:
Out of the large number of states an atomic nucleus can exist in,
those close to the yrast states (the lowest energetic level of
a given spin) are of particular interest as they are in general
due to fundamental excitation modes of the nucleus and thus display
most clearly the basic properties of a finite system of nuclear
matter. On the other hand, the levels in the yrast region are just
those accessible in γ-ray experiments; they decay predominantly
via the electromagnetic interaction, i.e. by emission of γ rays
and - less likely - of conversion electrons, as their decay by

─────────
*Work supported by Bundesministerium für Forschung und Technologie,
 Bonn, F.R. Germany

emission of nuclear particles is either energetical forbidden or
strongly hindered by the Coulomb and centrifugal barrier. As, moreover, the electromagnetic interaction is well understood, the information obtained from γ-ray experiments can be solely used to determine the properties of the γ-decaying states. These are - beside
excitation energies, spins, and parities - in particular the dynamic
electromagnetic moments[1], which present together with the static
moments one of our most valuable clue we have to understand
nuclear structure. The main use of γ-ray spectroscopy in heavy
ion reaction studies is to investigate special aspects of the reaction
amplitudes by particle-γ correlation measurements (exploited e.g.
in Coulomb excitation experiments) and to determine the amount
of angular momentum and excitation energy tied up in the residual
reaction products. While the total intrinsic excitation energy can
be deduced from the sum of the energies of all γ rays, the γ-multi-
plicity is a measure of the total intrinsic spin as the γ-decay
proceeds, at least in deformed nuclei, predominantly via stretched
E2-cascades, i.e. each γ-ray removing two units of angular momentum.

    The important impact γ-ray spectroscopy had - and still has -
on these fields of nuclear research, however, was - and will be -
intimately connected to the successful development of suitable γ-
detection techniques. Keystones in these developments are the
invention of scintillation detectors in the early fifties and of
the high resolution Ge detectors in the sixties. In the seventies,
when it became possible to produce nuclei by (HI, xn) reactions
in the highest spin states still decaying by γ emission, experiments
were considerably improved by combining Ge detectors with two types
of NaI-detector arrangements: In one type of experiments arrays
of individual NaI detectors were used, so-called multiplicity
filters[2], which provided information on the γ multiplicity. The
other type of experiments involved so-called sum-energy spectro-
meters[3], large NaI detectors with the target positioned in its center
such that most of the energy carried by the γ rays is absorbed in
the crystal. The additional information obtained in these ways has
been used successfully to enhance the γ-ray cascades containing
the nuclear structure information of interest and to obtain at least
average values for the initial spins and excitation energies of
the residual nuclei. It became clear rather soon, however, that
further progress in this field of high-spin state and heavy-ion
reaction studies would require a new effort to improve the γ-detection techniques. The kind of instrument one likes to have for these
studies is easy to define: It should have a γ-detection efficiency
of 100% of 4π, a good energy, time, and angle resolution, and it
should be able to handle events involving a large number of coincident γ rays. Obviously, such an instrument would allow for the first
time to study all parameters characterizing a γ-ray cascade on an

event-by-event basis, moreover, it would be ideally suited to study
rare γ-decay modes.

At the end of the seventies several groups started to think
about such modularized detection systems, trying to find a compromise
between the ideal detector and the down-to-earth requirement to
be technically and financially realizable. Two types of detector
systems emerged which mainly differ in which of the properties of
the ideal detector have been sacrificed to achieve this compromise:
The first type involves as many Compton-shielded Ge detectors as
possible, the rest of the solid angle being covered by scintillation
detectors (TESSA2 (ref.4), OSIRIS (ref.5), LBL-array (ref.6)). Here
the emphasis is on the measurement of many-fold γ-coincidences with
very good energy resolution and low Compton background, while the
additional scintillator detectors mainly serve as a sum-energy and
multiplicity filter with moderate sum-energy and multiplicity resolutions. These instruments are tailored for measuring decay schemes
of high spin states (for more details the reader is referred to
the lecture by Frank Stephens[6]).
The alternative, rather complementary approach is the crystal ball
system consisting of a spherical shell of scintillator material,
which subtends the full solid angle and is subdivided into a large
number of individual detectors. Here one tries to optimize the
detection of all individual γ-rays emitted in an event, the total
efficiency, and the multiplicity and sum-energy resolution, the
sacrifice being the individual γ-energy resolution. Two systems
of this type have been realized so far, both using NaI as scintillator material: The Spin Spectrometer at Oak Ridge[7], consisting
of 72 individual detector modules forming a hollow sphere of 36 cm
inner diameter and a shell thickness of 18 cm, and the Heidelberg-
Darmstadt Crystal-Ball[8], consisting of 162 individual detector
elements forming a sphere with an inner diameter of 50 cm and a
shell thickness of 20 cm. These instruments are rather versatile
and can be used in many fields of γ-ray spectroscopy either as a
stand-alone detector or in combination with particle as well as
high-resolution, Compton-suppressed Ge-detectors.

As the general properties of the two crystal ball systems
realized so far are rather similar, in the present lecture the
Heidelberg-Darmstadt Crystal Ball will serve as an example to discuss
the properties of such modularized 4π detector systems (sect.2).
Furthermore, two recent experiments performed with the Heidelberg-
Darmstadt Crystal Ball will be discussed, one concerning the first
unique observation of the 2-γ decay in ^{40}Ca and ^{90}Zr, exemplifying
the possibilities opened up by the Crystal Ball in the investigation
of rare γ-decay modes (sect.3), the second one concerning the investigation of high-energetic γ-rays emitted in (HI,xn) reactions,
illustrating its use in studies involving high-spin states (sect.4).

2. THE HEIDELBERG-DARMSTADT CRYSTAL BALL

In this section some of the basic properties of the Heidelberg-Darmstadt Crystal Ball (CB) spectrometer will be discussed. The CB is a joint project of the Physikalisches Institut der Universität Heidelberg, the Max-Planck-Institut für Kernphysik, Heidelberg, and the Gesellschaft für Schwerionenforschung, Darmstadt. The instrument is presently located at the Tandem-postaccelerator facility of the Max-Planck Institut für Kernphysik.

As schematically shown in fig.1, the spectrometer consists of a spherical shell with an inner diameter of 50 cm and a thickness of 20 cm, which is built up by 162 mechanically separated and individually removable NaI detectors. The choice of the detector material, the number of detector elements and the dimensions of the total system reflect our compromise between the desired properties of the instrument and financial constraints.

One of our basic design goals was to built an instrument with a γ-detection efficiency ε of as close to 100% of 4π as possible. This does not allow for the use of Ge as the basic detector material because of technical difficulties to produce large enough Ge crystals. Therefore NaI was chosen for the detector material as of all large volume detectors known the best energy resolution can be obtained with NaI scintillation detectors. The thickness of the shell is then determined by the attenuation coefficient for γ rays in NaI; it was chosen to be 20 cm, corresponding to a transmission probability of 1.5% for 1 MeV γ rays.

The number of individual detector elements N is governed by the requirement to resolve the individual γ rays from a cascade of large multiplicity M_γ. The probability r that a single detector element is hitted by more than one γ-ray in an event is approximately given by $r \approx (M_\gamma-1)/(2N)$ assuming equal efficiencies for all N detector elements and $\varepsilon \approx 1$; thus to have $r \leq 0.1$ for a cascade with $M_\gamma = 30$, N is required to be larger than 150. Furthermore, if one wants to identify single γ-transition energies by requiring that none of the direct neighbours of a given detector element fires (thus suppressing the outscattering due to Compton scattering and pair production processes leading to an imperfect detection of the energy of the primary γ quant in the detector considered) the restrictions on N are even more severe. For N=150 and M_γ=30 only $n_{isol} \approx$ 5 of such isolated hits are expected [9] assuming a realistic outscattering probability of 0.3; this number decreases to $n_{isol} \approx 1.5$ for e.g. N=72. The number of elements actually chosen for the CB design is N=162. This odd number results from a solution of the interesting geometrical problem of subdividing a sphere into polyhedrons under the following restrictions: Each polyhedron should subtend equal solid angles, the number of different types of polyhedrons should be small, and the form of the detector

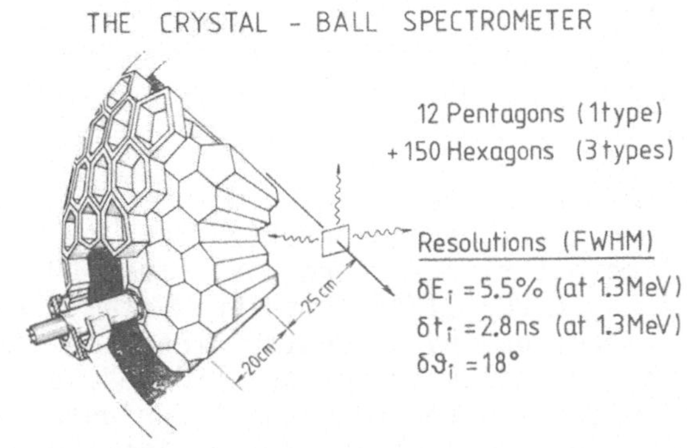

THE CRYSTAL - BALL SPECTROMETER

12 Pentagons (1 type)
+ 150 Hexagons (3 types)

Resolutions (FWHM)

$\delta E_\uparrow = 5.5\%$ (at 1.3 MeV)
$\delta t_\uparrow = 2.8$ ns (at 1.3 MeV)
$\delta \vartheta_\uparrow = 18°$

Fig. 1.   Schematic view of a section of the Crystal Ball displaying
the geometrical configuration of the modularized NaI shell.
The total solid angle of 4π is covered by 162 modules of
four different geometrical shapes but subtending equal
solid angles. Also given are the individual detector
resolutions for 1.3 MeV γ rays.

elements should be such as to allow for a good light collection.
Moreover, to minimize the outscattering probability as few as possible polyhedrons should meet at each corner and the ratio of front
face area to circumference should be a maximum. Acceptable solutions
of this problem - allowing only for polyhedrons with pentagonal
and hexagonal rather than triangular faces - lead to N=32, 42,
72, 92, 122, 132, 162, 192... . For the configuration selected
(N=162), 12 equal pentagonal polyhedrons and 3 different types
of hexagonal polyhedrons are needed to build up the detector shell.

The inner radius of the detector shell was chosen to be 25 cm
to maximize the front area of the individual modules, to allow
for a sufficient neutron separation by a time-of-flight technique,
and to provide enough room for a large scattering chamber for additional particle detector set-ups etc.

The 162 detector modules are canned separately in 0.6 mm thick
Al housings, produced without welding seams by a special hydroforming
technique, and are optically coupled to phototubes of 7.5 cm diameter.
The individual modules are mounted in a large spherical honey-comb
like structure, which allows to remove single modules or cluster
of modules as to make room for other detectors, the beam pipe etc.
The total ball is divided into two hemispheres which can be separated
to provide access to the central target region.

For an accepted event the energy and time signals (the time
relative to a common start being converted to a charge signal)
of all individual detectors are stored in a fast ADC system and
read out via CAMAC by a PDP 11/45 computer system. Two time signals
per module are derived, one by a constant fraction timing (CFT)
discriminator with thresholds as low as $\approx$ 80 keV, and one by a
leading edge (LE) discriminator running at a higher threshold.
The high voltage on the phototubes (controlling the gain of the
energy signals), the threshold and walk adjusts of the CFTs, the
leading edge thresholds, and the delays of the time signals can
be set and changed via CAMAC using an LSI 11/23 computer. The electronic settings can be stored on a floppy disk and reloaded for
later runs.

A CB specific fast trigger can be derived from (i) the OR of
all CFTs, (ii) the OR of all high-threshold signals, (iii) a gate
on the sum of all pulse heights, and (iv) a gate on the multiplicity
spectrum being proportional to the number of detectors which have
fired. This trigger signal can be combined with those from other
detectors built into the CB and with the beam pulse signal to select
a valid event. The event rate that can be handled presently by
the ADC and read-out system is about 200-300 cps with typical event
length of about 100 words. The total event rate of the ball - as
measured by the CFT-OR - is usually kept below 300-400 kcps to
avoid pile-up problems and electronic dead-times.

Many of the properties of the CB can be studied with radioactive
γ sources emitting two γ rays in coincidence. Detecting the full-
energy peak of one of these γ-rays, say γ', in one of the CB detectors (or e.g. a Compton-shielded Ge detector replacing one of the
NaI modules) the fate of the second γ ray in the CB can be studied
in detail. This procedure also allows to simulate the response of
the CB to γ-ray cascades of a given multiplicity M_γ by considering
during the play-back M_γ of such single-photon events together.

The energy and time resolutions of the individual NaI modules
guaranteed and reached by the supplier of the detectors (Harshaw
Comp.) are given in fig.1 and are close to what is technically
achievable even for considerably smaller crystals. The angular
resolution of $\pm9°$ is solely determined by the geometry.

In fig.2 the response function of the CB to γ rays with energy
E_γ = 898 keV ([88]Y source, $E_{\gamma'}$=1.84 MeV) are displayed. The spectra
were incremented subject to various conditions discussed below but
irrespective of which of the individual detectors fired in the event.
(a) Sum mode: Lineshape obtained by adding up the energy signals
of those detectors which fired in an event. The sum-peak resolution
and in particular the good peak-to-background ratio are the essential
ingredients which determine the properties of the CB as a sum-energy
spectrometer. (b) Normal mode: Spectrum obtained by incrementing
the individual energy signals of all detectors which fired in an
event. Note that in this mode a single event may contribute to the
spectrum more than once due to the cross talk between the detectors
(e.g. via Compton scattering or - at higher energies - via annihilation radiation). The average number of detectors firing is 1.3 at
E_γ = 0.9 MeV. (c) Single hit: Spectrum incremented by selecting
those events where only one detector fired. As this extreme anti-
scattering condition can only be used for events with M_γ=1, it is
important that already a considerably relaxed condition, which is
also useful for events with $M_\gamma \gg 1$, results in a sizable improvement
of the peak-to-background ratio as compared to the normal mode.
This is demonstrated by fig.2d, labelled as Anti-Compton, which
shows the spectrum obtained by incrementing the individual energy
signals of those detectors firing in an event, where none of the
next neighbour detectors (5 or 6, respectively) have responded.
(e) Compton mode: Sum-energy spectrum obtained for events where
only two next neighbour detectors fired. (This mode will be discussed
in more detail below.) It should be mentioned that in those cases
where the full-energy of the γ-quant is detected in one detector
module (b,c,d) the resolution of the full-energy peak amounts to
7.2%, which is slightly worse compared to the individual detector
resolution of 6.7% because of small gain mismatches between the
different (161!) detectors.

Absolute detection efficiencies of the complete (N=162) CB for
M_γ=1 events are displayed in fig.3 as a function of E_γ. Note in
particular, that the total detection efficiency ε_{CB}, i.e. the
probability that at least one detector is firing, amounts to e.g.
ε_{CB} = 97.5% for E_γ = 1 MeV. Of the 2.5% of the γ-rays remaining
undetected, ≈1.5% can be explained by the finite thickness of the
NaI shell. Thus the efficiency loss due to gaps between the detectors, absorption in the inactive housing material, and electronic
thresholds only amounts to ≈1%.

The sizable probability to detect a Compton scattered γ ray
in one of the next neighbours of the detector hitted by the primary
γ quant (see fig.2e, $\varepsilon_{com} \approx$ 15% for E_γ = 1MeV) allows for the use
of the CB as a 162-fold Compton polarimeter. Fig.4 displays the
measured polarization sensitivity $Q(E_\gamma)$. The two data points were
obtained by fitting the linear polarization angular correlation
functions $W(\theta_\gamma, \Psi_c)$ (ref.1) measured for the two γ rays emitted

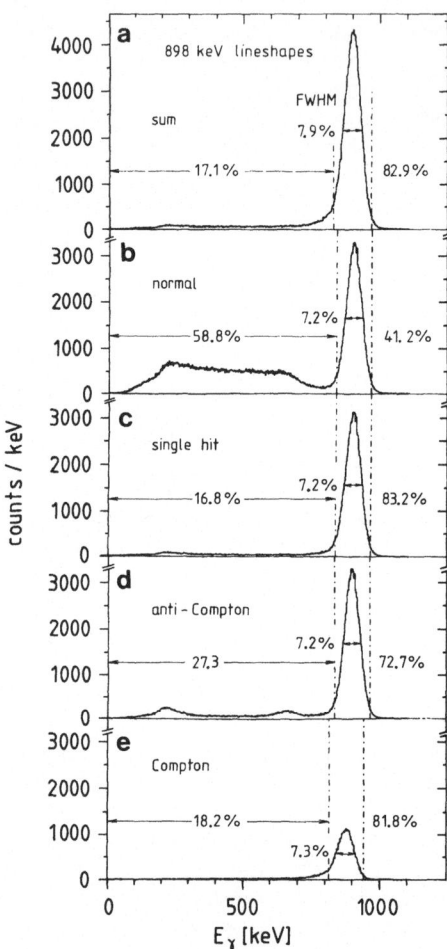

Fig. 2.   Response functions of the CB for γ rays of 898 keV obtained
under various conditions (see text).

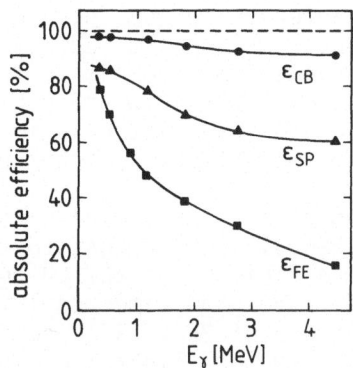

Fig. 3. Absolute efficiencies measured for the complete CB for
M_γ=1 events (ε_{CB}: Probability that at least one detector
fired. ε_{SP}: Probability to detect the full energy of the
γ-ray in the sum-energy peak (cf. fig.2a) ε_{FE}: Probability
to observe the full energy in one of the 162 individual
modules (cf. fig.2b,c,d))

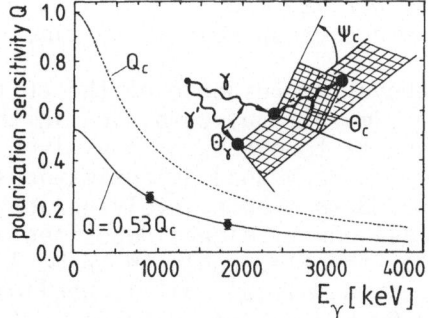

Fig. 4. Measured polarization sensitivity $Q(E_\gamma)$ for the two γ rays
of a ^{88}Y source (dots) when using the CB as a 162-fold
Compton polarimeter. The solid line is obtained by scaling
the sensitivity Q_c of an ideal Compton polarimeter for
θ_c=90° (dashed line).

by the ^{88}Y source (the angles θ_γ, ψ_C are defined in the insert of fig.4). The observed sensitivity amounts to 53% of that of an ideal Compton polarimeter with $\theta_c = 90°$. The polarization sensitivity of the CB is certainly inferior to what can be obtained in specifically designed Compton polarimeters (in fact, to optimize the multiplicity resolution we have tried to minimize the outscattering probability), however, the statistic which can be obtained for $W(\theta_\gamma, \psi_C)$ because we actually have 162 polarimeters covering the full solid angle of 4π outweights by far the reduced $Q(E_\gamma)$ values.

The time resolution of the individual detectors, measured relative to a plastic scintillator, amount to $\delta t(FWHM) = 2.8$ ns at $E_\gamma = 1.3$ MeV. As the efficiency of the CB for detecting neutrons is also large ($\epsilon_n \approx 85\%$), the good time resolution together with the target-detector distance of 25 cm is essential for separating neutrons and γ-rays by a time-of-flight technique. The separation which can be achieved by this method is illustrated in fig.5; it displays the time spectra - measured relative to the beam pulse - observed at three different angles with respect to the beam axis during the bombardment of ^{130}Te with a ^{34}S beam of 150 MeV. Neutrons with energies ≤ 7 MeV, that is the dominant part of the neutron spectrum emitted e.g. in (HI,xn) reactions, can be well distinguished in this way from prompt γ rays. Note, moreover, that a possible summing of a prompt γ-ray pulse with a pulse caused by a delayed neutron hitting the same detector is realized by the CFT discriminator if the neutron signal occurs within the internal delay time of the CFT (35 ns) as - due to the change of the rise time of the summed signal - the time output of the CFT will occur outside the prompt peak. The most important time range for summing is handled in this way; for larger delay times between the γ and n pulse, the additional time signal derived from the leading edge discriminator can be used to perform a corresponding rise-time analysis.

As pointed out above the response of the CB to higher multiplicity events can be simulated by combining in the off-line analysis multiplicity 1 events to events of any given multiplicity M_γ. This procedure results in very good approximations of the response characteristics of the CB to events involving long γ cascades; the probability that a small energy signal deposited in a detector module is overlooked by the electronic in an $M_\gamma=1$ but not in a real $M_\gamma > 1$ event because of a multiple hit is negligible small for CFT thresholds around 80 keV. As an example for this calibration procedure, fig.6 illustrates the characteristic quantities for the response of the CB to an event consisting of a cascade with M_γ transitions of equal transition energy $E_{\gamma i}$. Displayed are the centroids and FWHM values for the resulting N_γ and E_{CB} distributions for the complete CB, N_γ being the number of responding detectors and E_{CB} the sum-energy deposited in the ball. These distributions are well represented by Gaussians for $M_\gamma > 5$ and their FWHM values

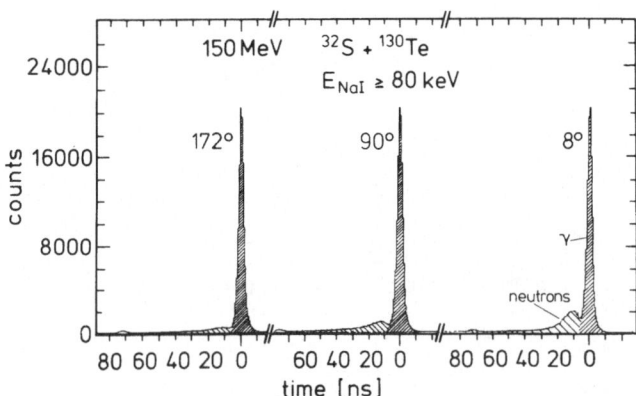

Fig. 5. Time spectra of events relative to the beam pulse, recorded with the CB at three different angles with respect to the beam axis.

are $\delta N_\gamma/\langle N_\gamma\rangle \lesssim .25$ and $\delta E_{CB}/\langle E_{CB}\rangle \lesssim .20$ for $M_\gamma \gtrsim 20$ and $E_{\gamma i} \lesssim 1$ MeV. For an incomplete ball (N<162) the multiplicity and in particular the sum-energy resolutions deteriorate; removing e.g. 5 detector modules, $\delta N_\gamma/\langle N_\gamma\rangle$ and $\delta E_{CB}/\langle E_{CB}\rangle$ increases by about 5% and 15%, respectively, for the M_γ and $E_{\gamma i}$ range given. For cascades involving different transition energies $E_{\gamma i}$, reasonable estimates of the centroids and resolutions can be obtained from the curves given in fig.6, replacing $E_{\gamma i}$ by the average transition energy $\langle E_{\gamma i}\rangle$.

In heavy-ion induced fusion reactions leading to the formation of evaporation residues, for example, the γ-sum energy E_{CB} and the number of (prompt) responding detectors N_γ can be directly converted to yield the excitation energy E_i of the entry state to the de-excitation cascade in the residue and the number M_γ of cascade transitions. Moreover, as rapidly rotating nuclei decay via cascades involving predominantly stretched ($\Delta I=2$) E2 transitions, the γ multiplicity M_γ can be further converted to yield the spin I_i of the entry state. As an illustrating example for the event-by-event measurement of E_{CB} (E_i) and $N_\gamma(I_i)$ by the CB, fig.7 displays the E_{CB} vs. N_γ correlations observed in the fusion reaction ^{130}Te(^{34}S, xn)$^{164-x}$Er at a beam energy of 160 MeV. The correlations shown for the 4n, 5n and 6n channel were obtained by gating on low-lying

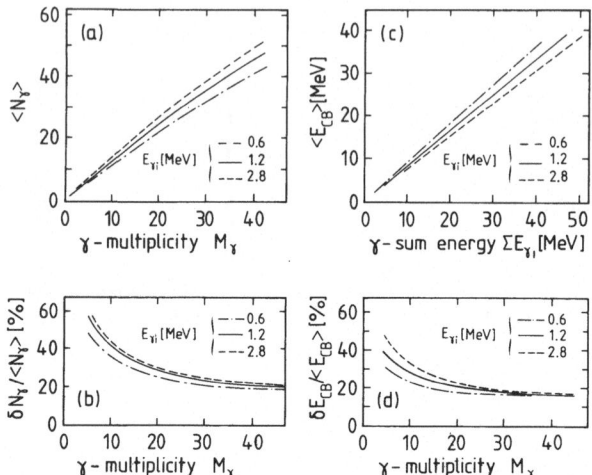

Fig. 6.   Response of the CB to an event consisting of a cascade
          with M_γ γ rays of equal energies E_{γ_i} = 0.6, 1.2, 2.8 MeV,
          respectively. (a) Average number $<N_\gamma>$ of responding
          detectors vs. multiplicity M_γ. (b) FWHM values of the N_γ
          distributions vs. M_γ. (c) Average detected sum-energy
          $<E_{CB}>$ vs. the total γ-ray energy ΣE_{γ_i} (d) FWHM values of
          the E_{CB} distributions vs. M_γ.

characteristic γ lines of the corresponding residual nuclei, which
were observed in a Ge detector operated in coincidence with the
ball; they reflect the distributions of the entry states (E_i, I_i)
to the cascades of the respective final nuclei and show the familiar
fractionation of the cross-section in the E_i vs. I_i plane. It
should be noted that e.g. in the 4n channel (^{160}Er) entry states
up to spins $I_i \approx 2 \ (M_\gamma(N_\gamma)-3) \approx 55$ (c.f. ref.1 and sect.4) are
populated but that the intensities of the transitions between
the yrast states as seen in the Ge detector by gating on these

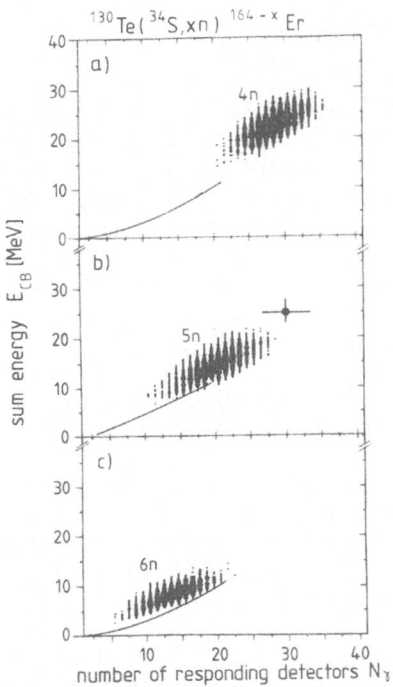

Fig. 7. E_{CB} - N_γ correlation observed in coincidence with a Ge
detector during the bombardment of a ^{130}Te target with
160 MeV ^{34}S ions, gated (a) with the 6-4 and 4-2 transitions in ^{160}Er, (b) with the 17/2-13/2 transition in ^{159}Er,
and (c) with the 6-2 and 4-2 transitions in ^{158}Er. The
solid line represents the calculated centroids of the E_{CB}-N_γ
correlation in case the cascades would start from the yrast
line known from discrete line spectroscopy. The cross indicates the expected resolution for E_{CB}= 25 MeV and N_γ=30.

entry states still decrease rapidly when approaching $I_{yrast} \approx 30$.
This indicates that either the γ decay in the high spin region of
^{160}Er proceeds via many different bands above the yrast line (pre-
venting the high-spin yrast transitions from collecting enough
intensity to be recognized in the Ge detector) or/and that at high
spins the yrast line is no longer a smooth line due to the loss of
collectivity of the nucleus (leading to an irregular line pattern
hard to identify in the Ge spectrum).

Not only the magnitude I_i of the spin of the entry state can
be determined, it is also possible - at least for high spins I_i -
to determine the direction of $\vec{I}_i$ on an event-by-event basis by
analysing the spatial distribution of the M_γ γ-rays. As for high
I_i values all stretched E2 transitions occurring in the decay
cascade obey an $1-\cos^4\xi$ correlation with respect to the direction
of the entry spin[1], ξ being the polar angle between $\vec{I}_i$ and the
direction of the γ-ray, $\vec{I}_i/I$ can be determined as the normal to
the ring of detectors containing the largest number of E2 transitions.
Moreover, for evaporation residues formed in (HI,xn) reactions, $\vec{I}_i$
has to lie approximately in the plane perpendicular to the beam
axis; thus only the direction of $\vec{I}_i$ in this plane has to be determined. The outlined procedure was tested using the reaction
^{128}Te$(^{34}$S,xn$)^{162-x}$Er at $E(^{34}$S$)=155$ MeV. In fig.8 the intensity
of characteristic stretched E2 transitions, which are observed in
two Ge detectors operated in coincidence with the CB, are displayed
as a function of the direction of $\vec{I}_i$ (measured by the angle θ defined
in the insert). According to the $1-\cos^4\xi$ correlation pattern,
for the detector positioned close to the beam (GeI) the intensity
is found to be independent of θ, while it is a maximum for detector
II, positioned in the spin plane, if $\vec{I}_i$ is perpendicular to the
detector axis (clearly, no polarization of $\vec{I}_i$ is observable, i.e.
θ is only defined mod.π). The resolution observed for θ is of
the order of $\Delta\theta = \pm35°$; it is found that it cannot be improved
considerably by using more fancy methods to deduce θ. Nevertheless,
the resolutions reached for the direction and the magnitude of $\vec{I}_i$
on an event-by-event basis are sufficient to allow for novel investigations of the angular momentum degree of freedom in heavy ion
reactions; several experiments concerning these questions are
presently under way.

To demonstrate the usefulness of the Anti-Compton mode discussed above (compare fig.2d) in cases where the events involve γ cascades of high multiplicity, fig.9a displays the probability for the
number of isolated hits per event for γ cascades observed during
the bombardment of a ^{128}Te target with ^{34}S ions of 155 MeV. Although
the average number of responding detectors per event is 25, there
are on the average 6-7 isolated hits per event. Fig.9b shows the
resultant γ-ray spectrum when incrementing only the energy signals
of isolated detectors; moreover, only those events were selected
which involve the 4n channel by gating on the known[10] yrast tran-

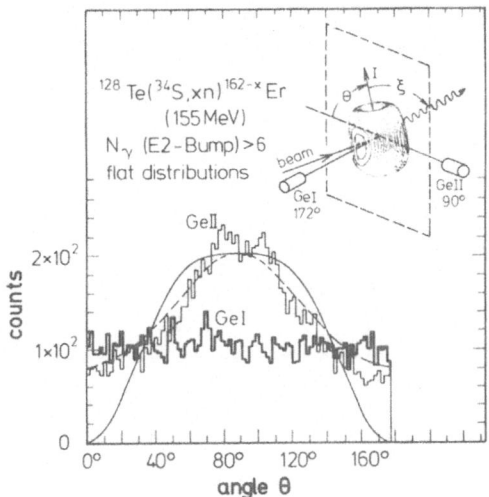

Fig. 8. Angular correlation of stretched E2 transitions observed
in two Ge detectors as a function of the spin direction
θ determined on an event-by-event basis by the CB. The
solid curve represents the theoretical correlation expected
for GeII, while the dashed curve is obtained allowing for
a resolution of Δθ = ±35°.

sitions of ^{158}Er recorded in a coincident Ge detector. Although
the energy resolution is certainly considerably inferior to that
obtained by using Ge detectors, the spectrum is of high quality
for NaI detectors; all the characteristic features of the known
yrast line in ^{158}Er are clearly observed.

The CB spectrometer can be supplemented by additional detectors,
which are installed either in a scattering chamber of 48 cm diameter
or which replace individual CB detector modules. In particular, a
modularized parallel-plate detector has been developed for heavy
reaction products, which subtends 82% of 4π (ref.11). For the detection of evaporation residues, a parallel plate ring detector is
available covering an angular range of 3° ≤ ϑ ≤ 12° with respect
to the beam[12]. For the same purpose, an electrostatic separator
has been developed and successfully tested[13]. Moreover, there are
presently 4 Ge detectors with NaI Anti-Compton shields available
which may replace anyone of the 12 pentagonal modules of the CB.

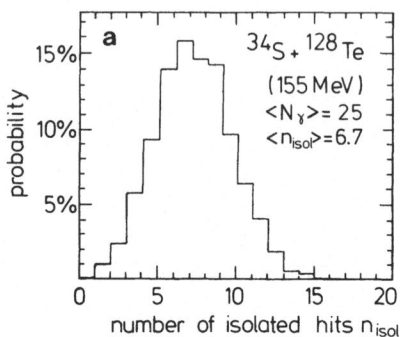

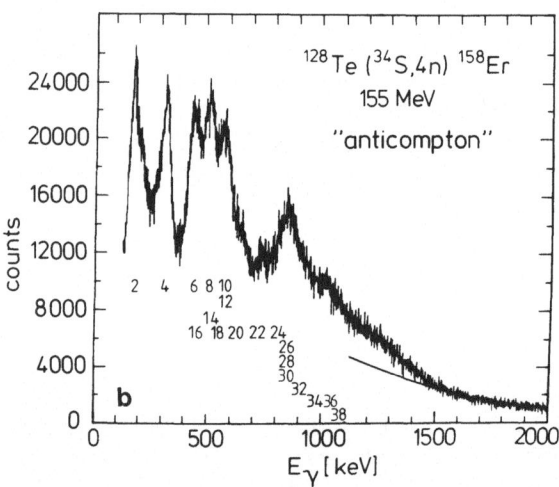

Fig. 9.   (a) Probability for the number of isolated hits per event
          (number of responding detectors where none of the next
          neighbours have fired) when bombarding a ^{128}Te target
          with ^{34}S ions of 155 MeV. (b) Anti-Compton spectrum obtained
          by incrementing the energy signals of the isolated detectors
          and selecting events belonging to the 4n channel by gating
          on the yrast transitions in ^{158}Er observed in a coincident
          Ge detector. The spins I label the known yrast transitions
          I → I-2 in ^{158}Er.

The Anti-Compton shields are necessary to shield the next neighbours against the large outscattering rates of the Ge detectors and to maintain a large total efficiency of the ball. Of course, the shields are also used to improve the peak-to-background ratio of the Ge detector; average Compton suppression factors of ≈ 3 are obtained.

Having described the basic properties of the CB we shall now turn to the discussion of two recent experiments to illustrate some of the new possibilities opened up by this instrument.

3. DOUBLE-GAMMA DECAY OF $^{40}Ca(0_2^+)$ AND $^{90}Zr(0_2^+)$

An excited nuclear state may decay by simultaneous emission of two γ quanta, each having a continuous energy spectra but summing up to the total transition energy. Such a double-gamma decay, being a second order process, is very slow compared to the usual one-quantum decay. However, as for $0^+ \to 0^+$ transitions the emission of a single γ ray is strictly forbidden, a radiative transition must proceed in these cases by emission of at least two γ rays, and the competing decay modes are thus restricted to the (in this case hindered) internal conversion and, if sufficient energy is available, internal pair conversion. Seven stable nuclei (^{16}O, ^{40}Ca, ^{72}Ge, ^{90}Zr, ^{96}Zr, ^{98}Mo) are known to have a ground and first excited state with spin and parity 0^+ and since about three decades many experiments[14] have aimed at proving the mere existence of the double-gamma decay in one of these nuclei. The CB is obviously tailored to takle this problem not only because of its large efficiency (the branching ratios are expected to be of the order $\Gamma_{\gamma\gamma}/\Gamma_{tot} \approx 10^{-4}$); its high granularity together with the good lineshape obtained in the single-hit mode and the possibility to use the ball as a Compton polarimeter should allow for a detailed investigation of this rare decay mode. Two cases have been studied so far[15], namely ^{40}Ca and ^{90}Zr.

The 0_2^+ first excited states[16] of ^{40}Ca (E_0 = 3.35 MeV, $T_{1/2}$ = 2.1 ns) and ^{90}Zr (E_0 = 1.76 MeV, $T_{1/2}$ = 62 ns) were populated via the (p,p') reaction at E_p = 5.08 MeV and E_p = 7.30 MeV, respectively, using a pulsed proton beam and metallic targets of ^{40}Ca (0.3 mg/cm^2, 99.97% enriched) and ^{90}Zr (0.7 mg/cm^2, 97% enriched). Backscattered protons were measured in four silicon detectors at ϑ_p = 130° with respect to the beam. Special care has been taken to be able to suppress the perturbing γ background originating from the decay of the 0_2^+ states via internal pair conversion and the subsequent positron annihilation in flight (PAF). For these events the γ-ray energies $E_{\gamma 1}$, $E_{\gamma 2}$ and their relative angle θ_{12} are kinematically related by

$$1/E_{\gamma_1} + 1/E_{\gamma_2} = (1-\cos\theta_{12})/m_e c^2 \qquad (3.1)$$

whereas for the 2γ decay all relative angles are allowed. Thus the perturbing PAF events with $E_{\gamma 1} + E_{\gamma 2} = E_{e^+} + 2m_e c^2 \approx E_0$ can be identified by means of eq(3.1). However, as θ_{12} is measured correctly only if the positron annihilates close to the CB center, we surrounded the target and silicon detectors by a 5 mm thick lucite housing. Positrons passing through the lucite are no longer troublesome as their kinetic energy E_{e^+} is sufficiently reduced. In the analysis of the data we accepted only γ rays which appeared simultaneously but delayed relative to the beam pulse and the inelastically scattered proton p'(0_2^+), thus using the 0_2^+ half-life to reject any prompt background. Furthermore, by requiring that two and only two CB detectors fired it was ensured that each of the two γ rays deposit their energy in one detector only. On the other hand, for the measurement of the linear polarization of the two γ rays the required condition was that one of the two γ rays fired two neighbouring detector modules.

In Fig.10a the sum-energy spectrum observed for $^{90}Zr(0_2^+)$ is displayed, which was obtained as discussed above without any further conditions on θ_{12} and $E_{\gamma 1}-E_{\gamma 2}$. The spectrum is dominated by the strong line at 1022 keV, due to positron annihilation at rest, and a continuous distribution, due to PAF, which extends up to the energy of 1.76 MeV where the sum energy peak from the double γ decay is expected. Restricting the relative angle and the energy difference to $120° \leq \theta_{12} \leq 180°$ and $|E_{\gamma 1}-E_{\gamma 2}| \leq 500$ keV, respectively, the PAF contribution is strongly suppressed for higher sum energies (fig.10b) and a clean sum peak corresponding to the double γ-decay of $^{90}Zr(0_2^+)$ emerges.

The special conditions employed in fig.10b can be explained with the help of fig.11, which displays the correlation between the energy $E_{\gamma 1}$ of one of the two γ quanta and their relative angle θ_{12} for two different sum-energy windows. Gating on a sum-energy region just below the double-γ peak (cf. window A in fig.10a), the upper matrix (fig.11a) is obtained, which is clearly dominated by PAF events obeying the correlation expressed by eq(3.1). Gating on sum energies around 1.76 MeV (cf. window B in fig.10b), in the resultant matrix (fig.11b) a strong additional yield is observed, which is due to the double-gamma decay of $^{90}Zr(0_2^+)$. The angle and energy conditions employed to obtain fig.10b just selected the region where the double-γ decay is concentrated. The projections of the matrix shown in fig.11b on the energy axis $E_{\gamma 1}$ and the θ_{12} axis are displayed in fig.12a and 12b, respectively, excluding the area contaminated by PAF. Moreover, the directional (θ_{12}) correlation has been corrected for the number of detector combinations contributing to a given θ_{12}, which corresponds to a division by $\sin\theta_{12}$. In addition, fig.12c shows the time distribution of the events attributed to the 2γ decay of $^{90}Zr(0_2^+)$; the observed decay time is in good agreement with the known[16] half-life of the 0_2^+ state of $T_{1/2}$=62 ns.

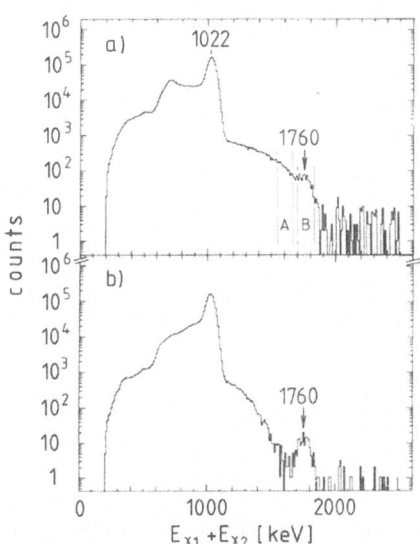

Fig.10 Sum-energy spectrum measured in delayed coincidence with protons populating the 1.76 MeV 0_2^+ level in ^{90}Zr (a) No restrictions on θ_{12} and $E_{\gamma_1}-E_{\gamma_2}$ (b) requiring $120° \leq \theta_{12} \leq 180°$ and $|E_{\gamma_1}-E_{\gamma_2}| \leq 500$ keV.

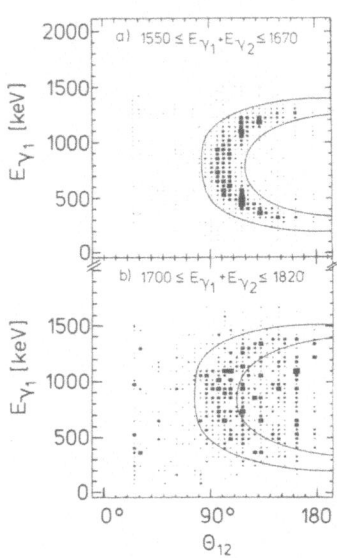

Fig.11   Correlation between the energy E_{γ_1} of one of the two γ
          rays against their relative angle θ_{12}, measured in delayed
          coincidence with protons populating the 1.76 MeV 0_2^+ level
          in ^{90}Zr. (a) Correlation observed by restricting the sum
          energy to 1.55 MeV $\leq E_{\gamma_1} + E_{\gamma_2} \leq$ 1.67 MeV (window A in
          fig.10a). The events are grouped around the kinematical
          curve expected for positron annihilation in flight (PAF).
          (b) Correlation observed for 1.70 MeV $\leq E_{\gamma_1} + E_{\gamma_2} \leq$ 1.82 MeV
          (window B in fig.10a), where the 2γ decay is expected
          to show up. The two curves enclose the area contaminated
          by PAF.

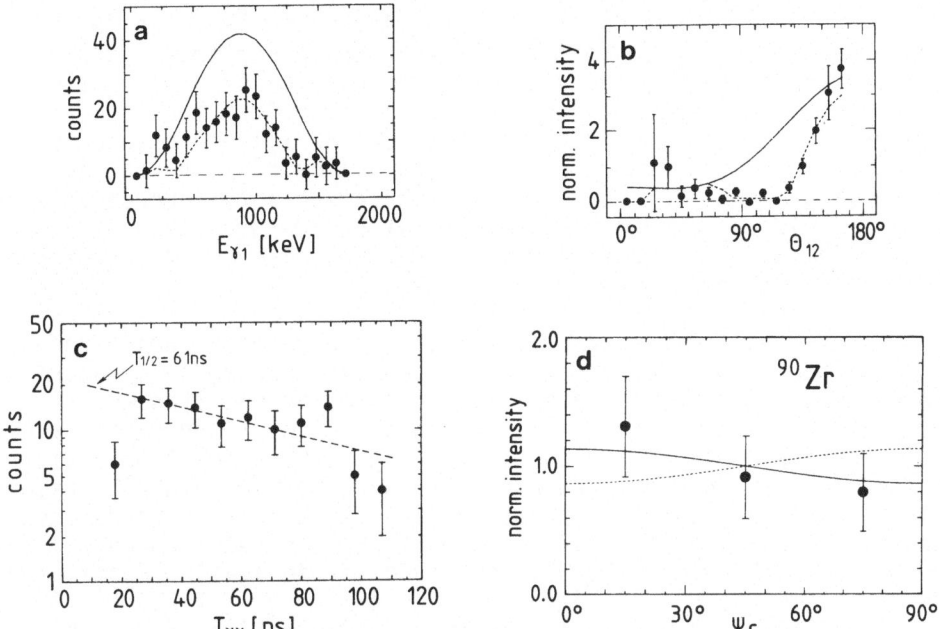

Fig.12   Correlations observed for the 2γ decay of ^{90}Zr(0_2^+).
(a) Projection of the matrix shown in fig.11b on the
energy axis excluding the region polluted by PAF. The
dashed and solid curve are the expected energy distributions
for double dipole decay with and without the PAF cut, respectively. (b) Projection of the matrix shown in fig.11b on
the θ_{12} axis, exclusive of the PAF region and corrected
for the number of detector combinations contributing to
a given θ_{12}. The dashed and solid lines are the calculated
directional γ-γ correlation functions with and without
PAF cut, respectively, after adjustment of the <2E1>/<2M1>
ratio. (c) Time correlation of the 2γ events relative to
the beam pulse. $T_{γγ}$ is the average time derived from the
signals of the two responding detectors. (d) Linear polarization correlation for the 2γ-decay events, excluding in
addition to the regions contaminated by PAF small and large
θ_{12} angles. The two lines display the calculated correlations for the two solutions of <2E1>/<2M1> (solid line:
+1.9; dashed line: $(+1.9)^{-1}$) obtained from the best fit
of the directional γ-γ correlation.

The most surprising result of our investigations of the 2γ decay of ^{90}Zr(0_2^+) and ^{40}Ca(0_2^+) is that the directional γ-γ correlations are found to be strongly asymmetric around $\theta_{12}=90°$. In most of the earlier experimental[14] and theoretical[17] works, double gamma decay between nuclear 0^+ states has been considered to be dominated by the 2E1 process, similar to what has been observed for the 2γ decay between atomic levels, where the E1 character of the two emitted photons has been verified experimentally as well as theoretically[18]. However, the assumption of a dominance of the 2E1 process leads to a γ-γ correlation function of $W(\theta_{12})$ α $1+\cos^2\theta_{12}$ in contrast to the observed asymmetry. Our result requires an interference between γ rays of different parities; indeed, it can be explained by a mixture of 2E1 and 2M1 transitions, which gives rise to an interference term linear in $\cos\theta_{12}$.

A detailed theoretical treatment of the 2γ decay using second order perturbation theory has been given by Grechukhin[17]. Regarding the A^2 term in the non-relativistic interaction Hamiltonian, Eichler[19] has shown that gauge invariance leads to an exact cancellation of its contribution to the 2E1 transition amplitude against part of the second order 2E1 amplitude such that Grechukhins result remains valid without any approximation. The effect of gauge invariance on the 2M1 amplitude is not clear; one can show, however, that the A^2 contribution to the 2M1 amplitude is negligible. Restricting the discussion to $0^+ \rightarrow 0^+$ decays and considering only 2E1 and 2M1 transitions (see also below), the spins and parities of the intermediate levels involved in the second-order decay process are limited to 1^- and 1^+. Moreover, as neither ^{40}Ca nor ^{90}Zr have 1^- or 1^+ states close to the 0_2^+ state, the 2γ decay is expected to proceed mainly through the corresponding high lying giant-resonance states. Including a linear polarization term, the 2γ correlation function is then found to be given by

$$\frac{dW(E_{\gamma_1},E_{\gamma_2},\theta_{12},\Psi_c)}{dE_{\gamma_1}\,d\Omega_{12}\,d\Psi_c} \quad \alpha \quad E_{\gamma_1}^{\;3}\,(E_o-E_{\gamma_1})^3$$

$$\left[1 + \cos^2\theta_{12} - \frac{4<2M1><2E1>}{<2M1>^2+<2E1>^2}\cos\theta_{12} + \right. \tag{3.2}$$

$$\left. + Q_c(E_{\gamma_1})\frac{<2E1>^2-<2M1>^2}{<2M1>^2+<2E1>^2}\sin^2\theta_{12}\cos 2\Psi_c \right]$$

Here Ψ_c denotes the angle between the Compton-scattering plane and the plane containing the two primary quanta (cf. also fig. 4), and the reduced matrix elements $<2\sigma1>$ are given by

$$<2\sigma 1> \equiv <0_1^+ ||\mathfrak{M}(2\sigma 1)||0_2^+>$$

$$\tag{3.3}$$

$$= -2\sum_n \frac{<0_1^+ ||i^{1-\Lambda(\sigma)}\mathfrak{M}(\sigma 1)||n><n||i^{1-\Lambda(\sigma)}\mathfrak{M}(\sigma 1)||0_2^+>}{E_n - E_0/2}$$

with $\Delta(E) = 0$, $\Delta(M) = 1$, where $\mathfrak{M}(\sigma 1)$ are the usual[1] multipole operators and E_n denotes the excitation energy of the intermediate state with spin 1 and parity $(-1)^{1-\Delta(\sigma)}$. Note that the quantities $<2\sigma 1>$ may be visualized as the off-diagonal counter parts to the $S_{-1}(\sigma 1)$ sum rule[20].

To determine the reduced matrix elements $<2\sigma 1>$ the measured γ-γ directional correlation has been fitted in a first step using eq.(3.2), integrating over $d\Psi_c$ and $dE_{\gamma 1}$, but excluding the region polluted by PAF and taking into account the energy dependence of the NaI peak efficiency as well as the granularity of the CB. The fit results in two equivalent solutions for the ratio $<2E1>/<2M1>$, one being the reciprocal of the other. For the measurement on ^{90}Zr (0_2^+), the fit obtained for the directional γ-γ correlation and the corresponding energy distribution, being just proportional to $E_{\gamma 1}^3(E_0 - E_{\gamma 1})^3$ for dipole transitions, are shown by the dashed line in fig.12b and 12a, respectively. To exclude one of the two solutions for $<2E1>/<2M1>$, the linear polarization of one of the two γ rays was measured. The observed linear polarization correlation for ^{90}Zr(0_2^+) is shown in fig.12d together with the two theoretical curves expected according to eq.(3.2) for the two reciprocal values of $<2E1>/<2M1>$. From this measurement and an identical investigation performed for ^{40}Ca(0_2^+) we conclude that the polarization data for ^{40}Ca and ^{90}Zr is consistent with a dominance of $<2M1>$ and $<2E1>$, respectively. Using the observed e^+ rate from the internal pair conversion of the 0_2^+ state to normalize the data, we finally obtain

$$^{40}\text{Ca}(0_2^+): \frac{\Gamma_{\gamma\gamma}}{\Gamma_{tot}} = (4.5\pm 1.0)10^{-4}, \quad \frac{<2E1>}{<2M1>} = +0.43^{+0.17}_{-0.12}$$

$$^{90}\text{Zr}(0_2^+): \frac{\Gamma_{\gamma\gamma}}{\Gamma_{tot}} = (1.8\pm 0.2)10^{-4}, \quad \frac{<2E1>}{<2M1>} = +1.9^{+0.7}_{-0.6}$$

From an analysis of the directional $\gamma\gamma$ correlation including 2E2 transitions, which leads to terms up to $\cos^4\theta_{12}$, an upper limit for the 2E2 contribution to the 2γ decay is obtained, namely

$\Gamma_{\gamma\gamma}(2E2)/[\Gamma_{\gamma\gamma}(2M1)+\Gamma_{\gamma\gamma}(2E1)] < 2\%$ for ^{40}Ca and $< 4\%$ for ^{90}Zr.

To explain first in a very qualitative way why the 2M1 transitions can successfully compete with the 2E1 transitions in $0_2^+ \rightarrow 0_1^+$ decays of spherical nuclei such as ^{40}Ca and ^{90}Zr, let us consider a simplified model where we describe the two 0^+ states as a mixture of two intrinsic shell-model states $|1>$ and $|2>$ i.e. $|0_1^+> = \alpha|1> + \beta|2>$ and $|0_2^+> = \beta|1> - \alpha|2>$. The $<2\sigma1>$ matrix elements are then given by

$$<2\sigma1> = \alpha\beta(<1||\mathfrak{M}(2\sigma1)||1>-<2||\mathfrak{M}(2\sigma1)||2>)$$

$$-(\alpha^2-\beta^2)<2||\mathfrak{M}(2\sigma1)||1> . \tag{3.4}$$

For ^{90}Zr, the case we shall consider here to explain the general argument, $|1>$ can be identified in a first approximation[21] with the state where all neutron and proton single-particle orbits are filled up to and including the $1g_{9/2}$ and $2p_{1/2}$ level, respectively, while in state $|2>$ the $2p_{1/2}$-proton pair is promoted to the $1g_{9/2}$ level, the neutron configuration remaining unchanged (cf. fig.13); moreover, the amplitudes α and β are roughly comparable. For these configurations, the matrix elements $<2||\mathfrak{M}(2\sigma1)||1>$ will be zero as the application of two single-particle dipol operators on state $|1>$ produces states having no overlap with state $|2>$. Moreover, the two 2E1 matrix elements $<1||\mathfrak{M}(2E1)||1>$ and $<2||\mathfrak{M}(2E1)||2>$ will be similar and very nearly cancel each other[22]; in fact, approximating the energies E_n of the intermediate 1^- states by the average energy of the giant dipole resonance, the two terms cancel exactly when harmonic oscillator wave functions are used. Thus the 2E1 transition rate is expected to be strongly hindered. On the other hand, the two 2M1 matrix elements $<1||\mathfrak{M}(2M1)||1>$ and $<2||\mathfrak{M}(2M1)||2>$ show a quite different behaviour. If we first consider the neutron contributions to the two matrix elements, we find $<1||\mathfrak{M}(2M1)||1> = <2||\mathfrak{M}(2M1)||2>$ as the neutron configurations of $|1>$ and $|2>$ are identical, i.e. the neutron contributions cancel each other exactly. In the proton subspace, however $<1||\mathfrak{M}(2M1)||1> = 0$ while $<2||\mathfrak{M}(2M1)||2> \neq 0$. This follows from the property of the M1 operator which allows only for spin-flip transitions between spin-orbit partners[1]. For state $|1>$, where all spin-orbit partners are either completely filled or empty (cf. fig.13), the Pauli principle excludes spin-flip transitions, while for state $|2>$, where the two $2p_{1/2}$ protons are promoted to the $1g_{9/2}$ orbit, the transitions $1g_{9/2} \rightarrow 1g_{7/2} \rightarrow 1g_{9/2}$ and $2p_{3/2} \rightarrow 2p_{1/2} \rightarrow 2p_{3/2}$ are allowed, leading to contributions to $<2||\mathfrak{M}(2M1)||2>$ being of typical single particle strength. In view of the strong hindrance of the 2E1 transition one thus has to expect, in accordance with our experimental findings, that the 2M1 transitions contribute significantly to the $0_2^+ \rightarrow 0_1^+$ 2γ decay in ^{90}Zr. Following the

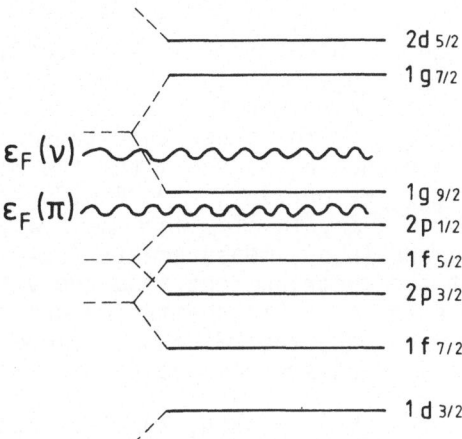

Fig.13   Single-particle states within the shell model together
          with the Fermi energies for protons [$\varepsilon_F(\pi)$] and neutrons
          [$\varepsilon_F(\nu)$] relevant for ^{90}Zr.

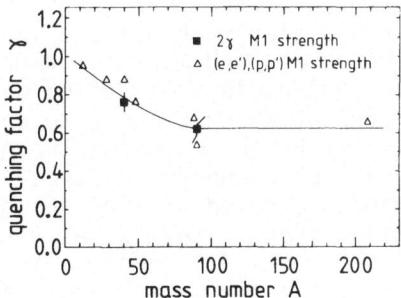

Fig.14   M1-quenching factors γ deduced from the measured 2M1 matrix
          elements in comparison with results deduced from (e,e')
          and (p,p') experiments (cf. ref.25).

general arguments given above, the same conclusion is obtained for the 2γ decay of $^{40}Ca(0_2^+)$.

For a quantitative assessment of the 2M1 transition matrix element we used calculated wave functions for the two 0^+ states in ^{40}Ca, employing an advanced shell-model code[23], and published[24] wave functions for ^{90}Zr together with M1 operators for bare nucleons and average experimental[25] M1 giant resonance energies. The resultant matrix elements <2M1>$_{theo}$ are larger than those observed experimentally; defining a quenching factor γ for the M1 strength by γ^2 = <2M1>$_{exp}$/<2M1>$_{theo}$ we obtain $\gamma = 0.76\pm0.05$ for ^{40}Ca and $\gamma = 0.67 \pm 0.09$ for ^{90}Zr, in nice agreement with those values deduced from (e,e') and (p,p') measurements (see ref.25 and fig.14). Note, however, that the contributions from the various 1^+ states to the 2M1 decay and to inelastic scattering are different. Because of this different weighting, the 2M1 matrix elements may thus help to differentiate the mechanisms responsible for the M1-quenching[26].

In contrast to the 2M1 strength, the 2E1 transition matrix elements are much more difficult to calculate in a quantitative way because of the strong cancellation effects. Within the two state model, used also above in our qualitative discussion, Bertsch[22] obtained estimates for the <2E1> matrix elements, which are of the right order of magnitude when compared with our experimental results. However, the relative signs of the <2E1> to the <2M1> matrix elements do not agree with experiment, indicating that further contributions to <2E1> exist which have to be taken into account.

4. HIGH-ENERGETIC γ-RAYS FOLLOWING (HI,xn) REACTIONS

In heavy-ion induced reactions compound nuclei can be formed at high excitation energies E^* and high spins I^*. For compound nuclei in the rare-earth region with not too large spins ($I^* < 70$), the subsequent decay proceeds predominantly along the following path[1] (cf. fig.15): The hot nucleus is cooled down in a first step by emitting neutrons each of which removes an energy of $B_n+\bar{e}_n \approx$ 10-12 MeV from the system, B_n and $\bar{e}_n$ being the neutron separation energy and the average kinetic energy, respectively, while the amount of angular momentum carried away by the neutron is small (~ 1 ħ). This evaporation process terminates when the remaining excitation energy E_i in the residual nucleus is too small to allow for a further emission of a neutron with small angular momentum. This limit is reached for excitation energies approximately given by $E_i(I_i) \lesssim E_y(I_i) + B_n$, where $E_y(I_i)$ is the energy of the yrast state at spin I_i, the value of I_i being close to I^* of the compound nucleus. The functional dependence of E_i on I_i is usually called the entry line, which is expected according to the above estimate to run roughly parallel to the yrast line. The number of neutrons emitted is thus given by $x \approx (E^*-E_y(I_i)-B_n)/(B_n+\bar{e}_n)$ and as $E_y(I_i)$

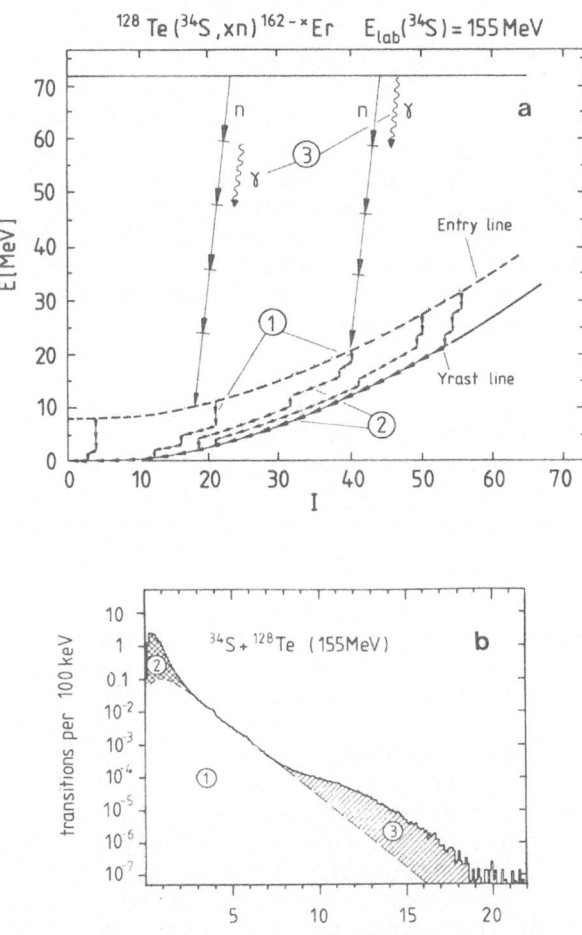

Fig. 15 (a) Decay paths of highly excited Er compound nuclei in the excitation energy vs. spin plane. (b) Gamma spectrum observed during the bombardment of ^{128}Te with 155 MeV ^{34}S. The three components contributing to the spectrum are indicated schematically.

is increasing with spin, this leads to the well-known fractionation
of the xn cross-section in the E_i vs. I_i plane exemplified e.g.
by fig.7. After the particle evaporation has terminated, the further
decay proceeds via the emission of γ rays. The resulting γ cascade
basically consists of two components: (i) Statistical transitions
towards the yrast line which mainly remove excitation energy but
not spin from the evaporation residue. These γ rays give rise to
an exponentially decreasing component in the γ-ray spectrum (labelled
by 1 in fig.15) containing an average number of ∿ 4 γ rays with
average energies of ∿ 2 MeV. (ii) Collective, stretched (ΔI=2) E2
transitions along or parallel to the yrast line removing the angular
momentum from the system. This component (labelled by 2 in fig.15)
contributes mainly to the low energetic part of the γ spectrum
(E_γ < 2 MeV) and is responsible for most of the γ intensity observed.
According to this picture, the total number of γ rays M_γ emitted
during the de-excitation of the evaporation residue can be approximately connected with the entry spin I_i by

$$I_i \approx 2 (M_\gamma - k) + I_0 \qquad (4.1)$$

where I_0 denotes the spin at which the γ cascade terminates (e.g.
the spin of the ground or an isomeric state of the residue) and k
is a number correcting for the fact that the statistical γ rays
remove only little angular momentum; k is expected to be of the
order of 3 in rare earth nuclei, in agreement with experimental
findings (see e.g. ref.2).

In 1981 Newton et al.[27] observed a third component in the γ
spectra following heavy-ion induced fusion reactions, leading to
an excess of high energetic γ rays with E_γ > 10 MeV (cf. the hatched
area labelled by 3 in fig.15); as the most likely explanation of
this enhancement they suggested, that with a probability of ∿ 10^{-3}
one of the neutrons in the early stage of the cooling-down process
is replaced by a γ transition of high energy arising from the decay
of the isovector giant dipole resonance (GDR), which is expected[28]
to be built on every excited state of a nucleus and to be largely
independent of the microscopic structure of the state. In fact,
an extrapolation of the hydrodynamical model, which accounts for
the shape of the GDR strength functions built on the ground and
low-lying states of heavy nuclei in terms of their deformation,
suggests that the study of these high energetic γ rays might yield
information on the average deformation of hot and rapidly rotating
nuclei.
Motivated by these exciting possibilities, several groups
started to perform more detailed experimental and theoretical
investigations and the reader is referred to the lecture given by
Jens Jørgen Gaardhøje at this school[29] for a summary of the experimental and theoretical work done so far. The following discussion
will be restricted to some specific aspects of the work performed
in this context at the CB. Again this instrument seems to be well

suited to takle the experimental problems connected with these
measurements. In particular, it allows to study in detail the spin
dependence and the directional angular correlation of these high-
energetic γ transitions.

In the experiments discussed below, Er compound nuclei were
produced using a pulsed beam of (a) 132 MeV ^{34}S on a ^{128}Te target.
(b) 50 MeV α particles on a ^{160}Dy target, and (c) 158 MeV ^{34}S on
a ^{130}Te target. Allowing for the energy loss of the beam in the
covering foils of the Te targets and in half of the target material,
in reaction (a) and (b) ^{164}Er compound nuclei are produced at the
same excitation energy of E* = 48 MeV, while via reaction (c) compound nuclei of ^{162}Er at E* = 71 MeV are formed with angular momenta
reaching up to I* $\approx$ 65. The dominant reaction channels are the 4n
channel in (a) and (b) and the 5n channel in (c). The bulk of the
data was taken requiring the energy signal of at least one of the
individual CB modules to surmount the leading edge discriminator
threshold adjusted to 6 MeV. For normalization purposes, however,
data were also taken without requiring the high energy-trigger.
Each measurement was accompanied by calibration runs using radioactive sources and the ^{12}C (p,p') ^{12}C* reaction at E_p = 20.5 MeV,
detecting the backscattered protons in two Si detectors at 140°
with respect to the beam.

In the analysis of the data care was taken (i) to obtain good
lineshape, efficiency, and energy calibrations, (ii) to suppress
background due to reactions on light target impurities such as C
and O, (iii) to eliminate events due to cosmic rays, and (iv) to
discriminate between gamma and neutron pulses and to suppress high
energy events caused by the summing of a prompt γ pulse with a pulse
produced by a slow neutron hitting the same detector module.

Lineshape, efficiency and energy calibration: Fig.16 displays
the experimental lineshapes observed for 4.4 MeV and 15.11 MeV γ-
rays by gating on the inelastically scattered protons leading to
the 4.4 and 15.11 MeV state of ^{12}C, respectively. The lineshapes
shown were obtained by incrementing only the largest energy signal
detected by an individual CB module, however, for pulse heights
> 2 MeV and > 3 MeV, respectively, these lineshapes are identical
to those observed in the normal mode (see sect.2) actually used
to analyse the data from the (HI,xn) reactions. The lineshapes were
normalized to the number of inelastically scattered protons detected
in the Si counters such that the integral over the lineshape yields
the total detection efficiency ϵ_{CB}. For γ-ray energies above 4 MeV
ϵ_{CB} was found to be rather constant with ϵ_{CB} = 0.95 ±0.03. Although
the lineshapes can be considerably improved by requiring Anti-Compton
conditions similar to those discussed in sect.2, the normal mode was
used in analysing the (HI,xn) data to avoid any multiplicity dependent distortions of the detection efficiency. Moreover, the resolution
obtained in the normal mode seems to be sufficient for the present

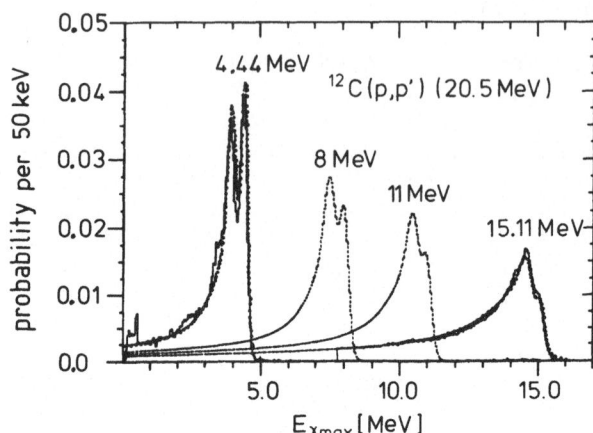

Fig. 16   Experimental γ lineshapes (solid curves) observed in coincidence with protons inelastically scattered from a ^{12}C target
(see also text). The dotted curves represent the calculated
lineshapes using an analytical expression.

purpose; for 15.11 MeV γ rays the width of the peak, centered approximately around (15.11-0.51) MeV, amounts to 11%. The experimental
lineshapes (4.44, 12.7 and 15.11 MeV) were described by an analytical expression, the parameters of which were interpolated to yield
the γ lineshapes for other energies (cf. fig.16). The energy calibrations for the individual CB modules were performed using the
positions of the full energy peaks of all calibration lines. The
accuracies of these calibration curves at energies above the highest
calibration point of 15.11 MeV were checked by means of a light
pulser; at $E_\gamma \approx 20$ MeV it was found to be $\lesssim 2\%$.

Background reactions: In the ^{34}S experiment performed at the
low bombarding energy, but also for the measurements carried out
at higher energies, background reactions on small amounts of light
target impurities (C,O) were found to considerably disturb the high
energy part of the γ-ray spectra, in particular when selecting low

multiplicity events ($N_\gamma \lesssim 15$). Using auxiliary measurements performed by bombarding ^{12}C targets with ^{34}S ions, a method has been developed which allows to suppress quantitatively the perturbing background due to reactions on light target contaminations. The method exploits the different kinematics of the neutrons emitted from light and heavy reaction products; both types of events are well separated in a plane spanned by the number N_γ of prompt γ rays and the number N_{del} of delayed neutron pulses if the time windows used for defining N_{del} are carefully selected (see ref.30 for more details).

Cosmic-ray background: Requiring prompt coincidences with the beam pulse, contributions from cosmic ray events distorting in particular the high energy part of the γ spectra were suppressed by a factor of ≈ 20. Moreover, as the CB is rather effective in detecting a cosmic-ray shower, an additional suppression of more than a factor of 10 could be obtained by requiring that none of the individual CB detectors responded with a pulse height in excess of 25 MeV. Finally, by limiting the total energy detected by the CB to values smaller than the excitation energy E^* of the compound nucleus, the remaining background due to cosmic ray events is found to be negligibly small (cf. fig.17).

Neutron discrimination: To discriminate γ rays from neutrons the time-of-flight technique described in sect. 2 has been used. Moreover, to recognize summing events due to the detection of a prompt γ ray and a slow neutron in the same detector module, a rise-time analysis has been performed as discussed in sect.2. The γ-n summing contribution reaches a maximum ($\sim 25\%$) around pulse-heights corresponding to $E_\gamma \sim 7$ MeV (the γ energy released in an (n,γ) reaction on ^{128}I is 6.8 MeV) but is found to be negligibly small for energies above $E_\gamma \approx 10$ Mev.

The γ-ray spectra observed with the reaction (c) - taking care of the precautions discussed above and correcting for Doppler-shift effects - are displayed in fig.17 as a function of the number of prompt responding detectors N_γ. The spectra are not corrected for the γ-ray lineshapes and for summing effects due to multiple γ hits. For a given N_γ window the high energy parts of the spectra were normalized to the corresponding total γ spectra measured without high-energy threshold, which in turn were normalized to the corresponding average number $\langle N_\gamma \rangle$; the scale (transitions per 100 keV) applies to the spectral regions with $E_\gamma > 3$ MeV only as for smaller energies the spectra are distorted by the cross-talk between the detectors. The average numbers $\langle N_\gamma \rangle$ were converted to yield the average spin values of the entry states by using the calibration curves shown in fig.6 together with eq.(4.1) using k=3 and I_0=0. As discussed above, I_t is also an approximate measure of the spin I^* of the compound nucleus.

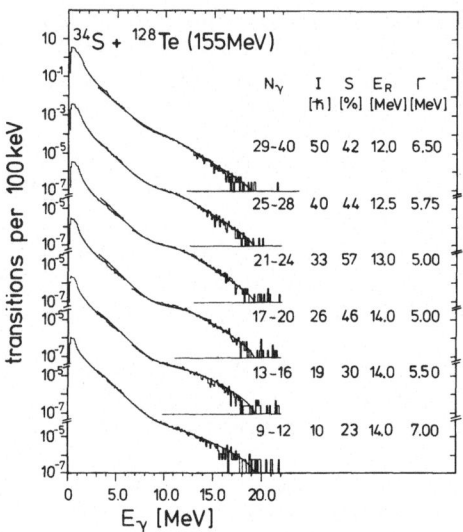

Fig. 17   Gamma-ray spectra observed in the bombardment of ^{128}Te with 155 MeV ^{34}S ions as a function of the number of responding CB detectors N_γ, i.e. of the spin I of the entry states. The solid curves were calculated as described in the main text, using a single Lorentzian to represent the giant dipole resonance strength function and adjusting the strength (S), the centroid position (E_R) and the width (Γ_R) as to obtain the best fit to the high energy portion of the spectra.

The high energy portions of the γ spectra, which are clearly dominated for $E_\gamma > 9$ MeV by the additional component (3), have been fitted with the results obtained from calculations using the CASCADE[31] code, which treats the decay of a compound nucleus within the statistical model. The calculations were performed assuming an E1 giant dipole resonance strength function of the form[33]

$$f(E_\gamma) \equiv \Gamma_\gamma^I \rho_I(E) E_\gamma^{-3} =$$

$$= 3.32 \cdot 10^{-6} (1+0.8\chi) \frac{NZ}{A} \cdot S \frac{\Gamma_R E_\gamma}{(E_\gamma^2 - E_R^2)^2 + E_\gamma^2 \Gamma_R^2} \qquad (4.2)$$

where Γ_γ^I is the E1 radiative width between individual levels for a decay $I \rightarrow I \pm \Delta I$ with $\Delta I = 0, \pm 1$, $\rho_I(E)$ is the density of initial states, E_R and Γ_R are the energy and width of the GDR, and S represents the GDR strength in units of the classical dipole sum-rule corrected for exchange force contributions ($\chi \approx 0.4$ for $A \approx 160$, ref.34). As discussed further below, the assumption of a pure E1 character for the high energetic γ rays is consistent with the observed γ-angular distributions. The spectra, calculated as a function of I*, were folded with the γ lineshapes and corrected for summing effects caused by possible double hits of the individual detectors by two γ rays from the same event. For the measurement shown in fig.17 the calculated spectra obtained by adjusting S, E_R and Γ_R as to achieve the best fit are shown by the solid curves, assuming a single Lorentzian (4.2) to describe the $\Delta I = 0, \pm 1$ transitions. Note, however, that the energy spectra are not sensitive to a possible splitting of the GDR transitions into e.g. two components because of the limited statistics.

A compilation of the results for S, E_R and Γ_R deduced from the present three experiments is given in fig.18. The most surprising result is the spin dependence observed for the centroid energy E_R; while at small spins E_R (and Γ_R) agrees with the average position (and width) of the known GDR build on the ground state of stable Er nuclei, E_R is found to decrease steadily ($dE_R/dI \approx -0.05$ MeV $\hbar^{-1}$) with increasing spin, reaching a value of ≈ 12 MeV at $I \approx 50$. This feature of the data is stable against reasonable variations of other input parameters to the CASCADE calculation (e.g. changing the level density parameter $a = A/9$ MeV^{-1} by ±15%, assuming a spin-independent width Γ_R etc.), and it is at least consistent with the observations of other groups for spin-averaged values of the position of the GDR energy: Haas et al.[35] reported a measurement in ^{144}Gd, using the 10^+ isomer to select the reaction channel, which resulted in $\bar{E}_R \approx 12.2$ MeV for $I > 30$ instead of 14.9 MeV expected from the systematic of the ground state GDR positions. The measurement of Gaardhøje et al.[36] on ^{166}Er compound nuclei results for $\langle I \rangle \approx 20$ in $\bar{E}_R = 14.9$ MeV using a single Lorentzian and in $E_R \approx$ 14.1 MeV if two Lorentzians are used, which has to be compared to

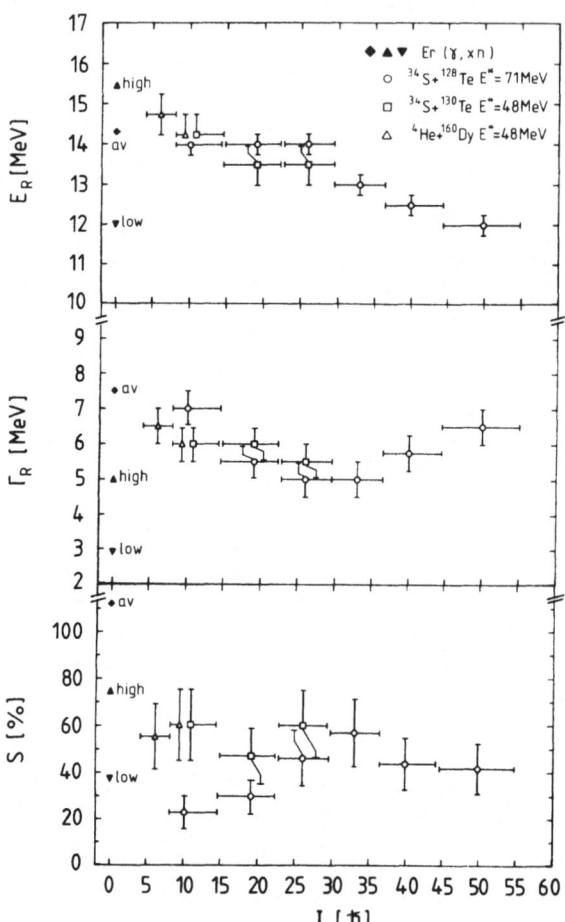

Fig. 18   Centroid position E_R, width Γ_R and strength S (in units
          of the E1 sum-rule) of the GDR strength function of Er
          nuclei as a function of the entry state spin I. Open sym-
          bols: Present experiments. Closed symbols: GDR from
          Er(γ,xn) (ref. 32) using one (av.) or two (high,low)
          Lorentzians to describe the shape.

E_R = 13.8±0.5 read off from fig.18. It should be noted, however, that - based on the theoretical arguments given so far (see e.g. ref.37,38) - one does not expect any sizable shift of the centroid energy of the GDR with increasing spin. Further noticable results of our investigation are that within the statistical accuracy we do not observe a dependence of the GDR parameters on the temperature of the compound nuclei or a dependence on the entrance channel, the latter result being expected in case the γ emission results from decays of equilibrated compound nuclei. The average strength of the GDR is found to be only ≈ 0.5 of the observed strength of the GDR built on the ground state of stable Er isotopes, in approximate agreement with results obtained by other groups (see e.g. ref. 36). Note, however, that S is usually subject to rather large systematic uncertainties caused by the normalization procedure (≈ ±0.2 in our case), that S is measured only relative to the total neutron decay width, and that eq.(4.2) assumes that the GDR decays only statistically. The average width Γ_R of the GDR is found to be around 6 ± 1 MeV as compared to the width of 7.5 MeV observed for the GDR built on the ground state.

Fig.19 displays the measured angular correlation coefficients A_2 and A_4 as a function of E_γ, selecting events with $16 \leq N_\gamma \leq 40$. The coefficients were obtained from the angular distributions measured relative to the beam - transformed into the rest-frame of the γ-decaying nucleus - using $W(\theta_\gamma) \propto 1 + A_2Q_2P_2(\cos \theta_\gamma) + A_4Q_4P_4(\cos \theta_\gamma)$, ($Q_k$ being the geometrical attenuation coefficient) and restricting θ_γ to $80° \leq \theta_\gamma \leq 180°$ as to be on the safe side regarding any conceivable problems with unrecognized neutron events. The angular correlation coefficients deduced at low energies are consistent with a dominance of stretched (ΔI=2) E2 transitions, expected due to the dominance of the component (2) in this energy region (c.f. fig.15), while for energies around 4-5 MeV, governed by component (1), isotropy is observed as expected for a statistical mixture of stretched and unstretched transitions. At energies above 7 MeV the data is consistent with A_4=0 in agreement with the assumed E1 character of these γ-rays. However, A_2 is clearly observed to be negative for E_γ < 13 MeV, which calls for a dominance of stretched ΔI = ±1 transitions in this energy region, while A_2 values around zero are found for energies E_γ ≈ 13-14 MeV. These results are corroborated by the γ-angular correlations measured relative to the spin direction (c.f. sect.2 and fig.8) although the statistical accuracy obtained in this case for the correlation coefficients is somewhat inferior because of the selection criteria which have to be employed in order to get reasonable resolutions for $\vec{I}_i$.

The clear anisotropy of the γ rays in the energy region $5 < E_\gamma \lesssim 13$ MeV and the isotropy observed around $E_\gamma \approx 13$-14 MeV is an exciting result. Within a consistent interpretation of the high-energy component (3) as being due to the E1 decay of the giant dipole resonances these findings imply that the resonance is energet-

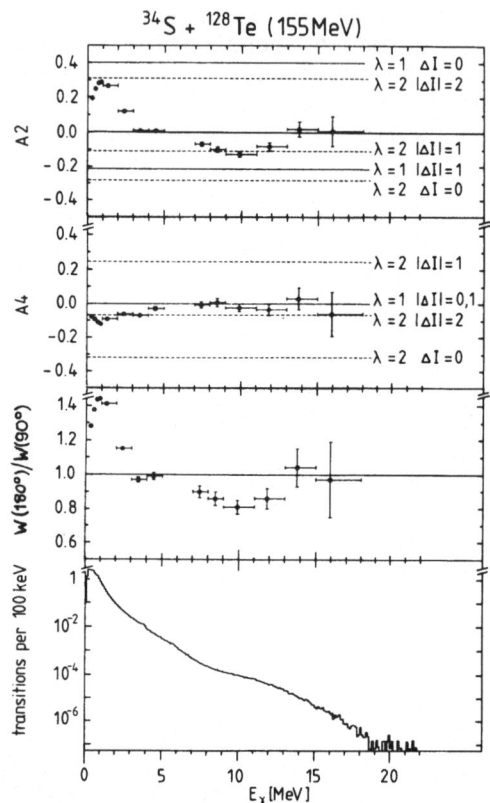

Fig. 19 Angular correlation coefficients and 180° - 90° anisotropies
         as a function of γ-ray energy observed in the ^{128}Te + ^{34}S
         (155 MeV) experiment. The values were deduced from the
         directional γ-angular distributions measured relative to
         the beam direction restricting N_γ to $16 \leq N_\gamma \leq 40$. Also
         shown are the expected correlation coefficients for pure
         transitions assuming a dealignment factor of $\beta_2 = 0.8$
         and $\beta_4 = 0.5$, respectively.

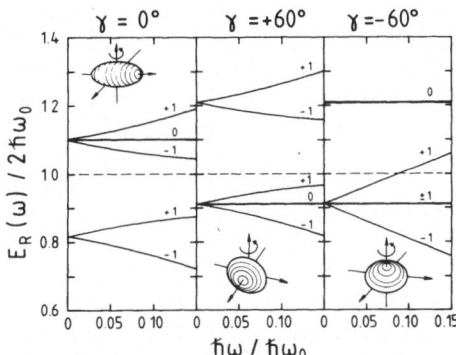

Fig. 20   Giant dipole resonance energies $E_R(\omega)$ as a function of
the rotational frequency $\hbar\omega$ assuming a β deformation of
0.3 and three different γ deformations as indicated
(Er: $\hbar\omega_0 \approx 7.2$ MeV; $\hbar\omega/\hbar\omega_0 \approx 0.055$ for I = 30). The states
labelled with 0,+1,-1 lead to $\Delta I = 0,+1,-1$ E1 transitions.

ically splitted with the centroid energy of the $\Delta I=0$ component being
shifted to higher energies as compared to the centroid energy of
the $\Delta I = \pm 1$ components. Such an energy splitting is expected if
the states the GDR is built on are rotational states which have
either a prolate deformation with a rotational angular momentum
perpendicular to the symmetry axis or an oblate deformation with
the "rotational angular momentum" along the symmetry axis. This
is schematically shown in fig.20, which displays the centroid energies of the five main components of the giant dipole resonance in
rotating nuclei for three specific deformations; they were calculated within the framework of a cranked deformed oscillator model
in which dipole-dipole two-body forces are taken into account[37,39,
29]. At rotational frequencies $\hbar\omega = 0$ the splitting of the GDR is
determined by the deformation only. In fast rotating nuclei, the
two fundamental modes corresponding to vibrations along the two
axes perpendicular to the rotational axis are further splitted,
giving rise to $\Delta I = \pm 1$ excitations, while the vibration along the
rotational axis, giving rise to the $\Delta I = 0$ excitation, is not

effected. Note, however, that for realistic rotational frequencies the rotational splitting is still small compared to the splitting caused by the deformation; even more, the difference between the centroids of the $\Delta I = \pm 1$ and the $\Delta I = 0$ components are still only determined by the deformation. Although one expects some modifications of this simplified picture e.g. due to fluctuations of the spin direction with respect to the principle axes of the deformation elipsoid, our data show that the anisotropy is still large enough to yield information on the deformation of hot, rotating nuclei. A preliminary analysis of the angular distribution measurements shown in fig.19 (^{162}Er,E* = 71 MeV, 20 $\lesssim$ I $\lesssim$ 65) results in $\beta \approx 0.2$ for $\gamma = 0°$ or $\beta \approx 0.1$ for $\gamma = -60°$. Obviously, a more precise angular distribution measurement in the energy region 14 MeV $\lesssim E_\gamma \lesssim$ 18 MeV is needed, in particular to distinguish between these two limiting cases; corresponding measurements are presently under way.

Combining our results with those obtained by other groups e.g. on the energy distributions[29,36] one might indeed claim that we are on the threshold to a γ-ray spectroscopy of states far-off the yrast line.

5. OUTLOOK

Although we have discussed only two types of experiments presently performed at the Heidelberg-Darmstadt Crystal Ball, we hope we were able to demonstrate the versability of this unique instrument for nuclear γ-ray spectroscopy. It seems to us that the nuclear physics community has taken with the crystal balls and the multi-Ge arrays a decisive experimental step towards a more detailed and deeper understanding of the atomic nucleus.

REFERENCES

1. D. Pelte and D. Schwalm, In-Beam Gamma-Ray Spectroscopy with Heavy Ions, in: "Heavy Ion Collisions", R. Bock, ed., North-Holland Publ. Comp., Amsterdam (1982), 3:3
2. D. L. Hillis, J.D. Garrett, O. Christensen, B. Fernandez, G.B. Hagemann, B. Herskind, B. B. Back, and F. Folkmann, Nucl. Phys. A325:216 (1979)
3. P. O. Tjøm, I. Espe, G. B. Hagemann, B. Herskind, and D. L. Hillis, Phys. Lett. 72B:439 (1978)
4. see e.g. P. J. Twin, P. J. Nolan, R. Aryaeinejad, D.J.G. Love, A. H. Nelson, and A. Kirwan, Nucl. Phys. A409:343c (1983)
5. R. M. Lieder, P. Kleinheinz, K. H. Maier, P. v. Brentano, J. Eberth, and H. Hübel, "OSIRIS", KFA-Jülich, Report (1983)
6. F. S. Stephens, these proceedings

7. M. Jääskeläinen, D. G. Sarantites, R. Woodward, F. A. Dilmanian, J. T. Hood, R. Jääskeläinen, D. C. Hensley, M. L. Halbert, and J. H. Barker, NIM 204:385 (1983)

8. D. Habs, F. S. Stephens, and R. M. Diamond, LBL Report PUB-5020 (1979)
   V. Metag, D. Habs, K. Helmer, U.v. Helmolt, H. W. Heyng, B. Kolb, D. Pelte, D. Schwalm, W. Hennerici, H. J. Hennrich, G. Himmele, E. Jaeschke, R. Repnow, W. Wahl, R. S. Simon, R. Albrecht, The Darmstadt-Heidelberg Crystal Ball, in: "Detectors in Heavy Ion Reactions", W. v. Oertzen, ed., Springer, New York (1983)

9. R. S. Simon, J. de Physique C10:281 (1980)

10. J. Burde, E.L. Dines, R. M. Diamond, J. E. Draper, K. H. Lindenberger, C. Schück, and F. S. Stephens, Phys. Rev. Lett. 48:530 (1982)

11. R. Schmidt-Fabian, P. Glässel, D. v. Harrach, MPI-Annual Report, Heidelberg (1981), 42

12. R. D. Fischer, D. Habs, R. Kroth, W. Kühn, V. Metag, R. Mühlhans, R. Novotny, R. Repnow, R. S. Simon, D. Schwalm, and H. Ströher, MPI-Annual Report, Heidelberg (1983), 29

13. W. Bonin, M. Dahlinger, D. Habs, B. Schwartz, and R. S. Simon, MPI-Annual Report, Heidelberg (1983), 31

14. S. Gorodetzky, D. Sutter, R. Armbruster, P. Chevallier, P. Menrath, F. Scheibling and J. Yoccoz, Phys. Rev. Lett. 7:170 (1961)
    G. Sutter, Ann. de Phys. 8:323 (1963)
    E. Beardsworth, R. Hensler, J.W. Tape, N. Benczer-Koller, W. Darcey and J.R. MacDonald, Phys. Rev. C8:216 (1973)
    Y. Asano and C.S. Wu, Nucl. Phys. A125:557 (1973)
    B.A.Watson, T. T. Bardin, J. A. Becker and T. R. Fisher, Phys. Rev. Lett. 35:1333 (1975)

15. J. Schirmer, D. Habs, R. Kroth, N. Kwong, D. Schwalm, M. Zirnbauer, and C. Broude, Phys. Rev. Lett. 53:1897 (1984)

16. C. M. Lederer and V. S. Shirley, "Table of Isotopes VII", Wiley, New York (1978)

17. D.P. Grechukhin, Nucl. Phys. 35:98 (1962)
    D.P. Grechukhin, Nucl. Phys. 47:273 (1963)
    D.P. Grechukhin, Nucl. Phys. 62:273 (1965)

18. R. Marrus and P. J. Mohr, Forbidden Transitions in one- and two-electron atoms, in: "Advances in Atomic and Molecular Physics", D. R. Bates, ed., Academic Press, New York (1978) 14:181
    S. P. Goldman and G. W. F. Drake, Phys. Rev. A24:183 (1981)

19. J. Eichler, Phys. Rev. A9:1762 (1974)

20. P. Ring and P. Schuck, "The Nuclear Many-Body Problem", Springer, New York (1980)

21. I. Talmi and I. Unna, Nucl. Phys. 19:225 (1960)

22. G. F. Bertsch, Part. Nucl. 4:237 (1972)

23. The Oxford Shell Model Code, W.D.M. Rae (1982), unpublished

24. W. J. Courtney and H. T. Fortune, Phys. Lett. 41B:4 (1972)

25. A. Richter, Phys. Scripta T5:63 (1983)

26. H. Morinaga and T. Yamasaki, "In-Beam Gamma-Ray Spectroscopy",
    North-Holland Publ. Comp., Amsterdam (1976)
27. O. Newton, B. Herskind, R.M. Diamond, E.L. Dines, J. E. Draper,
    K. H. Lindenberger, C. Schück, S. Shih, and F. S. Stephens,
    Phys. Rev. Lett. 46:1383 (1981)
28. D. M. Brink, Ph.D. Thesis, Univ. of Oxford (1955)
29. J. J. Gaardhøje, these proceedings
30. W. Hennerici, Ph.D. Thesis, Univ. of Heidelberg (1984)
31. F. Pühlhofer, Nucl. Phys. A280:267 (1977)
32. R. Bergere, H. Beil, P. Carlos, A. Veyssiere, Nucl. Phys. A133:417
    (1969)
33. G. A. Bartholemew, E. D. Earle, A. J. Ferguson, J. W. Knowles,
    and M. A. Lone, Adv. in Nucl. Phys, 7:229 (1973)
34. B. L. Berman, S. C. Fultz, Rev. Mod. Phys. 47:713 (1975)
35. B. Haas, D. C. Radford, F. A. Beck, T. Byrski, C. Gehringer,
    J. C. Merdinger, A. Nourredine, Y. Schutz, and J. P. Vivien,
    Phys. Lett. 120B:79 (1983)
36. J. J. Gaardhøje, C. Ellegaard, B. Herskind, and S. G. Steadman,
    Phys. Rev. Lett. 53:148 (1984)
37. K. Neergaard, Phys. Lett. 110B:7 (1982)
38. J. L. Egido, H. J. Mang and P. Ring, Nucl. Phys. A339:390 (1980)
    J. L. Egido, and P. Ring, Nucl. Phys. A338:19 (1982) and Phys.
    Rev. C25:3339 (1982)
    M. Faber, J. L. Egido and P. Ring, Phys. Lett. 127B:5 (1983)
    P. Ring, L. M. Robledo, J. L. Egido and M. Faber (to be published)
39. R. R. Hilton, Z. f. Phys. A309:233 (1983).

HIGH SPIN PHYSICS

F. S. Stephens

Nuclear Science Division, Lawrence Berkeley Laboratory
University of California
Berkeley, California   94720

Nuclei can be studied from their ground states ($\sim 0\hbar$) up to angular momenta of order $100\hbar$, where they are literally pulled apart by centrifugal effects. This range of angular momenta can be viewed as resulting from "cranking" the nucleus around a "rotation" axis, where the critical variable is the cranking velocity. The calculated response of nuclei to such an imposed angular velocity corresponds well with recent observations, and includes a rich and varied interplay of collective and single-particle phenomena.

This work was supported by the Director, Office of Energy Research, Division of Nuclear Physics of the Office of High Energy and Nuclear Physics of the U.S. Department of Energy under Contract DE-AC03-76SF00098.

The information on high-spin states comes mostly from studies of γ rays; thus there is a strong incentive to develop more powerful γ-ray detection systems. The high-spin field is currently giving rise to a new generation of such data acquisition systems. This will be the subject of Section 3 of this report. Fortunately, the theoretical developments have kept pace with the experimental. Single-particle motion can be calculated in deformed potentials of several types, and, further, these can be cranked about various axes to simulate the rotation. Such calculations, though still approximate, can be an excellent guide. Virtually all the properties now measured can also be calculated, and comparison provides a stimulus both to interpreting the experimental data and to improving the calculations. In the lower-spin region some rather detailed comparisons will be made in Section 1; whereas Section 2 will discuss the less detailed information on the very highest spin states.

1. The New Spectroscopy

Nuclei can generate high angular momentum either by alignment along a common axis of the angular momentum of several individual nucleons or by a collective rotation of the nucleus as a whole. Recent developments in this field have been centered on understanding the competition of these two modes. This can be illustrated in fig. 1, where level schemes of ^{158}Er and ^{147}Gd are shown[1,2]). The ^{158}Er scheme is quite regular and the dominant behavior is collective rotation of a prolate-deformed nucleus as is illustrated at the left of fig. 1. The ^{147}Gd scheme is quite irregular, with complicated decay pathways and isomeric states (dark levels). Its dominant behavior is certainly single-particle alignment, as is illustrated at the right of fig. 1. Yet both of these schemes contain elements of the other type of behavior. There are irregularities in the ^{158}Er

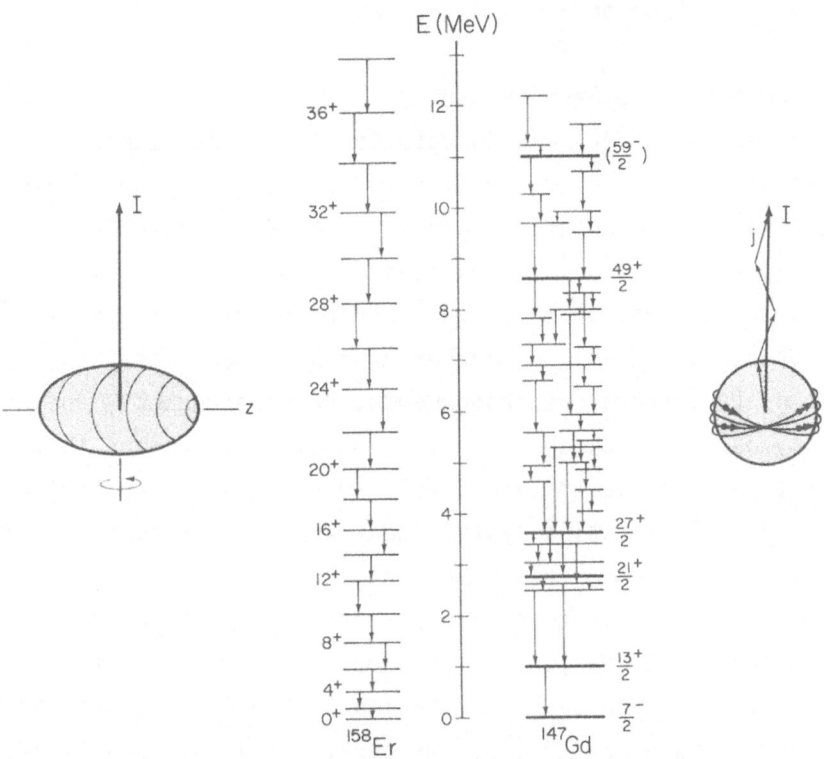

Fig. 1. Level scheme for ^{158}Er and ^{147}Gd, together with illustrations of the dominant source of angular momentum for each case.

rotational pattern at spins around 16 and 26, which correspond to
single-particle alignments, and the 49/2+ isomer at 8.6 MeV in
^{147}Gd has a quadrupole moment that suggests that the aligned
particles are polarizing the core so a collective oblate shape is
developing. In this talk I will discuss our present understanding of the interplay of these two types of behavior. I will
do so starting from good rotational nuclei.

1.1. Alignment in collective nuclei

The lack of smoothness in rotational spectra came as a
surprise in 1971 when a discontinuity was found in the energies
of the ground-state rotatonal bands of several rare-earth
nuclei[3]). As the nucleus de-excites from a high initial spin,
the regular increase in rotational period (slowing down) is
interrupted occasionally by rather sizeable decreases. These
correspond to internal rearrangements, "nuclearquakes", and are
generally called "backbends". It is amusing to compare them with
another type of quake—"starquakes". Neutron stars or "pulsars"
are also rapidly rotating systems that are slowing down. Occasionally they too have sudden internal rearrangements that
decrease the moment of inertia and therefore speed up the rotation (called "glitches"). An earthquake is a similar phenomenon,
but the change in the earth's rotation from even the largest
earthquake is much too small to measure. It is quite common that
rapidly rotating objects modify their internal structure to produce larger moments of inertia, and these modifications revert
back, often in sudden jumps, as the system slows down. The
interesting question for each system has to do with the nature of
the internal modification. The slowing down of the nucleus
^{158}Er below spin 20 is compared with the pulsar Vela in fig.
2. The behaviors are quite similar, though the percentage change
in the nuclear case is much larger. The pulsar glitches are not
too well understood at present—early explanations had to do with

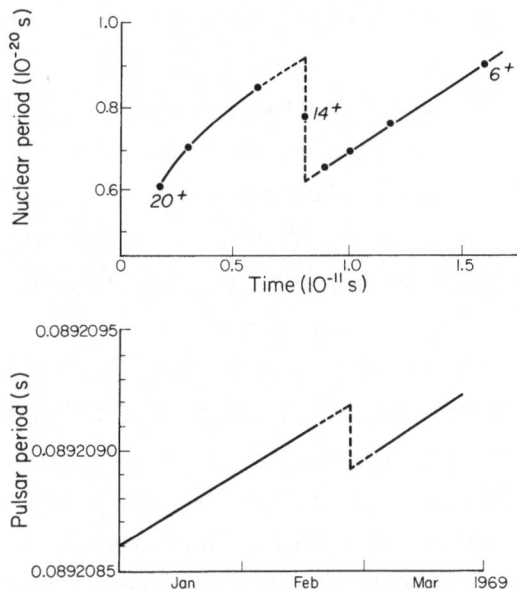

Fig. 2. Plots of the rotational period <u>vs</u> time for the nucleus ^{158}Er (top) and the pulsar Vela (bottom).

a sudden breaking of the solid crust on the neutron star, but
more recent ones involve vortices in the flow pattern. The
nuclear glitch is much better understood and is due to the sudden
pairing of two high-j particles. In the case of this first backbend in ^{158}Er, the particles are $i_{13/2}$ neutrons. Above I ~
14 this pair of aligned particles contributes 10ħ along the rotation axis, but this is lost below I ~ 14 when the particles
suddenly couple to spin nearly zero and begin to participate in
the pairing correlations. The angular momentum has to be made up
by the collective rotation, which must speed up, thereby decreasing the period.

Such behavior is now well studied in nuclei around ^{158}Er,
and the change described above corresponds to a crossing of two
rotational bands[4]). A band with two aligned $i_{13/2}$ neutrons
crosses the ground-state band, which has all particles participating in the pairing correlations (pairing vacuum). Thus the
discontinuity actually corresponds to a shift into another band,
though the mixing between these bands gives collective enhancement to the transition connecting the bands, often to the point
where they are stronger than the "in-band" transitions at the
crossing. The energy of the aligned band relative to the ground
band gradually decreases with increasing spin because of the
Coriolis interaction. Just as a toy gyroscope will attempt to
align its rotation axis with that of its rotating frame, so a
pair of high-j particles tends to align its rotation axis (angular momentum) with that of the rotating nucleus, thereby decreasing its energy relative to a band without such alignment. This
shift in angular momentum between the orbital motion of individual particles and the collective rotation of the nucleus is
illustrated in fig. 3, where the top figure is the moment of
inertia plotted against angular frequency ($\hbar\omega = E_\gamma/2$) for the
nucleus ^{158}Er. The sharp increases in $\mathcal{J}$ due to the alignments
are apparent, the first one giving rise to a "backbending" as the

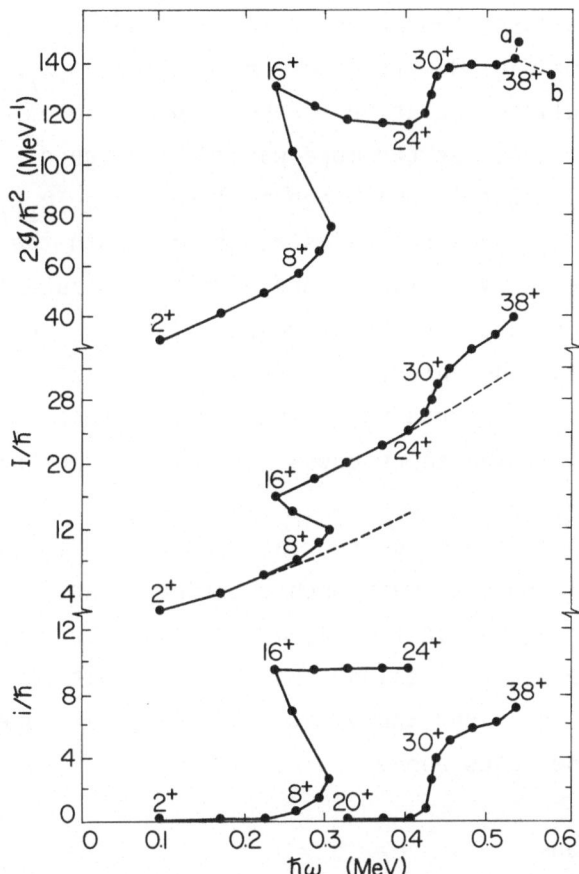

Fig. 3.   Plots of the moment of inertia (top), spin (middle),
and spin alignment (bottom) vs the rotational frequency for the
yrast sequence in ^{158}Er.   From ref. [1]).

sequence shifts bands and the second to an "upbend". In the
center of fig. 3, spin is plotted against angular frequency. The
members of the three different bands fall rather clearly on separate lines, and the difference in spin between the lines at a
given frequency represents the difference in aligned angular
momentum, Δi, between the bands at that frequency. This is shown
at the bottom of fig. 3. It is clear that the $(i_{13/2})^2$ band
has about $10\hbar$ aligned relative to the ground band of ^{158}Er.
The next higher band has two more particles aligned
(4-quasiparticle state), which are believed to be $h_{11/2}$
protons, and the additional Δi is about $7\hbar$. Both the spin and
the angular frequency in fig. 3 are directly measurable quantities, another of which is the interaction of the two bands as
they cross. A strong interaction means heavy mixing of the bands
and a "smoothed-out" crossing, whereas weak interactions are
associated with sudden sharp crossings. The three quantities,
frequency, change in alignment, and the interaction strength, (ω,
Δi, V), characterize a band crossing, and there is now developing
a spectroscopy connected with such crossings. An indication of
this new area is given in fig. 4, which shows the energies of
levels of ^{160}Yb plotted against spin[6]). In addition to the
two band crossings along the yrast sequence, there are many
occurring in the bands above.

1.2. Calculations

There is curently much interest in relating these crossings
to calculated ones. The deformation requires a modification of
the single-particle level spectrum from the usual shell model, in
that this spectrum must be calculated in the appropriate non-
spherical potential. A number of such calculations have since
been made using an anisotropic hormonic-oscillator (Nilsson)
potential[7]), a Woods-Saxon potential[8]), and most recently a
Hartree-Fock type approach towards a self-consistent

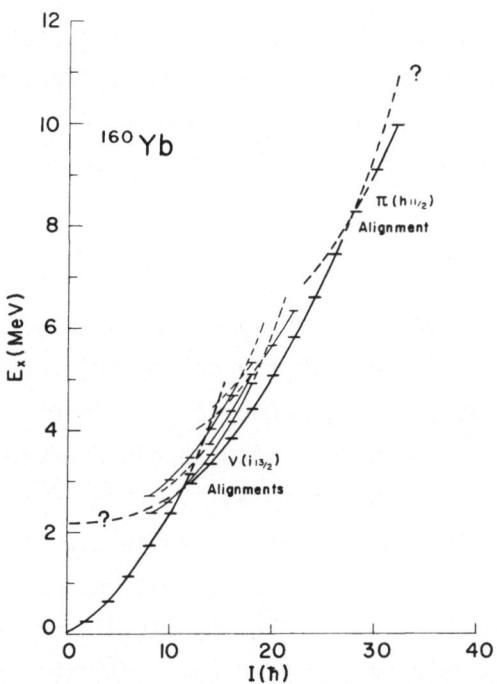

Fig. 4.   Rotational band trajectories on an E vs I plot for the
levels of [160]Yb. The observed levels are indicated by the horizontal marks. From ref. [11]).

potential[9]). These give rather similar results, one of which
is shown in fig. 5. The potential there is the Nilsson one:

$$V = M \left[\omega_3^2 x_3^2 + \omega_\perp^2 (x_1^2 + x_2^2) \right] + V_{\ell\ell} \hbar\omega \, (\ell^2 - \langle\ell^2\rangle N) + V_{\ell s} \hbar\omega_0 \, \ell \cdot s \; , \qquad (1)$$

where M is the nuclear mass, $-V_{\ell\ell} \sim 0.03$ and is added to "flat-
ten" the harmonic oscillator potential, $\langle\ell^2\rangle N = 1/2 \, N(N + 3)$
keeps the average shell energies unchanged, $-V_{\ell s} \sim 0.13$, and
$\hbar\omega_0 \sim 41 \, A^{-1/3}$ MeV. The $(2\omega_\perp + \omega_3)$ oscillator frequencies
can define a deformation, $\delta = 3(\omega_\perp - \omega_3)/(2\omega_\perp + \omega_3)$, which is to
lowest order $\Delta R/R$ and leads to a nuclear shape,

$$R(\theta) = R_0[1 + \frac{2}{3} \, \delta P_2(\cos \, \theta)] \; , \qquad (2)$$

having quadrupole character with a symmetry axis (3) and a
reflection plane (1,2) through the center perpendicular to the
symmetry axis. The 2j+1 degeneracy of the spherical shell model
orbits is broken, and the resulting levels are characterized by
their projection on the symmetry axis, Ω, their signature, α, and
their parity, π. They are two-fold degenerate corresponding to
time reversal symmetry of the nucleon motion. It is perhaps
worth emphasizing that in this fourth oscillator shell shown in
fig. 5, the $g_{9/2}$ orbit has dropped down into the next lower
shell, and the $h_{11/2}$ orbit has intruded from the shell above
due to the large spin-orbit splitting. These high-j "intruder"
orbits are very important for the high spin states. In general
the potentials are not restricted to axial symmetry nor to purely
quadrupole shapes. There is a broad spectroscopy based on identifying observed levels with those calculated, and validity of
this kind of approach is now established beyond question.

The next step is to include the rotation. If the problem is
simplified to cranking the nuclear potential around the x-axis
with a constant frequency ω (the "cranking model"), then the

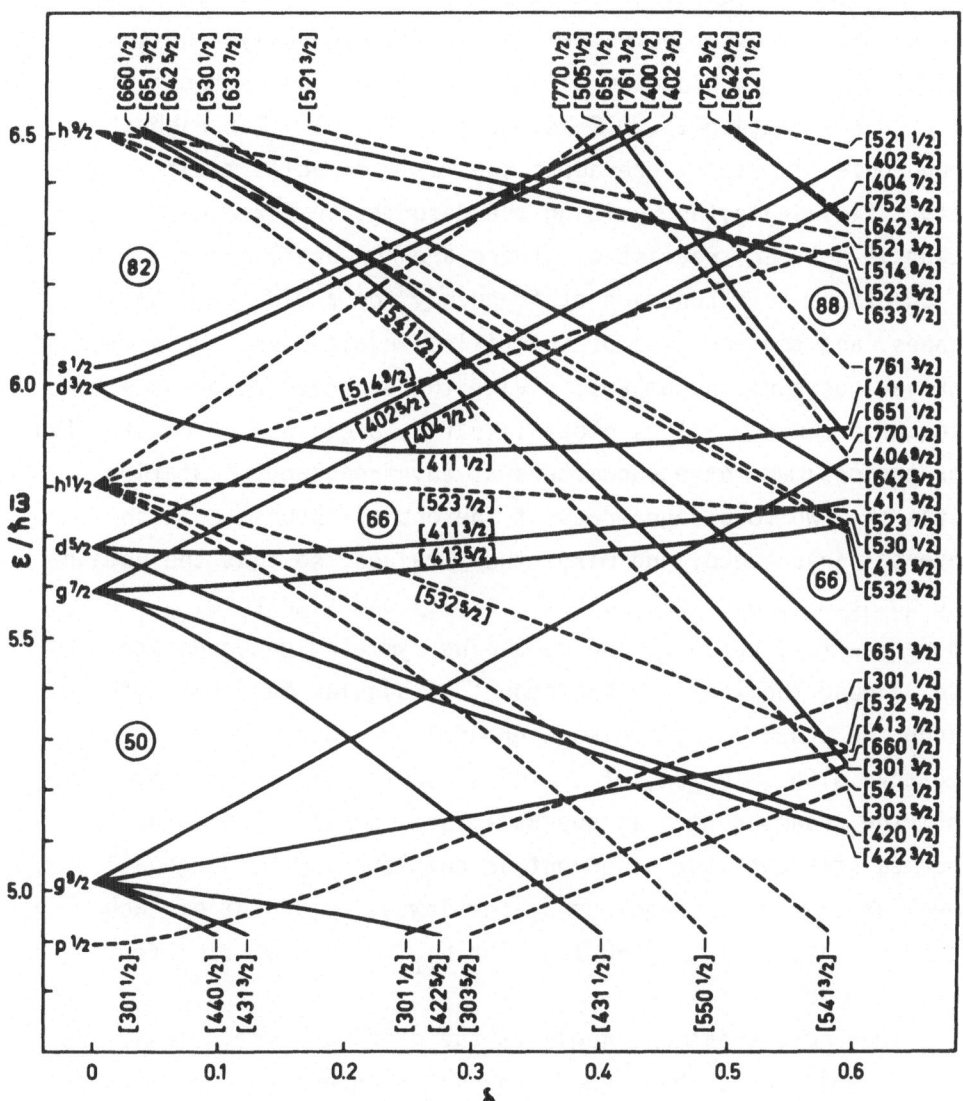

Fig. 5.   Energy levels as a function of prolate deformation for protons in the range 50 ≤ Z ≤ 82.

rotation corresponds simply to an additional term, $-\omega j_x$, in the particle Hamiltonian, where j_x is the component of particle angular momentum along the x-axis. This problem has been solved for all the types of potential mentioned previously, and the result[10]) for cranking the Nilsson potential for protons (fig. 5) at a deformation $\epsilon = 0.2$ is shown in fig. 6. The highest frequencies in fig. 6 are about the maximum nuclei might acquire before fissioning, but holding the deformation fixed up to this point is not very realistic. There are now available calculations as a function of frequency for a wide variety of shapes, and the energies of the filled levels have been summed for various nucleon numbers, giving the expected shape as a function of frequency. Thus plots like fig. 6 can now be constructed for the optimum shape parameters at any frequency. Rotation lifts the two fold degeneracy of the orbits associated with time reversal invariance, and mixes the Ω values, so that the remaining quantum numbers are just the parity and signature, π and α, where $\alpha = \pm 1/2$ reflects the remaining symmetry under rotation by 180° around the x-axis. The amount of angular momentum aligned along the x-axis, j_x, for each orbital is proportional to the negative slope, $-d\epsilon/d\omega$, of the lines on fig. 6, as is apparent from the cranking term in the Hamiltonian, $-\omega j_x$. When the line becomes straight, the alignment is complete (j_x is diagonal). The high alignments acquired by the lowest component of each high-j orbital, e.g., [550 1/2] for $h_{11/2}$ and [660 1/2] for $i_{13/2}$ are apparent.

The above treatment neglects the pairing correlations, and this is believed to be a good approximation above spins of 30 or 40, where the correlations are thought to be quenched. However, for lower spins one must solve the cranking calculations with pairing. This can be done either by subsequently solving the BCS equations for a given set of levels (e.g. from fig. 6) or by adding the pairing into the original diagonalization as in the

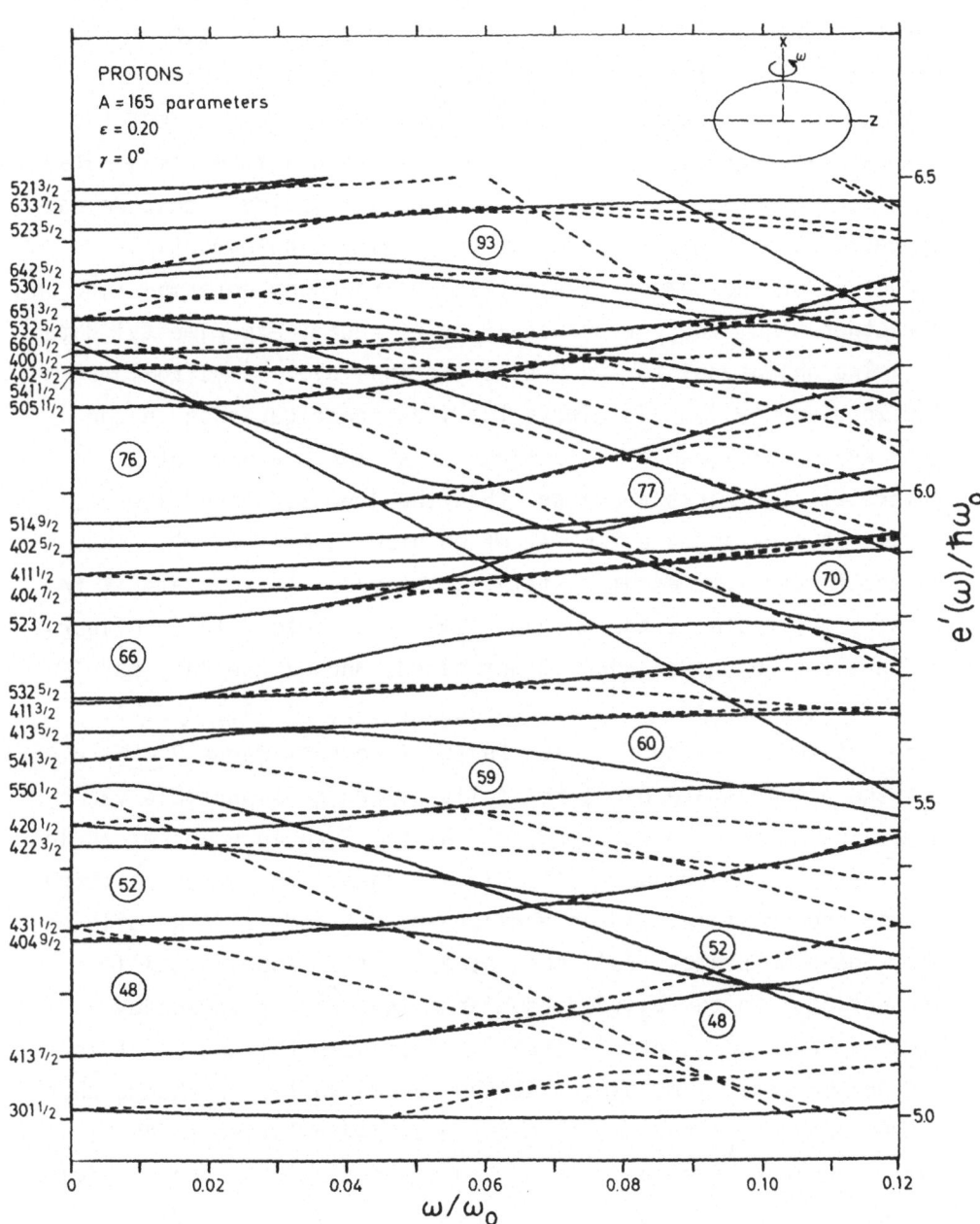

Fig. 6.   Energy levels as a function of rotational frequency for protons in a prolate nucleus having ε = 0.2 and mass number 165. The solid and dashed lines correspond to states having different symmetry with respect to $e^{i\pi j}x$ (different "signatures").   From ref. [10]).

Hartree-Fock-Bogolyubov method. In either case one gets
results[11]) like those shown in fig. 7 for the levels in
^{160}Yb. This diagram covers only a very restricted region, the
five states nearest the Fermi level. It is a little complicated
to understand fig. 7; however, it is only this level that we can
appreciate, on the one hand, the underlying single-particle
structure of the competition between pairing and rotation and, on
the other hand, the current analysis of recent experimental data.

The philosophy for understanding this plot is as follows.
The zero-energy line is always the yrast configuration in the
even-even nuclei, which is defined to be the "vacuum" of
interest. The lines above are then the lowest, second lowest,
etc. states having a given set of quantum numbers (α, π). Thus A
is the first $(+,+)$ excited state, B the first $(-,+)$ state, E the
first $(+,-)$ state, C the second $(+,+)$ state, etc. The configuration along any given line is not fixed, and we use "A", "B",
etc. to refer to the initial configurations, and A,B etc. to
refer to the lines in fig. 7. In the frequency range $0 \leq \hbar\omega \leq$
ω_1, the ground state band, the fully paired quasiparticle vacuum, lies at zero energy (yrast). As the frequency increases in
this range, the various excited levels change in energy relative
to the ground state, and in particular, "A" and "B" (the aligning
$i_{13/2}$ neutron states) fall most steeply (the slope is still
proportional to the aligned angular momentum). At frequency
$\hbar\omega_1$, the two-quasiparticle state "AB" crosses the ground state
and become yrast, implying that the curves in fig. 7 change character; above $\hbar\omega_1$, the energies are measured relative to "AB".
Thus a nucleus originally in any configuration "X" (not involving
"A" or "B") will cross over smoothly at $\hbar\omega_1$ into the state
"ABX", having two unpaired and aligned $i_{13/2}$ quasineutrons,
i.e., there will be a band crossing in this configuration. This
prediction of alignments in many bands at a given frequency will
be very important in studies of the highest spin states. The

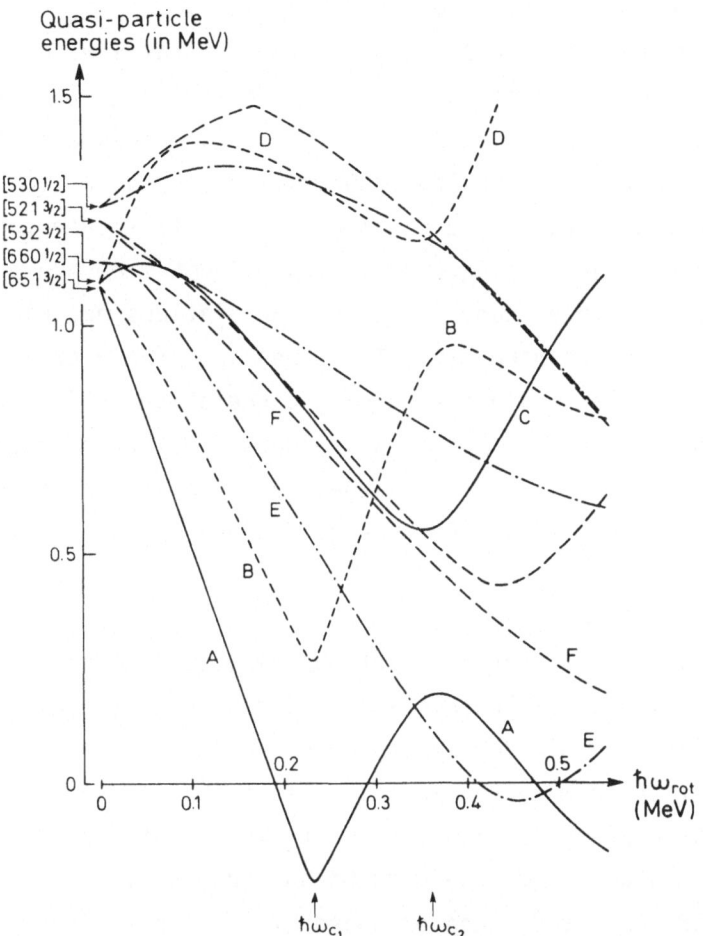

Fig. 7.   Quasiparticle spectra for neutrons as a function of
rotational frequency, which are calculated from a cranked modified–oscillator model for the fixed parameters suitable for the
low–lying states of ^{160}Yb. The quasiparticle orbits with the
quantum numbers, (α=+,π=+), (α=−,π=+), (α=+,π=−), and (α=−,π=−),
are denoted by solid lines, short–dashed lines, dash–dotted
lines, and long–dashed lines, respectively.

states A and B (or more complicated ones involving "A" or "B")
have this crossing blocked, and they do not change configuration
at $\hbar\omega_1$. Therefore their energies change slope, reflecting the
new "vacuum" (or reference) configuration. States with the same
quantum numbers (α,π) cannot actually cross on fig. 7 so that at
$\hbar\omega_2$ the state A (configuration still "A") undergoes a virtual
crossing with the state C (configuration "ABC" above $\hbar\omega_1$). The
same thing occurs for state B and D at about this same frequency. Since the pairing correlations cannot make use of the
two aligned particles above $\hbar\omega_1$, the correlations are dimin-
ished, as is the pairing gap. This breaking of one or a few
pairs of particles while the pairing correlations still exist for
the remaining pairs is a process analogous to that of gapless
superconductivity, where the aligned particles play the role of
the dilute impurities in the superconductor.

1.3. Pairing effects

The nuclear pairing correlations play an important role up
to spins around $30\hbar$. The nucleon orbitals in a static deformed
potential are twofold degenerate, corresponding to a time
reversal of their motion. This situation for an axially symmetric prolate nucleus is illustrated at the top of fig. 8. The
angular momentum, j, of the nucleon has projections, $\pm\Omega$, along
the symmetry axis and, when occupied by two nucleons, results in
total angular momentum zero. Every orbital, characterized by
j,Ω, can give rise to such a spin-zero pair. The nucleons in a
filled orbital near the Fermi level can scatter <u>as a pair</u> into a
nearby empty orbital, and the coherent scattering pattern that
develops comprises the nuclear pairing correlations. It is
interesting that these correlations are closely analogous to
those in superconductors or superfluids. In fact, the equations
of Bardeen, Cooper, and Schreifer[12]) (BCS) that first gave an
explanation of superconductivity are taken over exactly into the

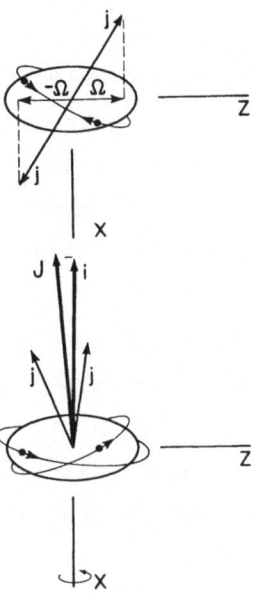

Fig. 8. The two important coupling schemes in deformed nuclei. In the absence of rotation (top) particles with angular momentum, j, are in time-reversed orbits with projections $\pm\Omega$ along the symmetry (Z) axis. At high rotational frequencies the particles couple to a J, aligned as well as possible with the rotation (X) axis, along which they have projection, i.

nuclear case[13-15]) and give nearly correctly both the systematic mass difference between even-even nuclei (all paired, zero
quasiparticles) and odd-mass nuclei (one quasiparticle) and the
~2 MeV level gap in even-even nuclei (zero to two quasiparticle
energy).

These pairing correlations affect the ability of the nucleus
to generate angular momentum. It is easy to see that this is
plausible since, insofar as the pairs are coupled to spin zero,
they can contribute nothing toward generating angular momentum.
This causes a factor of two or three reduction in the nuclear
moment of inertia, which is given reasonably well by the BCS wave
functions. It follows that angular momentum will tend to weaken
the pairing correlations, thus increasing the moment of inertia
and reducing the rotational energy. The mechanism of this weak-
ening is the Coriolis force, which acts oppositely on the two
members of the pair, lifting their degeneracy. Ultimately the
Coriolis force wants to align the particle angular momentum as
well as possible with the rotation axis, as illustrated at the
bottom of fig. 8. This process is analogous to the effect of a
magnetic field on the paired electrons in a superconductor, where
there is a sudden change back to the normal state when a critical
magnetic field is reached. There were initial thoughts that a
nucleus might behave similarly when a critical angular momentum
is reached, but the nucleus differs from the superconductor in at
least two respects: 1) rather than approximately Avogadro's
number of electrons, there are ~100 nucleons in the nucleus, of
which only 10-20% are near the Fermi level and thus participating
in the pairing correlations, and 2) the nucleons have a wide
spread of j values ranging from 1/2 to ~13/2 (for mass around
100). The result is that the nuclear phase transition is not
sharp but broad—i.e. gradual—as evidenced by a gradual rise in
the moment of inertia for spins up to ~20-30$\hbar$. But, within this
gradual rise is occasionally a large irregularity that corre-

sponds to the rather complete alignment of a particular pair of
high-j nucleons. This comes about because the Coriolis force is
proportional to j and thus affects high-j particles most
strongly, so that at some point the nucleus finds it energetically most favorable to align such a pair rather completely while
keeping the pairing correlations among the lower-j nucleons.
This is what causes the nuclearquakes discussed in connection
with fig. 2. In ^{158}Er and many other nuclei of that region, it
is the sudden alignment of a pair of $i_{13/2}$ ($\ell = 6$, $s = +1/2$)
neutrons that causes the large irregularity at frequencies around
0.25 MeV (I ~ 16). This was discussed in connection with fig. 3.

There is a point of view developing that angular velocity or
frequency is an important dimension along which nuclear properties can be measured. In this picture a rotational band is just
a sequence of snapshots of the same configuration at increasing
rotational frequencies, and there are a number of properties that
can be readily observed in these snapshots. The alignments discussed above stand out in the various bands like mileposts, and
the present discussion will be limited to these mileposts and,
further, to just the critical frequency, $\hbar\omega_c$, at which they
occur. Much of the information so far available is on the alignment of two $i_{13/2}$ neutrons in nuclei around mass 160. In fig.
9 the aligned angular momentum, i, for this pair (measured by the
difference in angular momentum between the band under consideration and a reference band) is plotted against rotational frequency (approximately half the rotational γ-ray energy) for three
bands. The solid line is for the lowest-energy band in the
even-even nucleus ^{162}Yb, which is much the same as the band in
^{158}Er plotted in fig. 3. The critical frequency is about
0.26 MeV and the aligned angular momentum is ~10$\hbar$ (12$\hbar$ is the
maximum for two $i_{13/2}$ neutrons). There are methods to evaluate
both these quantities more accurately, but that is not necessary

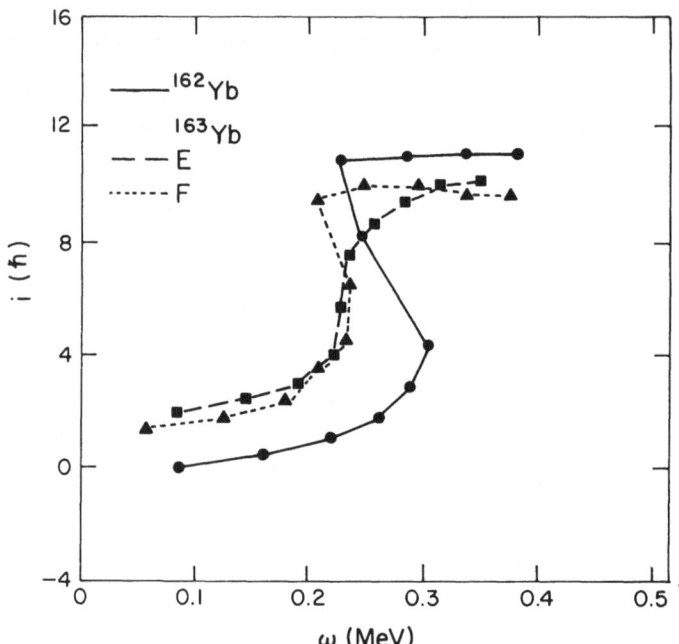

Fig. 9.   The aligned angular momentum, i, is plotted against
rotational frequency, ω, for the first backbend ($i_{13/2}$ alignment) region of the lowest-lying (yrast) sequence in ^{162}Yb and
for two bands in ^{163}Yb (labeled E and F). The midpoint of the
sharp rise is approximately the crossing frequency. From ref.
16).

here. The dashed lines are for two bands in the nucleus ^{163}Yb
with one additional neutron located in an orbital labeled either
E or F. These orbitals comprise a time-reversed pair at zero
rotational frequency and are not very pure shell-model states,
though their dominant component is $h_{9/2}$. In the even-even
nucleus ^{162}Yb, this pair of states (E,F) is available for the
pairing correlations, and, in particular, a pair of $i_{13/2}$
neutrons can scatter into it. On the other hand, in ^{163}Yb it
is blocked by the odd nucleon for the bands based on either E or
F. The pairing correlations are thereby weaker--in general and
in particular for a pair of $i_{13/2}$ neutrons. It is then easier
to unpair and align the $i_{13/2}$ neutrons, and this occurs at a
lower rotational frequency, ~0.22 MeV, as seen in fig. 9. The
shift is clear and closely reproducible in other nearby nuclei[17]). This shift can be related through calculations to the
change in the pairing correlations involved and turns out to correspond to a 20-30% reduction in pairing. As discussed in Section 1.2 the calculation of such properties is done by cranking a
deformed shell-model potential around an axis perpendicular to
the symmetry axis and can relate this frequency shift, $\delta \hbar \omega_c$,
to a change in the pairing gap, $\delta \Delta$, with reasonable confidence.
Thus we learn that blocking just one orbital near the Fermi level
reduces the pairing correlations appreciably, a result that is
confirmed by other kinds of experiments--transfers of pairs of
nucleons and directly from the odd-even mass difference. The
pairing correlations in nuclei are marginal, and three or four
blocked levels of either type (protons or neutrons) are enough to
destroy the correlations for that nucleon type. But the analysis
of data like that shown in fig. 9 can be carried considerably
further.

The discussion so far has involved blocking one particular
pair of orbitals, E and F. Others can be blocked, and to date
most of the calculations of nuclear pairing effects assume iden-

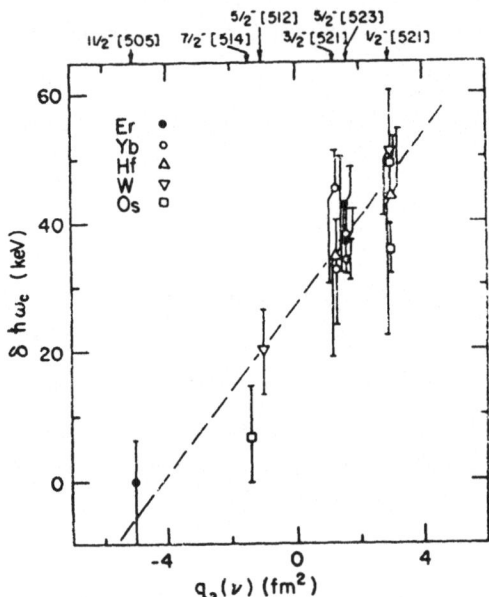

Fig. 10. The shift in crossing frequency, $\delta\hbar\omega$, between the odd-neutron and neighboring even-even nucleus is plotted against the quadrupole moment of the odd-neutron orbit, $q_2(\nu)$. The configuration of the odd-neutron orbit is given by the Nilsson quantum numbers at the top. The dashed line illustrates the correlation between $\delta\hbar\omega$ and $q_2(\nu)$.

tical results for blocking any orbital equally distant from the
Fermi level (called "monopole pairing" for reasons that will
become apparent). This is, in fact, not very reasonable, since
the aligning neutrons in this case are $i_{13/2}$ (with a specific
orientation, implied by their alignment) and some orbitals will
have better spatial overlap with these than others. It seems
likely that the more similar ones will affect the pairing more
(larger $\delta \hbar\omega_c$), but no experimental information previously
existed on such detailed properties. One measure of the shape of
an orbit is its quadrupole moment relative to the nuclear symmetry axis (this is the lowest-order useful moment since nucleon
electric dipole moments vanish due to parity conservation). Figure 10 is a plot of the shift in the alignment frequency for two
$i_{13/2}$ neutrons, $\delta \hbar\omega_c$, vs the quadrupole moment of the
blocked orbital[18]). The aligning neutrons have a large positive (prolate) quadrupole moment ($\sim+4$ fm^2), and the magnitude
of $\delta \hbar\omega_c$ is reasonably clearly correlated with the similarity
of the quadrupole moment of the blocked state to this value. In
fact, the $h_{11/2}$ (11/2$^-$ [505]) orbital is strongly oblate
(very different), and blocking it produces no difference in the
pairing behavior of the $i_{13/2}$ neutrons ($\delta \hbar\omega_c \sim 0$). Such
higher order effects are referred to as "quadrupole pairing."
Their appearance results from the few-body nature of the nucleus,
and gives us some information about pairing phenomena which do
not occur in macroscopic systems such as superconductors.

Exploitation of the rotational-frequency dimension has just
begun. Studies of the type outlined above can be extended to 1)
additional blocked orbitals and 2) other aligning pairs. Also,
it is apparent in fig. 9 that the amount of aligned angular
momentum, i, varies between the even (0-2 quasiparticle) and the
odd (1-3 quasiparticle) systems. This is probably also a pairing
effect but is not yet so well understood as the "critical-
frequency" effects discussed above. There are still other

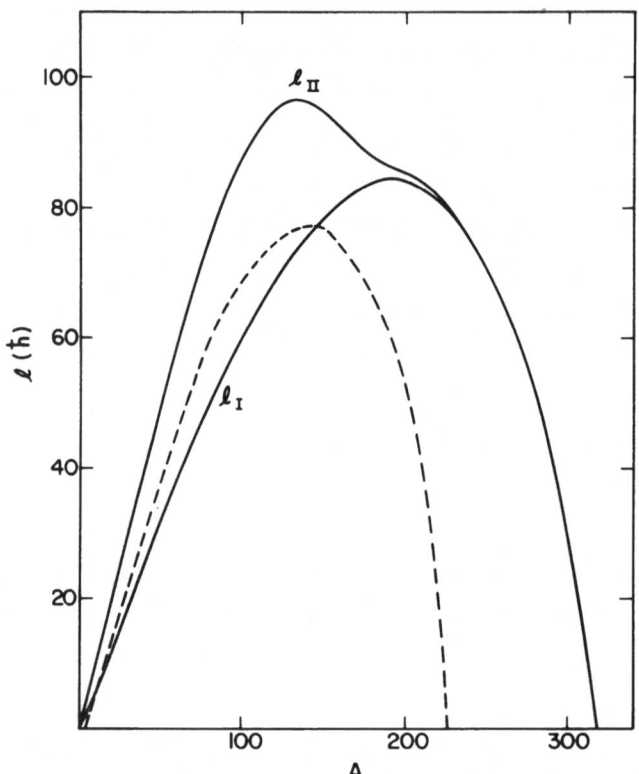

Fig. 11. A plot of angular momentum in a nucleus vs the nuclear
mass number. The curve ℓ_{II} traces out the points where the
fission barrier is zero, according to the liquid-drop model.
Below the curve ℓ_I the equilibrium shape is an oblate spheroid
(Maclaurin); between ℓ_I and ℓ_{II} it is an ellipsoid, generally
triaxial (Jacobi). The dashed line indicates a fission barrier
of 8 MeV. From ref. [10]).

properties to study as a function of frequency. The fact that E and F are split by the rotation is due to their different symmetry properties. Such "signature splitting" appears to be sensitive to details of the shape and may thus give us more insight into the shapes of nuclei and how they change with frequency (and other properties). In addition, the process of quenching the pairing correlations by the presence of several quasiparticles (blocking) and a high rotational frequency (Coriolis effects) is under rather intensive study at this time. Thus far it has not been possible to obtain quantitative measures of this quenching, though it is clearly rather large in some cases.

This "new spectroscopy" is just beginning and seems likely to be quite exciting. It is in addition to the very interesting question of what happens to the nucleus at still higher frequencies, which will be taken up in the next section.

2. The Highest Spins

2.1. Angular momentum limitations

The amount of angular momentum a nucleus can hold is limited. The usual limitation has to do with fission, and it is easy to understand that the centrifugal force associated with rotation will tend to encourage this process. Without any barrier, fission is rapid—occurring in around 10^{-20} seconds. Thus to "exist" (i.e., longer than ~10^{-20} seconds) the nucleus must have a barrier against fission. Such barriers can be estimated using the liquid-drop model[19]), and the angular momenta at which they just vanish are shown in fig. 11 by curve ℓ_{II} as a function of nuclear mass (along the valley of beta stability). The maximum is about $100\hbar$ for A ~ 130. The curve falls off sharply at higher mass because of the increased Coulomb repulsion due to the additional protons. (Nuclei in the actinide region have measurable spontaneous fission lifetimes even at spin

zero.) It falls off at lower mass because the nuclear moment of
inertia (proportional to $A^{5/3}$) becomes smaller, so that for a
given spin the rotational frequency is larger, thereby increasing
the centrifugal force. If cold nuclei could be produced corresponding to the curve ℓ_{II}, spectroscopic studies might be possible up to this limit of angular momentum. But the heavy-ion
fusion reactions that bring in this much angular momentum to the
compound nucleus also bring in several tens of MeV of excitation
energy, greatly increasing the fission probability. To prevent
fission, another process must successfully compete to de-excite
the nucleus, and at such excitation energies this can only be
particle evaporation. The time scale for this particle evaporation is 10^{-17}-10^{-18} s, and in order to slow the fission down
to such times, a fission barrier of the order of the neutron
binding energy (~8 MeV) is required. The dashed line in fig. 11
corresponds to an 8-MeV fission barrier, below which particle
evaporation should dominate.

There is another interesting aspect shown in fig. 11. The
curve ℓ_{II} is determined by two different types of nuclear shape
in different mass regions. At the highest masses the shape where
the fission barrier just vanishes is spheroidal, more specifically, oblate. In fact, this is the stable shape everywhere below
the curve ℓ_I. In the region between ℓ_I and ℓ_{II} the equilibrium shape is ellipsoidal, generally triaxial, and elongates
with increasing spin. These shapes are well known in the general
context of equilibrium shapes of rotating objects. If one just
reverses the sign of the Coulomb energy term in the liquid-drop
model, the equations describe a rotating gravitational object
with surface tension (i.e., a liquid). To have a surface energy
comparable to its gravitational energy (a nucleus has Coulomb and
surface energies comparable), the object would have to be of
order 10 meters in diameter (density taken as 5, surface tension

like water). A rotational period of about an hour would bring the object to the shape boundary analogous to the line ℓ_I in fig. 11. If asteroids go through molten stages, they would be an example of such a system. The limit of negligible surface tension was applied to calculations of the shape of astronomical bodies (first the earth) by Maclaurin[20]) in 1742. The spheroidal shapes below ℓ_I in fig. 11 bear his name, whereas the ellipsoidal ones above were found about a hundred years later by Jacobi[21]). If, on the other hand, one increases the surface energy greatly relative to the gravitational energy, one describes weightless rotating droplets (as were examined by the Apollo astronauts using blobs of water). One can even consider negative masses (moments of inertia) and describe bubbles in ordinary fluids (or nuclei, for that matter). It is apparent that the nuclear properties depicted in fig. 11 are related to a very widespread type of behavior. Since the dashed line in fig. 11 includes some area above ℓ_I, there is a hope to find some of these elongated shapes in nuclei at very high spins. They go under the name "superdeformed nuclei" and will be energetically favored by shell effects for some nuclei and disfavored for others. There is, at present, a great interest in identifying such shapes, as they would give interesting information about the applicability of the liquid drop model at the very highest spins. Some evidence for them will be discussed later in this section.

2.2. Population of high-spin states

A schematic illustration of the decay modes for a nucleus of mass ~160 is shown in the "phase-like diagram" of fig. 12. The coordinates are nuclear excitation, E*, and spin, I. The heavy lines divide the E*-I space into regions of different decay modes. The yrast line is the locus of states of lowest energy for a given spin, so that no states exist in the nucleus below

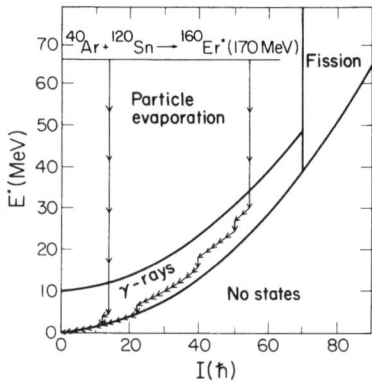

Fig. 12. The heavy lines constitute a phase-like diagram for the
decay modes of a nucleus having mass number about 160 as a function of excitation energy and angular momentum. The lighter horizontal line indicates the range of angular momentum brought in
by a typical heavy-ion reaction, following which two of the many
possible decay pathways are shown (longer arrows represent neutron evaporations and shorter ones γ-ray emissions).

this line. A typical heavy-ion fusion reaction might lead to an initial excitation energy of ~65 MeV and a spin distribution ranging from 0 to ~65ħ as indicated by the light line in fig. 12. As long as the nucleus has sufficient energy above the yrast line to emit nucleons (~10 MeV), it usually does so; γ-ray emission is too slow to compete well with particle evaporation. But at excitations below the nucleon binding energy, γ-ray emission takes over and de-excites the nucleus to its ground state. The angular momentum removed by the particle evaporation is small if neutrons or protons are involved (~1ħ per particle and only a few particles). In fig. 12 it can be seen that the highest spins (longest γ-ray cascades) will be associated with the fewest neutrons emitted.

It is now known that these γ-ray cascades have two principal types of transitions. The "statistical" transitions carry off energy but little angular momentum and so cool the nucleus towards the yrast line. The "yrast-like" transitions follow paths roughly parallel to the yrast line and remove the angular momentum of the system. These latter are sometimes collective rotational transitions, and sometimes not. There are an enormous number of pathways from the beginnings of a high-spin γ cascade until a region near the yrast line is reached, with the result that no single transition has enough intensity to stand up in the spectrum (with present techniques). This is the origin of the name "continuum" γ-ray spectrum, though this is not a true continuum. When the nucleus has cooled sufficiently the population condenses into a few pathways and the transitions in these pathways stand up in the spectrum and are resolved. This typically happens in the spin range 30-40ħ for masses around 160, and it provides a logical division in high-spin studies. In the lower-spin region, one can employ all the techniques of conventional γ-ray spectroscopy and develop detailed information on the nature of the transitions and states involved, as discussed in

Section 1. At the higher spins where the population is spread
out too much to permit the study of individual transitions, new
techniques are providing a picture of the average nuclear behavior. These are described in the present section.

2.3. Moments of inertia

One of the most important factors in determining the physics
of high-spin states is simply the rotational behavior of rigid
classical objects. In fig. 13 the moment of inertia of such an
object (solid lines) is compared with that of a rigid sphere for
a variety of shapes and rotational axes. The shape and axis is
defined by γ, which varies from $-120°$ to $60°$ as the object varies
from a prolate shape rotating about its symmetry axis, through
oblate and prolate shapes rotating about axes perpendicular to
the symmetry axis, to an oblate shape rotating about its symmetry
axis. These axially symmetric shapes are shown by small drawings
in fig. 13, and the regions between correspond to shapes with all
three axes different--triaxial. The deformation is given in
terms of a quantity ϵ, which is to lowest order just $\Delta R/R$, the
difference in radii divided by the average radius. Values of ϵ
around 0.3 are typical for the familiar deformed rare-earth and
actinide nuclei, and 0.6 corresponds to an axis ratio of 2:1, the
largest known in nuclei. The largest moments of inertia, and
therefore the lowest rotational energies, occur for the range of
shapes between $\gamma = 0°$ and $60°$. The very largest moment of inertia for moderate deformations is for $\gamma = 60°$, an oblate shape
rotating around its symmetry axis, and it is for this reason the
earth has such a shape. The full liquid-drop model (LDM) treatment of a rotating nucleus[19]) including surface and Coulomb
energies in addition to these geometrical shape considerations is
shown by the dots in fig. 3. It is apparent that there is no
strong shape preference in these additional LDM terms, so that
simple geometry determines the liquid-drop shapes. This is

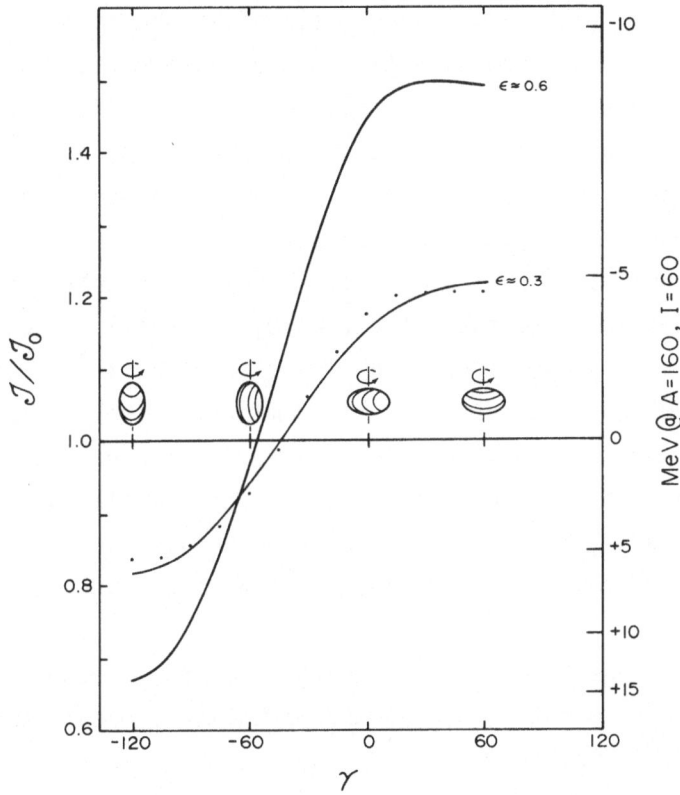

Fig. 13. This figure plots the ratio of the moment of inertia of
a rigid ellipsoid to that of a rigid sphere as a function of the
ellipsoidal shape parameter, γ. The two curves are for deformations (~ΔR/R) of ε = 0.3 and 0.6. The parameter γ defines a
rotation axis, as well as a shape, and the four axially-symmetric
shapes (two prolate and two oblate) are shown by the small drawings. The scale on the right converts the moment-of-inertia
ratios into energy differences for a mass number around 160 and a
spin of 60ℏ. The dots indicate the energy differences from the
full liquid-drop model (surface and Coulomb energies in addition
to the rotational energy implied by the moment of inertia.)

important, since the LDM is our best guide to such macroscopic
nuclear properties and is even the average limit to which some of
the microscopic (individual particle) models are normalized.

In order to see how significant these geometrical shape
effects are, one must choose a mass and spin, and for A = 160 and
I = 60, an energy scale is given on the right side of fig. 3.
The variation for ϵ = 0.3 of about 10 MeV is larger than typical
shell effects (~3 MeV) so that for this spin the shape effects
considered here should be dominant. However, the rotational
energy varies as I^2 so that, for I = 30, the shell effects and
these classical shape effects should be about equivalent, and
below I ~ 20 the shell effects will dominate. The arguments made
here would seem to apply only for collective nuclear rotations,
and even then only if the nuclear moment of inertia has the
rigid-body value, neither of which is obviously the case. In
fact, however, most people do believe that rotating nuclei at
high spins will, on average, have the rigid-body moment of inertia, and this has been shown to be the case for independent particle motion in a rotating anisotropic harmonic-oscillator potential[22]). (The smaller moments of inertia observed at low spins
are due largely to the pairing correlations, which should be
quenched by the Coriolis force above ~30ℏ, as will be
described.) Furthermore, even in noncollective cases, it has
been shown (for a Fermi gas) that the trajectory of lowest levels
follows that given by rotating the appropriately shaped rigid
body[23]) Thus, these geometrical arguments are expected to be
valid, and shapes in the γ = 0-60° range should dominate at high
spin, i.e., above ~30ℏ in the A = 160 region. There are a number
of detailed microscopic calculations that agree with these
expectations.

There is a further aspect of these shapes that is important. The rotation of a nucleus about an axis perpendicular to
the symmetry axis is a collective rotation with smooth bands--

$E(I) \propto I(I + 1)$—and strongly enhanced E2 transitions connecting the levels. There are many beautiful examples of such rotors in the region around mass 160, one of which was shown in the left part of fig. 1. This is the lowest-lying sequence of levels in ^{158}Er. The odd spins are missing, indicating that the two ends of the nucleus are indistinguishable (symmetry under rotation of 180° about the x or y axis—like homonuclear diatomic molecules). Also the electric quadrupole transitions connecting the levels generally vary smoothly in energy and are enhanced by about 150 times over that expected for a single proton, indicating a collective quadrupole shape as schematically shown on the far left of fig. 1. On the other hand a quantal system like the nucleus cannot rotate around a symmetry axis. This degree of freedom is contained in the single-particle motions. Thus a nucleus with $\gamma = 60°$ builds up its angular momentum by aligning that of one or more individual nucleons with the symmetry axis, like spherical closed-shell nuclei. An example of this behavior was shown on the right side of fig. 1. The ^{147}Gd nucleus is nearly spherical in its ground state and builds its angular momentum by aligning particles as schematically illustrated at the far right of fig. 1. The aligned particles give the system an oblate shape that "effectively rotates about its symmetry axis"—$\gamma = 60°$. The motion is almost completely noncollective and the transitions in the ^{147}Gd scheme are quite irregular in energy and are not enhanced. Most nuclei combine these types of behavior, leading to triaxial shapes between 0° and 60°. This tendency was seen in the ^{158}Er scheme, where there are irregularities around spins 16 and 26, which correspond to single-particle alignments. Our understanding of this alignment process was a major step for high-spin studies in the last decade and was described in the previous section.

As a consequence of the interplay between collective and

single-particle motions, there are a variety of moments of inertia one can measure and compare with detailed nuclear model calculations. The first distinction to make is between kinematic and dynamic values. The equation for the rotational energies of a symmetrical top is:

$$E(I) = \frac{\hbar^2}{2\mathcal{J}} I(I + 1) ,$$ (3)

where $\mathcal{J}$ is the moment of inertia. One sees that a moment of inertia may be defined from the first derivative of the energy with respect to spin:

$$\frac{\mathcal{J}^{(1)}}{\hbar^2} = I \left(\frac{dE}{dI}\right)^{-1} = \frac{I}{\hbar\omega} ,$$ (4)

where $^{(1)}$ is called the "kinematic" moment of inertia because it has to do with the motion of the system—the ratio of angular momentum to angular frequency. It is also apparent that the second derivative leads to a definition:

$$\frac{\mathcal{J}^{(2)}}{\hbar^2} = \left(\frac{d^2E}{dI^2}\right)^{-1} = \frac{dI}{\hbar d\omega} ,$$ (5)

where $\mathcal{J}^{(2)}$ is called the "dynamic" moment of inertia because it has to do with the way the system will respond to a force. If there is only the kinetic energy term as given in eq. (3), these are equal; but, in general, when there are additional I-dependent terms in the Hamiltonian these two moments of inertia will differ. In the present case, the Coriolis force perturbs the internal nuclear structure, giving rise, in lowest order, to an $(I \cdot j)$ term, so that $\mathcal{J}^{(1)} \neq \mathcal{J}^{(2)}$. This situation is not uncommon in other branches of physics. The arguments carry over into translational motion, where $p^2/2m$ is analogous to $I^2/2\mathcal{J}$, and additional momentum-dependent terms in the Hamiltonian give rise to two observed masses. An electron moving in a crystal lattice is a close analog[24]), where the kinematic mass determines the level density and related statistical mechanical properties; whereas the response of the electron to an external force depends on a different, dynamic mass. In cases where the extra (angular) momentum-dependent term(s) depend on $(I^2) p^2$ (or so long as they can be expanded in lowest order as such) they can be taken together with the kinetic term to give a renormalized (moment of inertia) mass.

These two moments of inertia can be defined in principle for any sequence of states desired, but certain ones occur rather naturally in the decay processes. If the particle configuration is frozen, so that one is confined to a collective rotational band, the appropriate moments of inertia are $\mathcal{J}_{band}^{(1)}$ and $\mathcal{J}_{band}^{(2)}$. When there is no perturbation (alignment, shape change, etc.) of the internal structure along this band, these correspond to tne true "collective" values, and this is an approximation often made. In general, however, a single decay pathway involves a sequence of bands having different alignments or shapes. It is then natural to define "effective" moments of inertia $\mathcal{J}_{eff}^{(1)}$ and $\mathcal{J}_{eff}^{(2)}$, which include both the collective contribution and, in addition, contributions caused by changes in particle alignment and shape. For the unresolved spectra from the highest spin states, the population is spread over many bands in many decay sequences. Nevertheless, average values for these moments of inertia can be determined in the following ways.

The γ-ray spectrum from a rotational nucleus is highly correlated in time, spatial distribution, and energy. For a perfect rotor, this spectrum is composed of equally spaced lines, up to some maximum energy corresponding to the decay of the state with highest angular momentum. One aspect of this distribution is

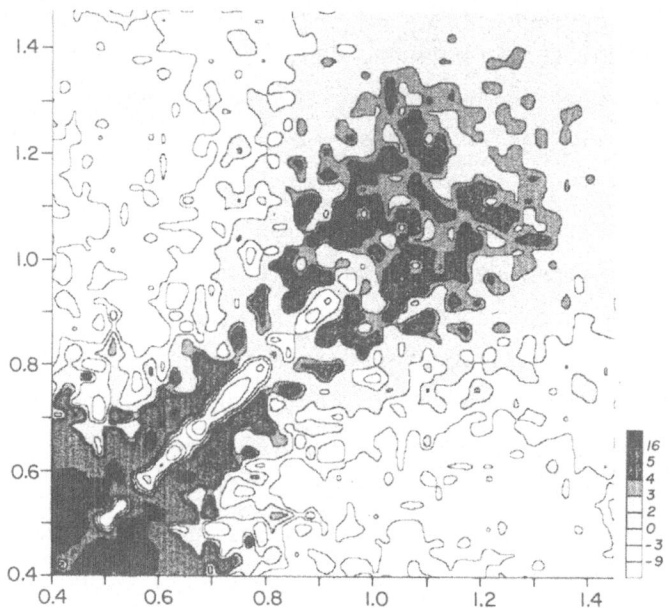

Fig. 14.   A plot showing the correlations between two γ-ray energies following the reaction ^{124}Sn (^{40}Ar, xn) $^{164-x}$Er at 185
MeV.   The data were taken on Ge(Li) detectors and uncorrelated
events were subtracted by the method of ref. [26].   The plot
shows contours of equal numbers of correlated events, according
to the scale at the right.   From ref. [25].

that no two γ rays have the same energy. If plotted on a two-
dimensional diagram of $E_\gamma^{(1)}$ vs $E_\gamma^{(2)}$, such energies give
a pattern with no points along the diagonal and a series of
ridges parallel to it. The width of the "valley" along the diagonal is determined by the difference between rotational γ-ray
energies and thus gives values for $\mathcal{J}_{band}^{(2)}$. The important
point is that the spectrum need not be resolved to determine the
valley width. All that is required is that the populated bands
have somewhat similar moments of inertia at a given frequency
(γ-ray energy).

The data in fig. 14 come mainly from ^{159,160}Er nuclei
formed by bombarding ^{124}Sn with ^{40}Ar at sufficient energy
(185 MeV) to bring into the fused system all the angular momentum
the nucleus can hold (~70ħ). The data have been "symmetrized"
around the diagonal in order to improve the statistics and have
an "uncorrelated" background subtracted. A valley is clear up to
energies ~1 MeV, and again probably from 1.1 to 1.2 MeV.
Resolved lines have been seen in this case only up to ~0.8 MeV.
The width of the valley in both the upper and lower region is
about the same and can be evaluated to give $\mathcal{J}_{band}^{(2)}/\hbar^2$
50 MeV^{-1}, around two-thirds of the rigid-body value. Together
with the effective moments of inertia, this is an important clue
as to the nuclear structure in this region, as will be seen.

There are other important features in these correlation
plots. For example, the valley is sometimes filled by irregularities in the bands, such as alignments. These produce several
transitions in the same energy region, and not only fill the
valley but produce "stripes" of higher coincidence intensity at
these γ-ray energies. The analysis of these features is not yet
very far advanced. There are also two or three recent
cases[27-29]) where the $\mathcal{J}_{band}^{(2)}$ values measured in this way
are very constant and near the full rigid-body value. This is a
puzzle, as it seems to leave no room for alignment effects which

are expected, as discussed below. There is clearly more
information in these correlation plots than is at present
understood.

The effective moments of inertia are simpler in some
respects. They involve only relating a collective γ-ray energy
with a spin or measuring the number of γ rays in an energy interval. The former gives $\mathfrak{J}_{eff}^{(1)}$ values and has been measured
several different ways, originally by relating the maximum γ-ray
energy in a spectrum with the estimated maximum spin input[30]).
However, $\mathfrak{J}_{eff}^{(2)}$ is much more sensitive to the nuclear structure. It is possible to measure $\mathfrak{J}_{eff}^{(2)}$ because, in a spectrum consisting only of "stretched" electric quadrupole ($I \rightarrow I-2$)
transitions (which is known to be a good approximation in regions
of rotational behavior), the number of transitions in a given
γ-ray energy interval is just half the spin removed by that
interval. If one knows the fraction of the observed population
that goes through the interval, then the height of the spectrum
gives directly $\mathfrak{J}_{eff}^{(2)}(\omega)$. This had been recognized earlier,
but the difficulty was to find the feeding as a function of
spin. Recently a method was developed[31]) using the spectra
from two similar but slightly shifted spin distributions, whose
difference is generally proportional to the feeding curve.

Figure 15(top) shows a spectrum due mainly to ^{160}Er decaying from a rather broad distribution of spins centered at ~55$\hbar$.
The spectrum of statistical γ rays, whose high-energy tail is
seen above ~2 MeV, is subtracted leaving essentially pure collective transitions, and the $\mathfrak{J}_{eff}^{(2)}$ values shown by the solid
line in fig. 15(bottom) result from correcting this for feeding.
Two other cases, ^{162}Yb and ^{166}Yb, are also shown in fig.
15(bottom). The general rise at low frequencies in all these
nuclei is due to the quenching of the pairing correlations, and
the irregularities below $\omega \approx 0.3$ MeV result from partially
resolved individual γ-ray transitions and the known alignments

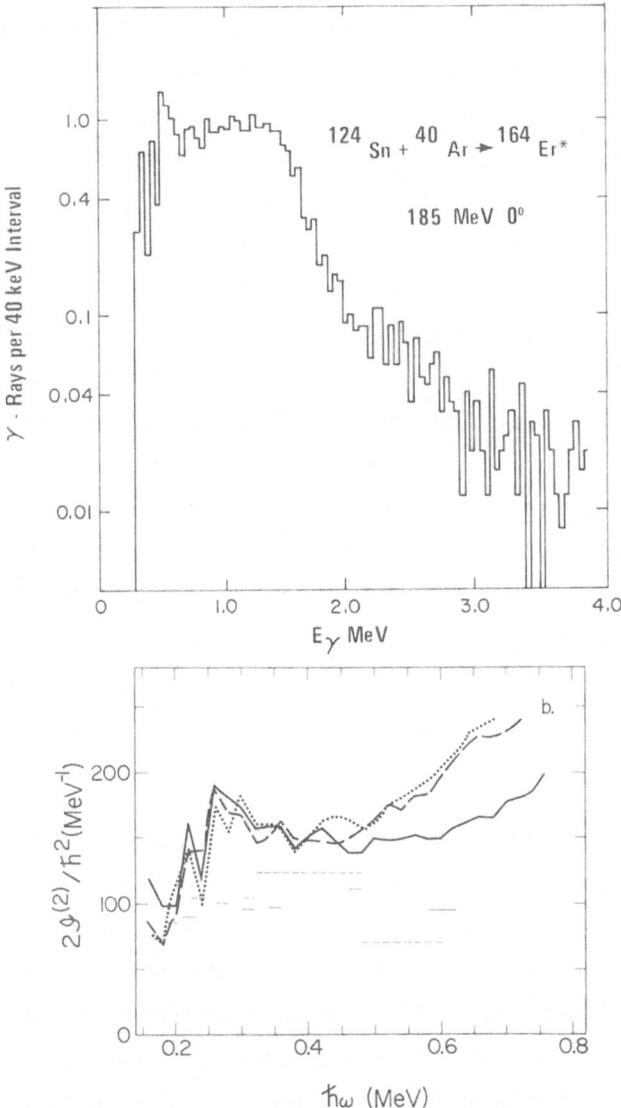

Fig. 15. The γ-ray spectrum (top) from the reaction indicated, taken with a 5" x 6" NaI crystal and corrected for its response function. The spectrum corresponds to a rather broad distribution of initial spins centered around 55ℏ (ref. [24]). The bottom plot is of $\mathcal{J}^{(2)}_{\text{eff}}$ vs ω as derived from the above data (heavy solid line). Also shown are similar plots for the systems ^{126}Te + ^{40}Ar (heavy dotted line) and ^{130}Te + ^{40}Ar (heavy dashed line) and some values for $\mathcal{J}^{(2)}_{\text{band}}$ for ^{124}Sn + ^{40}Ar (thin solid line) and ^{130}Te + ^{40}Ar (thin dashed line).

(backbends), which cause several transitions to pile up at the
same frequency. The band moments of inertia from the correlation
data are plotted as lighter lines in the regions where they have
been determined. The rise in the effective moments of inertia
above frequencies of 0.5 MeV seem to be associated with a drop in
the band values. This suggests that alignments are becoming more
important contributors of angular momentum. The higher values
for the Yb (Z = 72) nuclei compared with ^{160}Er (Z = 70) suggest
that protons play an important role here, which is in accord with
calculations that predict proton $h_{9/2}$ and $i_{13/2}$ alignments in
this frequency region.

     While such data do not give the detail obtained at lower
spins from studies of resolved lines, they are, nevertheless,
beginning to give important insights into the physics of these
highest spin states. No doubt techniques to study unresolved
spectra will be developed considerably further if we do not learn
how to resolve this continuum spectrum.

2.4. Superdeformations

     A special application of the E(γ)-E(γ) energy-correlation
spectra has to do with the search for superdeformations. The
data[32]) in fig. 16 come mainly from ^{152}Dy formed by
bombarding ^{108}Pd with ^{48}Ca at sufficient energy (205 MeV) to
bring into the fused system all the angular momentum the nucleus
can hold (~70$\hbar$). The data have been corrected for the response
function and efficiency of the Ge detectors, "symmetrized" around
the diagonal in order to improve the statistics, and have an
"uncorrelated" background subtracted. A valley is reasonably
clear between 0.8 and 1.35 MeV. (There are also horizontal and
vertical lines corresponding to coincidences with strong discrete
γ-ray lines from states below 40$\hbar$.) The width of the valley is
roughly constant and can be evaluated to give $\mathcal{J}_{band}^{(2)}/\hbar^2 =$
(85±2) MeV^{-1}, around 1.4 times larger than the rigid-sphere
value. Such a large band moment of inertia very likely implies a

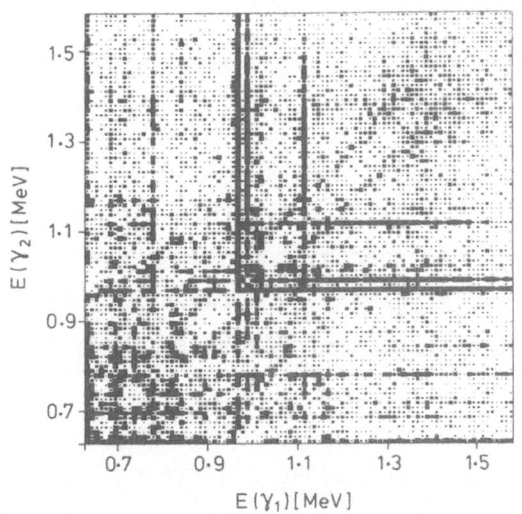

Fig. 16. A plot showing the $E(\gamma_1)-E(\gamma_2)$ correlations in ^{152}Dy following the reaction of 205 MeV ^{48}Ca on ^{108}Pd. The data were taken with Compton-suppressed Ge detectors, have been corrected for the response function and efficiency of the detectors, and have had uncorrelated events subtracted by an iterative procedure. From ref. [27]).

large deformation—close to an axis ratio 2:1. Thus this spectrum provids the best evidence to date for superdeformed shapes at
high spins (0.8-1.35 MeV suggests a spin range 35<I<60), and,
indeed, the microscopic calculations suggest a strong shell
effect favoring such shapes in the region of ^{152}Dy. This is a
powerful technique for identifying rotational bands in an unresolved spectrum, and will, no doubt, contribute greatly to our
knowledge of the highest spin states.

3.   The Frontier

In this section I would like first to discuss some of the
recent developments in γ-ray spectroscopy and then give some preliminary results from our partly completed Berkeley detector
array. The data I will present were taken with nine Compton-
suppressed Ge detectors, part of a planned array of twenty-one.

3.1. Ge balls

Germanium semiconductor detectors offer a unique combination
of energy resolution and efficiency for γ-ray spectroscpic studies. However, they cannot yet be made very large, so there is a
high probability that a Compton-scattered γ ray will escape from
the crystal, leaving a partial energy. This results in a poor
ratio of peak events (full energy) to total events (P/T ratio),
which is only about 0.16 for a 1.33 MeV γ ray in a moderately
large Ge detector. (The size implied here is about 2" diameter
by 2" thick, and is called "20%," where the 20% is relative to
the peak efficiency of a 3" x 3" NaI detector for γ rays of 1.33
MeV.) This means that in a γ-γ coincidence spectrum only ~2.5%
of the events are full energy—full energy (P^2), and in a γ-γ-γ
triple event (of particular interest in this section) only ~0.4%
of the events are useful (P^3). The remaining 97.5% of double
or 99.6% of triple events carry no information, and simply fill

up the storage systems with data that obscure the interesting
events. A partial solution to this problem has been known for
some time, and is to surround the Ge detector with a scintillat-
ing material, so that the γ rays scattered out of the Ge detector
can be detected in the scintillator and used to anti the partial-
energy pulse left in the Ge detector. With this technique one
can reasonably easily get a P/T of ~0.5 ; i.e., a P^2/T of 0.25
in the double events, and a P^3/T of 0.13 in the triple events.
These are improvements of ~3, 10, and 30, respectively, which are
extremely important, especially in the last two situations.
Compton-suppression techniques have been pioneered at Daresbury
and Copenhagen[33]), giving very impressive results with which I
am sure you are all familiar. In the Berkeley array[34]) we use
bismuth germanate (BGO) as the surrounding scintillator instead
of NaI as was used previously. The approximately 2.5 times
greater efficiency of BGO (a combination of density and higher
atomic number) means that the volume of such a shield is ~15
times smaller—a very signficant reduction. The Berkeley shields
are cylindrical and about 13 cm long and 13 cm in diameter.

The new feature we wanted to add in the Berkeley array was
the capability to go to very high rates and especially to be able
to use triple coincidences. There are two principal factors that
determine the rate of coincidence events. These are, the distance from the source to the detectors, and the number of detectors (the size of Ge detectors is not really variable over much
range, and furthermore is related to distance). The minimum
desirable distance is given by the point where summing in the
crystal (simultaneous arrival of two coincident γ rays in the
same detector) becomes excessive (~10% is the maximum summing
tolerable). For average multiplicities around 20, this gives an
upper limit for the efficiency of ~0.5%, or a minimum distance
for a 20% Ge detector of ~13 cm. It is important not to go much
farther away than this, since the rate for triple coincidences

falls off as the sixth power of the average distance. The
Berkeley detectors will be ~14 cm from the source, or about half
the distance of the Daresbury array, giving an improvement of
~2^6 = 64 for the triples rate.

Another obvious way to improve the rate is to increase the
number of detectors, and the small size of the Berkeley BGO sup-
pressors allows 21 detectors to fit rather easily around a source
at 14 cm distance. The rate for triples goes roughly as the cube
of the number of detectors so that the Berkeley array gains
another factor of ~50 over the Daresbury array (6 detectors), for
a total gain in triples rate of about 3000. A real study of
triple Ge events will be possible for the first time using the
Berkeley array.

The reason we are so interested in triple coincidences has
to do with the effective resolution. For a single detector the
energy resolution is defined as the reciprocal of the number of
resolvable points below a given peak. For a 1 MeV peak with 2
keV resolution, this becomes ~1/500 = 2×10^{-3}. For a two
dimensional matrix the number of resolvable points will be
$(500)^2$, for an "effective" resolution of 4×10^{-6}. One can
appreciate this higher resolution by realizing that two γ rays of
essentially identical energy (unresolved in one dimension) will,
in general, have completely different coincident relationships,
and thus be easily resolved in the two-dimensional coincidence
matrix. For triple coincidences the effective resolution becomes
much higher—8×10^{-9}. To illustrate this resolution, consider
a model, where one has N unbranched rotational bands, each with
25 γ rays spread over ~1 MeV. This gives ~$N(25)^3$ possible
γ-ray triplets (G). Since there are ~$(500)^3$ resolvable triplets (R), we have:

$$G/R = N/(20)^3 \sim N \times 10^{-4} \tag{6}$$

One thousand bands leaves this matrix 90% empty--a situation
which would result from only two bands in the one-dimensional
γ-ray spectrum. It is this enormous resolving power that inter-
ests us in triple coincidences.

A sketch of half of the Berkeley array is shown in fig. 17.
The 21 Compton-suppressed Ge detectors are arranged in three
rings of 7 each around a central ~4π BGO ball. The ball consists
of ~44 BGO elements arranged in three concentric cylinders, with
holes through which the Ge detectors view the target. It serves
to measure the number, angular distribution and total energy of
the γ rays emitted (as well as individual γ-ray energies, if
desired). This information can identify, or help to identify,
the product nucleus, its temperature and spin prior to γ-ray
emission, and the spin orientation. One can estimate, using a
simple model, which I won't take the time to describe here, that
this system will enable us to resolve about ten times more levels
than is now possible following a typical heavy-ion fusion reaction. The first results, which I will report now, come from nine
detector systems and no central ball, but already give double
events of unprecedented quantity and quality, and a first glimpse
of triple events.

3.2. Physics with the Berkeley ball

A problem with high-resolution in-beam γ-ray spectroscopy
has to do with Doppler broadening of the peaks. An approximate
expression for this width is:

$$\frac{\Delta E_o}{E_o} \approx \frac{v}{c} \sin \theta \sin \phi , \qquad (7)$$

where v is the velocity of the emitting nucleus, θ is the angle
between the recoil velocity and the γ-ray emission, and ϕ is the
opening angle of the γ-ray detector. In the worst case ($\theta = 90°$,

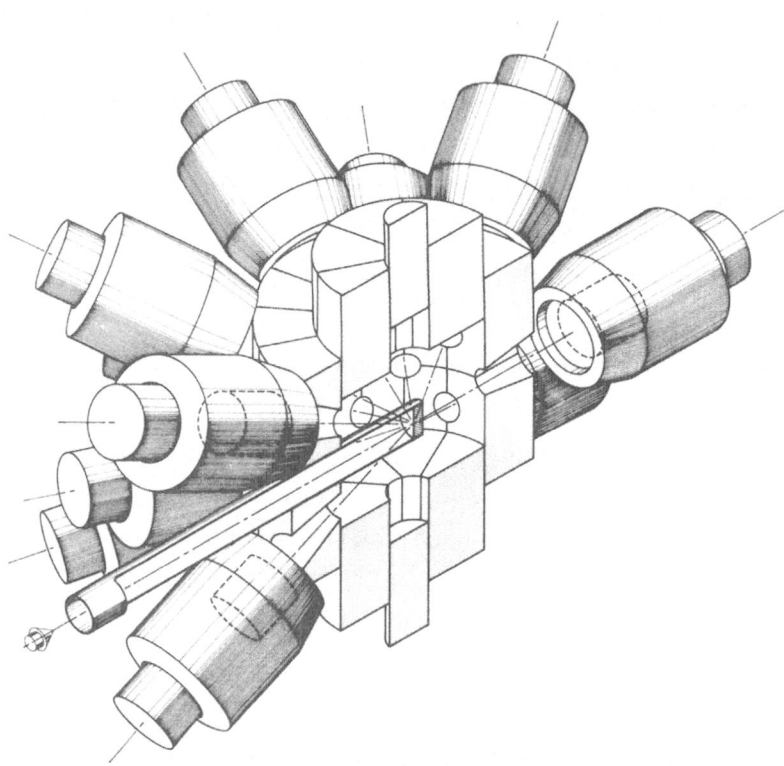

Fig. 17. Cutaway sketch of the Berkeley array. Phototubes which would attach to each element at the top and bottom of the central BGO ball are not shown. The Compton-suppressed Ge detectors are arranged in three rings of seven each. The beam entrance tube is shown from the left.

$\phi \sim 20°$) for a typical $v/c \sim 2.5\%$ (for projectiles like ^{40}Ar),
one finds: $\Delta E_0/E_0 \sim 1\%$, or $\Delta E_0 \sim 10$ keV at 1 MeV, which is
about five times the Ge resolution of 2 keV. This is a serious
loss, which, however, it is possible to avoid in certain cases.
A product nucleus recoiling into lead (or other high-density
material) stops in about 1 psec. If the γ rays are emitted after
this time, they have no Doppler broadening. There is a class of
nuclei, those near closed shells, which have long been known to
have deexcitation times long compared to 1 psec. In a preliminary survey of nuclei just above the $Z = 64$, $N = 82$ double closed
shell, we have found essentially all the discrete γ rays to be
emitted after stopping in Pb. The resolution is thus determined
by the Ge detector to be ~2 keV at 1 MeV. The nuclei in the survey were: ^{144,146,148,150}Gd, ^{148,150,152,154}Dy and
^{152,154,156,158}Er, from which we are currently studying
^{146}Gd, ^{148,150}Dy and ^{156}Er. I will present here some of
the results on ^{156}Er.

The ^{156}Er data consist of about 1.5×10^8 double and
~10^7 triple events. It was taken with nine detector systems
over a period of about two days. The quality of the doubles data
is shown in fig. 18 for one of the spin 28-26 transitions (884
keV), whose intensity is ~3%. One sees a rather clean decay down
to the ground state, and no continuation of the band above the
gating transition energy. An example of a triple spectrum is
shown in fig. 19. The $2 \to 0$ transition (344 keV) is weakly coincident with a transition of the same energy. The spectrum in
fig. 19 results from requiring two transitions of energy 344 keV
as a gate. It contains a number of interesting features, especially the transitions in the energy range 550-600 keV, which are
very weak in the full spectrum. Nevertheless, it has not yet
been possible to place this second 344 keV transition in the
level scheme, and we estimate that an order of magnitude more
events are needed to make full use of the higher resolution in

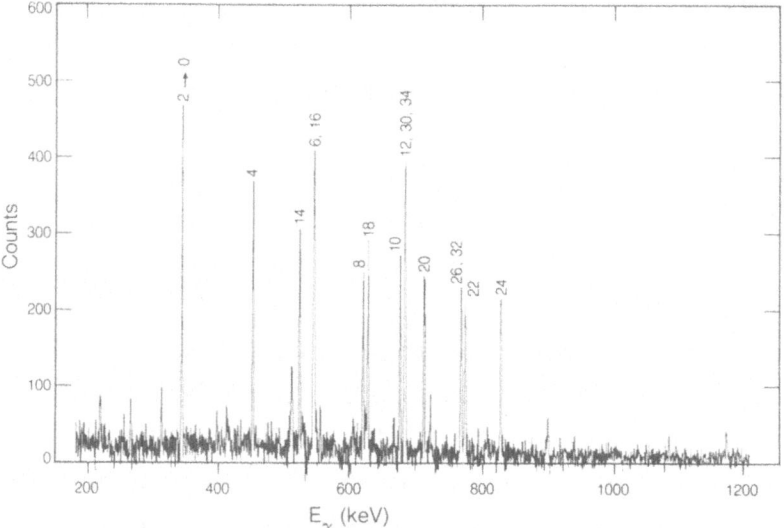

Fig. 18.   Spectrum in coincidence with the 884 keV (28 → 26)
transition.

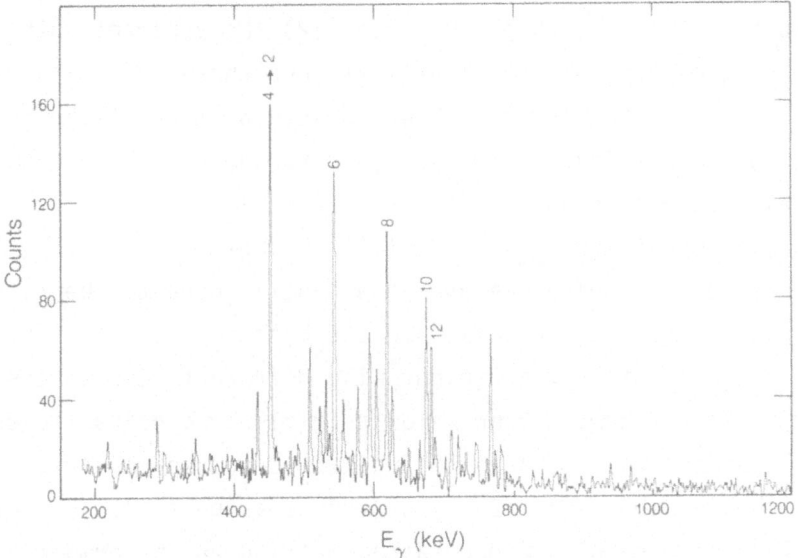

Fig. 19. Triple coincidence spectrum with a gate set on two 344 keV transitions. The spectrum has been smoothed using a simple three channel algorithm.

such spectra. The full array of 21 detectors will take triple
events about ten times faster.

The preliminary level scheme for ^{156}Er is shown in fig. 20.
The spins above about 20 are tentative and based on angular
distribution data only. The spin and parities of the right-most
band differ from those proposed for the lower part by Sunyar et
al.[35]), and are all tenatative, pending conversion electron
data. The interesting features of the scheme are: (1) the change
in character (to non-collective) of the left-most (even parity)
band at spin 30 (or perhaps even 26); (2) the apparent rather
smooth continuation of the right-most band above this dis-
ruption; and (3) the highest observed spin of 42ħ. Although the
results are preliminary, some analysis to support these conclusions will be presented.

A plot of energy _versus_ I(I + 1) is shown in fig. 21. The
lines are only to guide the eye, and can, in general, be rather
misleading. However, in this case the data do seem to divide
rather clearly into three regions with different slopes consisting of: (1) the ground band below spin ~10 (0 quasiparticles);
(2) all bands between spins ~10 and ~25 (2 quasiparticles); and
(3) all bands above spin ~25 (4 or more quasiparticles). This is
probably just a result of decreasing pairing as the number of
quasiparticles increases, but it is interesting that the bands
follow one another so closely in this respect, and that the slope
above spin ~25 corresponds to nearly the rigid-body moment of
inertia.

A spin _versus_ frequency plot is shown in fig. 22. The ground
band undergoes a rather sharp backbend around ħω ~ 0.3, with an
alignment of nearly 10ħ. This is rather typical for the first
$i_{13/2}$ alignment (AB) in this region. During the second backbend
(or upbend) in this sequence (around ħω = 0.4) the smooth
(collective) character ends. Whether this represents a
termination of the sequence in a γ ~ 60° configuration, or whe-

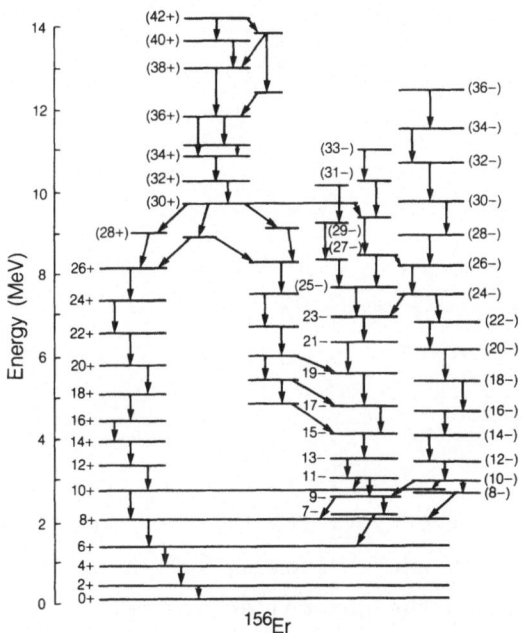

Fig. 20. Premliminary level scheme of ^{156}Er.

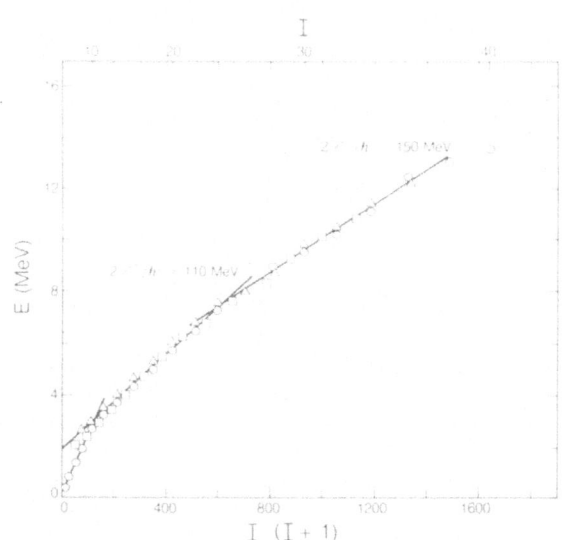

Fig. 21. Plot of energy vs I(I + 1) for the three main bands of ^{156}Er. The symbols are: circles, left-most (even-parity) sequence, including the non-collective region at the highest spins; triangles, the right-most (π, σ = –, 0) sequence; and squares, the center sequence (π, σ = –, 1).

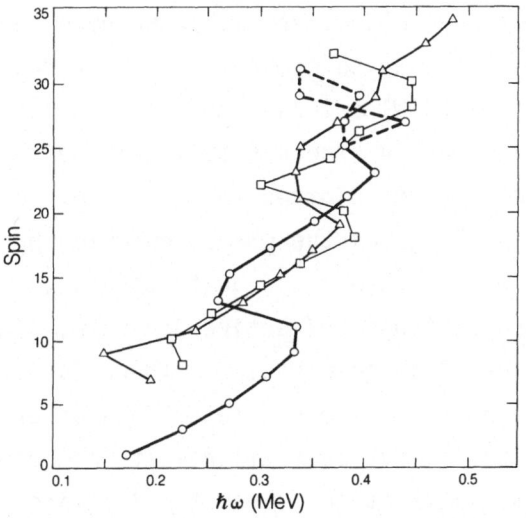

Fig. 22. Plot of spin <u>vs</u> ω. The bands have the same symbols as in Fig. 21.

ther this just means a shift in population to a different (non-collective) sequence is not clear at present.

The other two bands have similar alignments of 7 or 8ℏ at the lower frequencies (ℏω ~ 0.22--0.37). This would be consistent with the lowest configurations expected in the cranked shell model, AE and AF. The odd-spin (signature 1) band (AE) lies lower, as expected. The backbends in both of these bands at ℏω ≈ 0.35 have alignments around 7 or 8ℏ and would be consistent with the blocked $i_{13/2}$ backbend (BC). Above this frequency the situation is not very clear, except that neither of these bands becomes obviously non-collective, like the even-parity sequence.

Perhaps the best way to compare the ^{156}Er data with the cranked shell model (CSM) calculations is shown in fig. 23. Here the level energies have a smooth rotational energy (roughly rigid body) subtracted, and are plotted against spin. Figure 7a shows the data, and fig. 7b gives the CSM calculation by Dudek and Nazarewicz[36]). This agreement is quite striking, particularly the predicted onset of non-collective behavior. The very low-lying state at I = 42 is interesting. There is a very low-lying calculated state at 42ℏ. The discrete lines observed run up 42ℏ, and it seems possible that they all result from population collected into this I = 42 state. Even the lifetime observed (>1 psec) could arise from the decay of this state. Its configuration (not included in the information from Dudek and Nazarewicz) relative to the N = 82, Z = 64 closed shell is very probably: π, $(h_{11/2})^4$; ν, $(i_{13/2})^2(h_{9/2})^2(f_{7/2})^2$.

The present data on ^{156}Er are still preliminary, and the interpretation should probably not be pushed further at present. It will be interesting to study in some detail the transition to non-collective behavior (presumably $\gamma \approx 60°$) as a function of spin and configuration. A similar transition occurs in some of the other nuclei of this region (especially those with fewer neutrons), but has been difficult to study in much detail due to the

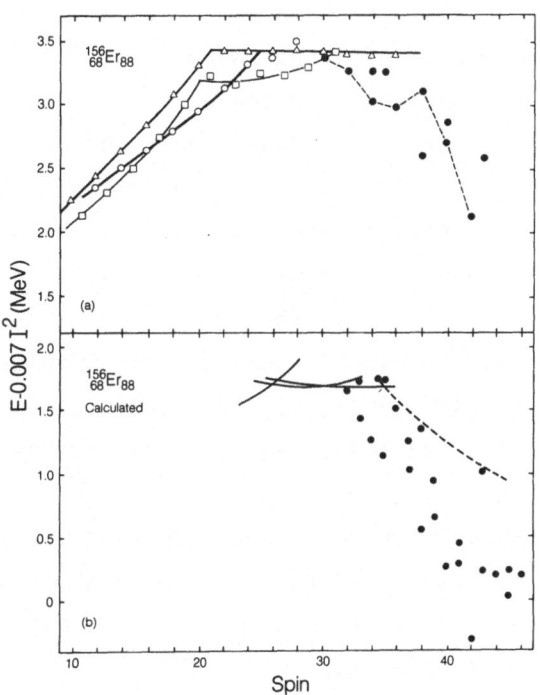

Fig. 23. Plot of energy (minus a smooth rotational energy—0.007 I^2) vs. spin. The top plot, a, is for the experiemental bands, with the same symbols as in Fig. 21. The lower plot, b, is for the calculated results of Dudek and Nazarewicz, where the lines indicate collective bands and the points indicate particle–hole (non-collective) states. The dashed line represents a band with large triaxial deformation.

complexity of the level schemes. Also in these cases the transition occurs at lower spin and the rotational structure of the
initial bands is not so well developed as in ^{156}Er.

In conclusion it appears that, large arrays of Compton-
suppressed germanium detectors will be extremely powerful tools
for γ-ray spectroscopy. In connection with studies of ^{156}Er,
the coincidence spectrum of a peak with 3% intensity was shown in
fig. 18. This is an intensity where a few years ago the peak
itself could barely have been identified. The quality of the
present coincidence spectrum is due to a combination of
Compton-suppression and very good statistics. We have chosen to
use the partially completed Berkeley array first in the region
just above N = 82 and Z = 64. The reasons for this are: (1) the
γ-ray lines from lead-backed targets are not Doppler broadened,
due to the long feeding times; (2) the level schemes are complex
and thus represent a challenge; and (3) the physics of the region
is interesting, involving many shape changes and the (theoretical) promise of superdeformed shapes at rather low spins. We
hope to be able to extend our knowledge of high spin states in
nuclei of this type in the way the Daresbury array has extended
our understanding of deformed nuclei.

This work was supported by the Director, Office of Energy
Research, Division of Nuclear Physics of the Office of High
Energy and Nuclear Physics of the U.S. Department of Energy under
Contract DE-AC03-76SF00098.

References

1) J. Burde, E. L. Dines, S. Shih, R. M. Diamond, J. E. Draper,
   K. H. Lindenberger, C. Schuck and F. S. Stephens, Phys. Rev.
   Lett. 48 (1982) 530
2) O. Hausser, H.-E. Mahnke, J. F. Sharpey-Schafer, M. C.

Swanson, P. Taras, D. Ward, H. R. Andrews and T. K. Alexander, Phys. Rev. Lett. 44 (1980) 132; O. Bakander, C. Baktash, J. Borggreen, J. B. Jensen, J. Kownacki, J. Pedersen, G. Sletten, D. Ward, H. R. Andrews, O. Hausser, P. Skensved and P. Taras, preprint (1982)

3) A. Johnson, H. Ryde and J. Sztarkier, Phys. Lett. 34B (1971) 605

4) F. S. Stephens and R. S. Simon, Nucl. Phys., A183 (1972) 257

5) J. Burde, E. L. Dines, S. Shih, R. M. Diamond, J. E. Draper, K. H. Lindenberger, C. Schuck and F. S. Stephens, Phys. Rev. Lett. 48 (1982) 530

6) B. Herskind, J. De Physique 41 (1980) C10-106

7) S. G. Nilsson, Mat. Fys. Medd. Dan. Vid. Selsk. 29, no. 16 (1955)

8) K. Neergaard, H. Toki, M. Ploszajczak and A. Faessler, Nucl. Phys. A287 (1977) 48

9) H. J. Mang, Phys. Rev. 18, 325 (1975); A. Faessler, K. R. S. Devi, F. Grummer, K. W. Schmid and R. R. Hilton, Nucl. Phys. A256 (1976) 106

10) C. G. Andersson, S. E. Larsson, G. Leander P. Moller, S. G. Nilsson, I. Ragnarsson, S. Aberg, R. Bengtsson, J. Dudek, B. Nerlo-Pomoska, K. Pomorski and Z. Szymanski, Nucl Phys. A268 (1976) 205

11) I. Hamamoto, Nordita preprint 81/28, (1981)

12) J. Bardeen, L. N. Cooper and J. R. Schrieffer, Phys. Rev. 106 (1957) 162; 108 (1957) 1175

13) A. Bohr, B. R. Mottelson and D. Pines, Phys. Rev. 110 (1958) 936

14) S. T. Belyaev, Mat. Fys. Medd. Dan. Vid. Selsk. 31, no. 11 (1959)

15) A. B. Midgal, Nucl. Phys. 13 (1959) 655

16) B. Herskind, private communication (1982)

17) J. D. Garrett, O. Andersen, J. J. Gaardhøje, G. B. Hagemann,

B. Herskind, J. Kownacki, J. C. Lisle, L. L. Riedinger, W. Walus, N. Roy, S. Jonsson, H. Ryde, M. Guttormsen and P. O. Tjøm, Phys. Rev. Lett. $\underline{47}$ (1981) 75

18) J. D. Garrett, G. B. Hagemann and B. Herskind, Nucl. Phys. $\underline{A400}$ (1983) 113c

19) S. Cohen, F. Plasil and W. J. Swiatecki, Ann. Phys. $\underline{82}$ (1974) 557

20) C. Maclaurin, A Treatise on Fluxions (1742)

21) C. G. J. Jacobi, Uber die Figur des Gleichgewichts, Poggendorff Annalen der Physik und Chemie $\underline{33}$ (1834) 229

22) A. Bohr and B. R. Mottelson, Mat. Fys. Medd. Dan. Vid. Selsk., $\underline{30}$, no. 2 (1955)

23) A. Bohr and B. R. Mottelson, Nuclear Structure, vol. 2, p. 80 (W. A. Benjamin, Inc., Reading, Mass.) 1975

24) A. Bohr and B. R. Mottelson, Phys. Scr. $\underline{24}$ (1981) 71

25) M. A. Deleplanque, F. S. Stephens, O. Andersen, C. Ellegaard, J. D. Garrett, B. Herskind, D. Fossan, M. Neiman, C. Roulet, D. C. Hillis, H. Kluge, R. M. Diamond and R. S. Simon, Phys. Rev. Lett. $\underline{45}$ (1980) 172

26) O. Andersen, J. D. Garrett, G. Bittagemann, B. Herskind, D. C. Hillis and L. L. Riedinger, Phys. Rev. Lett. $\underline{43}$ (1979) 687

27) H. G. Price, C. J. Lister, B. J. Varley, W. Gelletly and J. W. Olness, Phys. Rev. Lett. $\underline{51}$ (1983) 1842

28) C. Schuck, N. Bendjaballah, R. M. Diamond. Y. Ellis–Akovali, K. H. Lindenberger, J. O. Newton and F. S. Stephens, (to be published)

29) R. Chapman, J. C. Lisle, J. N. Mo, E. Paul, A. Simcock, J. C. Willmott, J. R. Leslie, H. G. Price, P. M. Walker, J. C. Bacelar, J. D. Garrett, G. B. Hagemann, B. Herskind, A. Holm and P. J. Nolan, Phys. Rev. Lett. $\underline{51}$ (1983) 2265

30) R. S. Simon, M. V. Banaschik, R. M. Diamond, J. O. Newton and F. S. Stephens, Nucl. Phys. $\underline{A290}$ (1977) 253

31) M. A. Deleplanque, H. J. Korner, H. Kluge, A. O.

Macchiavelli, N. Bendjaballah, R. M. Diamond and F. S.
Stephens, Phys. Rev. Lett. $\underline{50}$ (1983) 409

32) B. M. Nyako, J. R. Creswell, P. D. Forsyth, D. Howe, P. J.
     Nolan, M. A. Riley, J. F. Sharpey-Schafer, J. Simpson, N. J.
     Ward and P. F. Twin, Phys. Rev. Lett. $\underline{52}$ (1984) 507

33) P. J. Twin, P. J. Nolan, R. Aryäeinejad, D. J. G. Love, A.
     H. Nelson and A. Kirwan, Nucl. Phys. $\underline{A409}$ (1983) 343C

34) R. M. Diamond and F. S. Stephens, The High Resolution Ball,
     unpublished, Sept. 1981

35) A. W. Sunyar, E. Der Mateosian, O. C. Kistner, A. Johnson,
     A. H. Lumpkin and P. Thieberger, Phys. Lett. $\underline{62B}$ (1976) 382

36) J. Dudek and W. Nazarewicz, private communication (1984)

THE DYNAMICS OF NUCLEAR STRUCTURE

AT HIGH EXCITATION ENERGY

Jens Jørgen Gaardhøje

The Niels Bohr Institute
University of Copenhagen
Copenhagen, Denmark

1. INTRODUCTION

1.1 Giant Resonances in "cold" Nuclei

Collective excitations in nuclei have been known for more than 30 years and belong to the most studied phenomena in nuclear physics. Collective vibrations were first observed by Baldwin and Klaiber in 1947 [1]. They discovered a strong resonant behaviour of the nuclear photoabsorption cross section around 15 MeV, when bombarding stable nuclei with high energy gamma rays. The resonance turned out to be of electric dipole character and is called the Giant Dipole Resonance (GDR). It is now known to correspond to a vibration out of phase of protons against neutrons. Such a vibration can be generated by the oscillating electric field of the incident photons, which generates a force acting on the electrically charged protons causing them to move. Since the center of mass remains at rest, the neutrons move in the opposite direction. The restoring force of the resulting vibration is ultimately generated by the attractive nuclear force acting between the nucleons.

Since then the GDR has been studied intensively (see for instance the compilation from (γ,n) reactions by Berman & Fultz [2]), and it is found that the resonance is a property of all nuclei. The shape of the resonance is well described by a Lorentzian distribution (see fig.1) whose centroid (e.g. the excitation energy of the vibration) and width (related to the damping of the vibration) vary smoothly with the nuclear mass A. Experimentally the systematics [2] of the GDR centroid indicate that the energy varies as

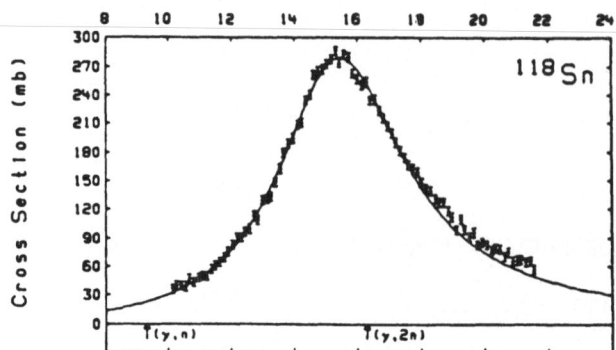

Fig.1. Total photo-neutron cross section for the spherical nucleus ^{118}Sn. The data are from Bermann [2].

$$E_{GDR} = 31.2 \ A^{-1/3} + 20.6 \ A^{-1/6} [\text{MeV}] \tag{1}$$

or simpler

$$E_{GDR} \simeq 79 \ A^{-1/3} [\text{MeV}] \tag{2}$$

the latter expression being a reasonable approximation for nuclei with A>100. Another aspect of the collective nature of the GDR is obtained from the measured strength. This can be related to the energy weighted sum rule for the dipole operator.

In general for an operator F_λ of multipolarity λ the sum rule $S(F_\lambda)$ is defined as

$$S(F_\lambda) \equiv \sum_a (E_a - E_o) |<a|F_\lambda|o>|^2 = \sum_a (E_a - E_o) B(F_\lambda; o \rightarrow a) \tag{3}$$

where $|a>$ and $|o>$ denote an excited state and the ground state, respectively [3]. In the case of the dipole

$$S(E1) = \frac{9}{4\pi} \ \frac{\hbar^2}{2m} \frac{NZ}{A} \ e^2 \simeq 14.8 \ \frac{NZ}{A} \ e^2 \ \text{fm}^2 \ \text{MeV} \tag{4}$$

and

$$\int \sigma_{TOT} dE = 6 \ \frac{NZ}{A} \ \text{MeV} \ \text{fm}^2 \tag{5}$$

It is found that the measured cross sections exhaust a large fraction of the calculated ones. These fingerprints indicate that the giant resonances involve a large fraction of the nucleons in highly coherent motion.

Several macroscopic models have been developped to describe the GDR. Goldhaber and Teller [4] and also Steinwedel and Jensen [5]described the nucleus in terms of a classical incompressible two-fluid liquid with a fixed surface. In this "hydrodynamical collective model" the restoring force is proportional to the volume symmetry energy of the semi-empirical mass formula (Bethe-Weizsäcker). This model yields a mass dependence $E_{GDR} \propto A^{-1/3}$ of the centroid energy (a standing wave in the nucleus will have a frequency inversely proportional to the radius). Another model uses a restoring force proportional to the surface symmetry energy. This yields a mass dependence $E_{GDR} \propto A^{-1/6}$.

The microscopic structure of the GDR can be qualitatively analyzed within the framework of the harmonic oscillator model. Fig.2 depicts such a simple potential. It is seen that dipole excitations (1^-) are created by particle-hole excitations promoting a particle from an orbital in the last filled shell into a free orbital of the next shell which has opposite parity. From such a scheme, one would

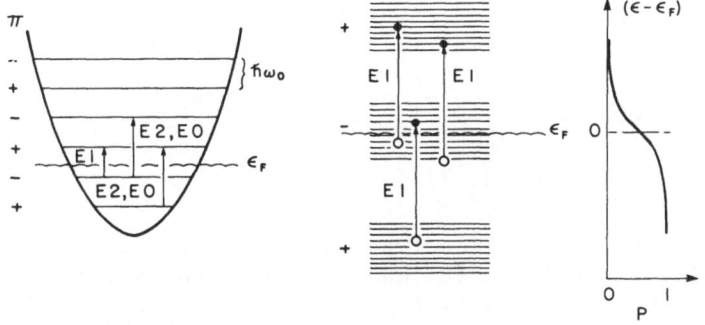

HARMONIC OSCILLATOR   REALISTIC POTENTIAL   TEMPERATURE

Fig.2. a) Schematic illustration of monopole (E0), dipole (E1) and quadrupole (E2) transitions in the pure harmonic oscillator. ε_F denotes the energy of the Fermi surface.
b) Partial view of allowed E1 transitions in a more realistic potential with the Fermi surface in a major shell.
c) The inclusion of temperature introduces a partial filling of levels above the Fermi surface as represented by this Fermi function.

predict an excitation energy for the GDR comparable to the inter-
shell distance, i.e. $E \simeq \hbar\omega_o = 41 \ A^{-1/3}$ (MeV). In contrast, the
observed energies are almost twice as large. The work by Elliot and
Flowers [6], Brink [7], and also Brown and Bolsterli [8] in the
late 1950's has shown that this discrepancy can be removed by introducing a residual particle-hole interaction. This gives rise to a
collective state of higher energy carrying essentially all the
strength. The collective state corresponds to a constructive super-
position of all possible p-h states of a given angular momentum
and parity.

In the past 15 years a number of other types of giant vibrations in nuclei have been identified and studied, primarily through
the use of inelastic scattering reactions such as (e,e'). An
example is the giant quadrupole vibration which corresponds to a
shape oscillation from one ellipsoidal shape to another of different orientation. This mode cannot be excited by gamma rays, since
the photon electric field only acts on the protons in one direction.
Particles scattering off the nucleus can, on the other hand, induce
more complex vibrations.

The various resonance modes can be classified according to
their multipolarity, by whether they involve the spin of the
nucleons (S=1: spin flip mode, such as the giant Gamow-Teller vibration) or not (S=0), and by whether the neutrons and protons
oscillate in phase (T=0: isoscalar modes) or out of phase (T=1:
isovector modes). This is schematically depicted in fig.3.

In general, the study of giant resonances can provide information on the effective nucleon-nucleon interaction (isospin and spin
dependent) for which the central part can schematically be written

$$V_{TS}(i) = V_{00}(r_{ij}) + V_{10}(r_{ij})\vec{\tau}_i \cdot \vec{\tau}_j + V_{01}(r_{ij})\vec{\sigma}_i \cdot \vec{\sigma}_j$$
$$+ V_{11}(r_{ij})\vec{\sigma}_i \cdot \vec{\sigma}_j \ \vec{\tau}_i \cdot \vec{\tau}_j \qquad (6)$$

By choosing suitable reaction types to interact with the nucleus
the individual terms in eq.(6) can be probed specifically. An
example of strong selectivity for the $\vec{\sigma}_i \cdot \vec{\sigma}_j \vec{\tau}_i \cdot \vec{\tau}_j$ term, is the use
of the (p,n) reaction.

1.2 The GDR in excited Nuclei

Most of the knowledge on giant resonances, which is available
today, has been obtained from the study of excitations, built on
the nuclear ground state, although there is no reason not to expect
giant resonances built on states of higher excitation energy E^*
and angular momentum I. In the case of the isovector GDR, that is
the main mode which will be discussed here, this proposition was

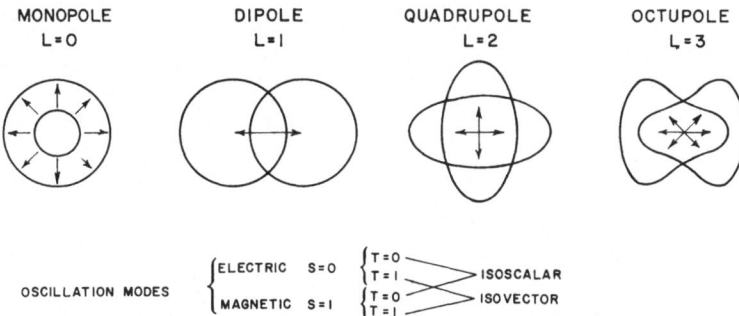

Fig.3. Lowest vibrational modes for nuclei. For the isoscalar modes
the neutrons and protons move in phase, for the isovector
modes, out of phase. Magnetic vibrations involve the spin
of the nucleons.

first stated by D. Brink in 1955 [7]. The situation may schematically be inferred from fig.2. In a more realistic potential (e.g.
a Nilsson potential) the degeneracy of the major shells, peculiar
to the harmonic oscillator potential, is lifted. The inclusion of
a finite nuclear temperature introduces a partial occupation of
orbitals above and below the fermi surface. For the temperatures
relevant here (T<3 MeV) only orbitals within the same shell are
mixed. The situation may then be visualised as a new "ground state"
upon which a collective resonance may be built. This collective
state then carries information about the structural properties of
the excited nucleus.

    While GDR's built on low-lying states have been studied in n
or p capture reactions [9], the first experimental observation of
resonances built on highly excited states of high spin was done
recently [10], with the measurement of gamma rays formed by heavy-
ion induced fusion reactions. In the latter case of complex particle capture, thermally equilibrated nuclei in which part of the
available energy is associated with GDR vibrations, are produced.
So far it seems that most of the GDR decay strength is statistical
(see section 3). The experimental information on this type of GR
is still scarce, but several groups [11,12,13,14,15,16,17,18,19]
have undertaken their study in varied regions of the nuclear chart.

    In fig.4 the decay of such a hot rotating nucleus is schema-
tically depicted. After compound formation, the highly excited
nucleus decays predominantly by evaporating particles (labelled 3),
each of which removes ≃10 MeV of excitation energy (for A>100).
When the remaining E* above the yrast line is less than the particle binding energy, gamma ray emission takes over with the emission of mostly statistical E1 radiation (labelled 2), and low

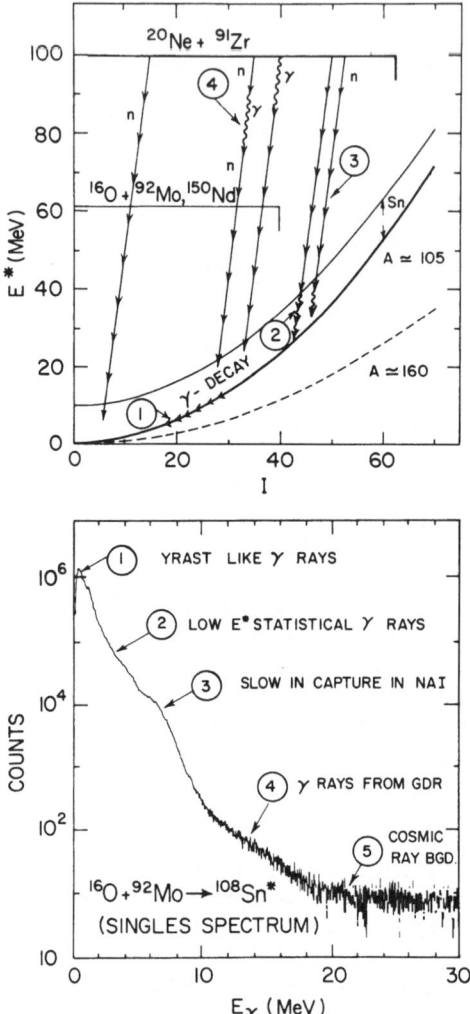

Fig.4. Top: Schematic diagram illustrating the decay of a highly excited compound system. Particle evaporation is indicated by straight arrows, gamma ray emission by wavy arrows. The approximate excitation energy and angular momentum ranges covered in the experiments discussed in this paper are also shown.

Bottom: Gamma ray spectrum ("singles") measured in a 5"x6" NaI detector from the decay of ^{108}Sn*. The various parts of the spectrum are identified and related to the decay of sequence.

energy quadrupole and dipole radiation from transitions of either
collective or non-collective character (labelled 1). These gamma
rays represent the largest part of the radiation detected in the
experiment as may be seen in the bottom part of fig.4. There is,
however, at $E_\gamma \approx 15$ MeV (for $A \approx 160$) an enhancement of the gamma ray
emission (labelled 4) over the extrapolation of the low E_γ region
of statistical transitions. This is due to gamma rays deexciting
giant dipole resonance modes built on excited states in the compound nucleus and its first few daughter nuclei. These gamma rays
are, in contrast to the bulk of the emitted gamma radiation, emitted in the unbound region in competition with particles. From statistical considerations it is expected that this competition is
strongest at the highest excitation energy. This is indeed substan-
tiated by experiment [12,16]. In the "singles" spectrum of fig.4
some of the experimental problems related to the study of these
gamma rays are apparent: i) the low cross section as compared to
the low energy (yrast-like) gamma rays (there are only $\approx 10^{-3}$ γ
rays above 10 MeV per decay cascade for $A \approx 160$ nuclei); ii) the
high background (labelled 5) originating from cosmic rays, which
limits the spectrum at high E_γ.

The study of giant resonances built on excited states is interesting because GR can be used as a probe of the nuclear structure
in the so far inaccessible region of high excitation energy.
Examples of phenomena available to study as a function of the two
new degrees of freedom (spin and temperature) are: deformation and
shape changes, disappearance of shell structure, damping of nuclear
motion, variations of the effective mass and restoring force.

The present discussion does not attempt to review the present
status of excited state giant resonances, but addresses instead a
few problems of current central interest, concentrating on experiments done at the Niels Bohr Institute and at the Lawrence Berkeley
Laboratory. In section 2 some of the experimental problems are
discussed – many of which are of quite general nature although
there are special problems related to the use of large detector
arrays in this field (see for example the lecture of D. Schwalm
at this school). In section 3, the applicability of the statistical model in the analysis of high energy gamma ray spectra is discussed. Section 4 deals with the possibility of studying the
nuclear shape at high excitation energy, while section 5 addresses
the attempts to identify higher multipoles of giant resonances
built on excited states.

The experiments discussed here were done together with C.
Ellegaard and B. Herskind in collaboration with O. Andersen, M.A.
Deleplanque, R.M. Diamond, G. Dines, L. Grodzins, A. Machiavelli,
S. Steadman, F.S. Stephens, Z. Sujkowski and P.M. Walker.
This work was supported by the Danish Natural Science Research
Council.

2. EXPERIMENTAL PROCEDURES

2.1 Studied Reactions

The experiments, which are discussed in detail in the following sections, were done with ^{16}O, ^{17}O and ^{12}C beams from the Niels Bohr Institute FN tandem accelerator and with a ^{20}Ne beam from the 88" cyclotron of the Lawrence Berkeley Laboratory. In order to give an idea of the energetics of the studied reactions, these are listed in table 1, with the corresponding values of the transferred excitation energy (calculated at the center of the target) and maximum angular momentum. The self-supporting metal targets were 1-2 mg/cm^2 (Nd, Mo and Ru) and $\simeq$5 mg/cm^2 (Zr) thick, respectively, and isotopically enriched to $\simeq$97% in the isotopes of interest.

Table 1. List of studied reactions. The maximum excitation energy and angular momentum transferred is in each case evaluated at the center of the target.

Reaction	Beam Energy (MeV)	E (MeV)	ℓ_{max}
$^{17}O+^{148}Nd\to^{165}Er*$	72	51	23
$^{16}O+^{150}Nd\to^{166}Er*$	84	61	36
$^{16}O+ ^{92}Mo\to^{108}Sn*$	84	61	39
	73	51	32
$^{12}C+ ^{96}Ru\to^{108}Sn*$	63	51	30
$^{20}Ne+^{91}Zr\to^{111}Sn*$	140	99	62
	100	66	41

2.2 Experimental Arrangements and Event Definition

Several experimental setups, continuously improved, were used, but in all cases the high energy gamma rays were detected in 7 to 12 5"x6" NaI(Tℓ) detectors located $\simeq$50 cm from the target. Absorbers of 6 mm Pb were placed in front of the detectors to reduce the count rate due to low energy transitions. Only gamma rays with $E_\gamma \gtrsim 5$ MeV were recorded in coincidence with a small-volume fast CsF detector located close to the target, and subtending $\simeq$10% of the total solid angle. A typical detector arrangement is shown in fig.5.

The use of the CsF detector and the long flight path to the NaI detectors allows an excellent discrimination against slow and fast neutrons. The influence of slow neutron capture in NaI is

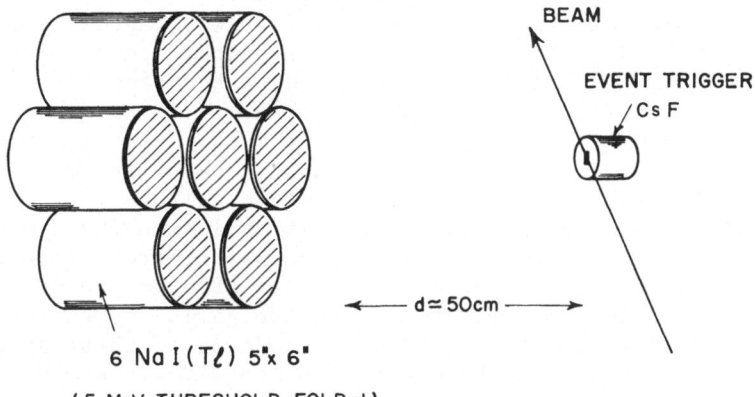

6 Na I(Tℓ) 5"x 6"

(5 MeV THRESHOLD, FOLD=1)

Fig.5. Typical experimental arrangement for the GR experiments
discussed in the text.

evident in the shoulder at 6-7 MeV (labelled 3) in the "singles"
spectrum of fig.4. This distortion of the gamma ray spectrum can
be removed completely, which is important for the reliability of
the statistical model fits to the data which are discussed in sections 3, 4 and 5. In addition, the participation of the CsF detector in the event definition favours the high multiplicity events of
interest since the background reactions (Coulomb excitation and
transfer) have low gamma-ray multiplicity and therefore are less
likely to trigger this detector.

The contribution from cosmic rays to the spectra, which is
the main limiting factor at high transition energies, was kept low
by requiring a high data rate (since the cosmic ray counting rate
is constant in time) and by employing fast coincidence timing.
This background can be further suppressed by arranging the detectors in a close geometry (as in fig.5) and by requiring that only
one gamma ray with $E_\gamma \geq 5$ MeV be detected in the array for each event.
This is due to the fact that there is a high probability for a
cosmic ray (f.ex. a highly energetic muon) which hits the detector
system to trigger several detectors, while the probability of detecting two high energy γ-rays from the target in the same event
is extremely low ($<10^{-13}$). Consequently, very few events of interest are lost by this procedure. This technique was used in the
Ne+Zr experiments.

2.3 Energy Calibration and Background Reactions

A difficulty with this type of experiments is to ensure pro-

proper gain stability of the energy recording NaI detectors.
Indeed, these detectors are quite sensitive to variations in count
rate and ambient temperature. Gain stability was carefully established by monitoring the constancy of the energy spectrum corresponding to slow neutron capture in NaI as a function of time.

The absolute energy calibration of the NaI detectors was de-
termined by the $E_\gamma \simeq 17.3$ MeV rays from the 3 MeV $^{11}B(p,\gamma)$ reaction
on a thick target. It is essential to obtain calibration points at
high transition energy since a simple extrapolation based on known
low energy lines from radioactive sources will introduce serious
errors. This is due to the drastic increase of pair production
processes in the detector material when the gamma ray energy increases. This is apparent in the spectra of fig.6. These spectra
have been calculated with the use of the Monte Carlo computer program EGS (Electron-Gamma Shower) [20] which simulates the propagation of gamma rays and leptons through matter. The spectra show
the response of a Pb shielded NaI (5"x6") detector to incident
monochromatic gamma rays of energies between 2.5 and 25.5 MeV. It
is seen that while at low E_γ the full energy peak dominates the
spectrum, the situation at high E_γ is drastically altered. The
spectrum is now increasingly dominated by the escape peaks and by
the long tails extending to low E_γ due to bremsstrahlung losses
from the produced e^- and e^+. The code was tested for the actual
detector geometry by simulating the gamma ray distributions originating from the $^{11}B(p,\gamma)$ reaction and from known radioactive sources (including ^{66}Ga). The calculated spectra were found to reproduce well the measured spectra.

Contamination from light target impurities (C,O,..) was also
investigated and found to be completely negligible in the region
of interest (10 MeV$\leq E_\gamma \leq$25 MeV).

3. STATISTICAL DESCRIPTION OF GIANT RESONANCES IN EXCITED NUCLEI

3.1 The Statistical Model

The now well established universality of GDR excitations in
thermally equilibrated nuclei suggests that the average GDR decay
is strongly statistical in nature. It is therefore natural to
attempt to analyse the experimental spectra within the framework
of the statistical model.

In this model, the decay probability from a system, formed
in complete statistical equilibrium with respect to all degrees of
freedom, into a particular channel (n,p,α or γ) is determined by
the barrier penetrability for the considered channel weighted by
the density of levels the decay may proceed to. The rate [see for
example Thomas in ref.21] $R_x dE_x$ for emitting a particle x from an
excited nucleus 1 (of excitation energy E1, spin J_1 and parity π_1)

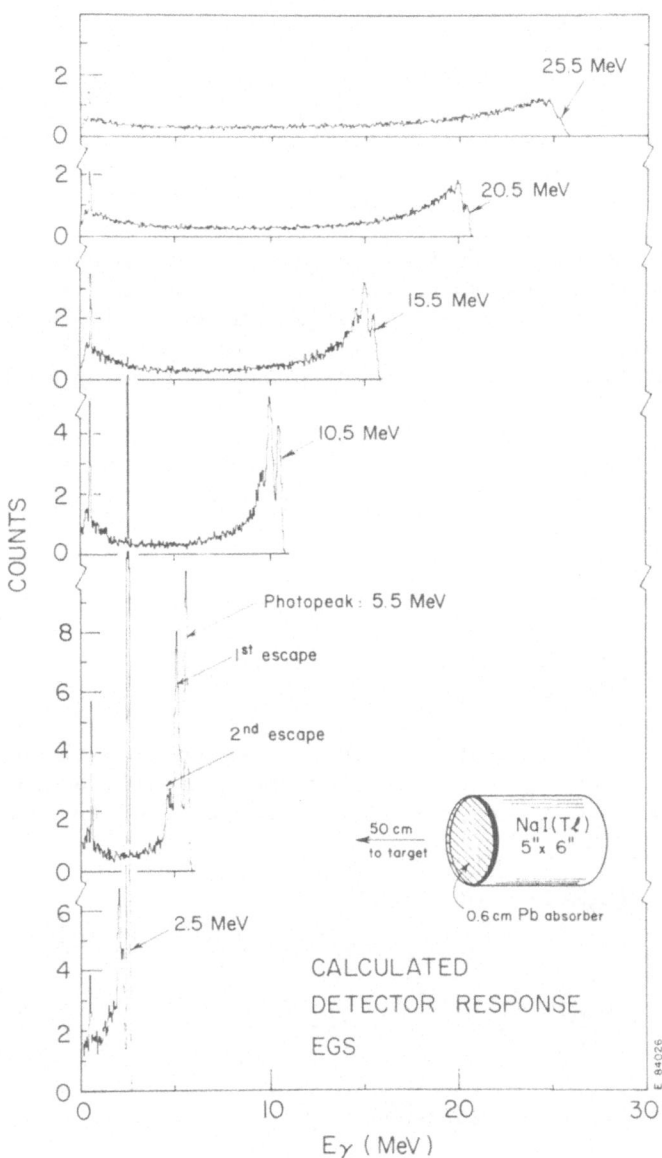

Fig.6. Calculated response of a 5"x6" NaI(Tℓ) detector, shielded with 0.6 cm Pb, to monochromatic gamma rays. The full energy peak, escape peaks and annihilation peak are recognized.

and thereby decaying into a nucleus 2 is

$$R_x dE_x = \frac{1}{\hbar} \Gamma_x(E_x)$$

(7)

$$= \frac{\rho_2(E_2,J_2,\pi_2)}{2\pi\hbar\rho_1(E_1,J_1,\pi_1)} \sum_{\substack{S=|J_2-s_x| \\ [\pi]}}^{J_2+s_x} \sum_{L=|J_1-s_x|}^{J_2+s_x} T_L^x(E_x)dE_x$$

Here E_x is the kinetic energy of the particle x, L and s_x its orbital angular momentum and spin, respectively, while $\vec{S} = \vec{J}_2 + \vec{s}_x$ is the channel spin. T_L^x denotes the transmission coefficient for the inverse problem, namely the scattering of the particle x on nucleus 2. The transmission coefficients are obtained from optical model calculations.

For the gamma decay a similar expression may be written up for a transition from a level 1 to a level 2:

$$R_\gamma dE_\gamma = \frac{1}{\hbar} \Gamma_\gamma(E_\gamma)$$

(8)

$$= \frac{\rho_1(E_2,J_2,\pi_2)}{2\pi\hbar\rho_1(E_1,J_1,\pi_1)} \sum_\lambda E_\gamma^{2\lambda+1} f^\lambda(E_\gamma)dE_\gamma$$

where λ is the multipolarity of the considered transition. The level densities appearing in these formulas are generally based on the Fermi gas model and are expressed as a function of the excitation energy E* and spin J

$$\rho(E_1 J) = \omega(E,M=J) - \omega(E,M=J+1)$$

(9)

with

$$\omega(E,M) = \omega(\frac{E-M^2}{aR}, 0)$$

(10)

$$\omega(E,0) = \frac{1}{12\sqrt{R} \, a^2 T^3} \exp(2\sqrt{aU})$$

(11)

using for the effective excitation energy

$$U = E-\Delta = aT^2 - \frac{3}{2} T$$

(12)

This formalism introduces a level density parameter a, which at high excitation energy may be estimated from the liquid drop model.

It is expected to be close to a $= \frac{A}{8}$ MeV^{-1}, where A is the nuclear mass number.

Several codes using this scheme exist. In most current studies of excited state GDR's, as in the present study, the code CASCADE written by Pühlhofer [22] was used. This code combines a reasonable execution time with the calculation of particle and gamma ray spectra for all the decay routes open to the considered system.

The gamma ray strength function $f^{\lambda}(\varepsilon_{\gamma})$ appearing in eq.(8) is generally determined from the photoabsorption cross section as suggested by Axel [23]

$$f_J^{\lambda}(E_{\gamma}) = 26 \; 10^{-8} \frac{\bar{\sigma}_J^{\lambda}(E_{\gamma})}{g_J E_{\gamma}^{2\lambda-1}} \quad (\text{MeV})^{-(2\lambda+1)} \tag{13}$$

On the other hand, this photoabsorption cross section can (for the GDR) be expressed in terms of the electric dipole sum rule (see eq.(3) and Hayward [24])

$$\bar{\sigma}^1 = 38 \frac{NZ}{A} (1+0.8x) \frac{\Gamma_G E_{\gamma}^2}{(E_{\gamma}^2 - E_G^2)^2 + \Gamma_G^2 E_{\gamma}^2} \; \text{mb} \tag{14}$$

The statistical factor g_J in eq.(13) depends on the spin of the initial and final levels, Brink [7] has shown that the cross section for spin J states may be related to a spin independent cross section $\bar{\sigma}(\lambda=1)$ by

$$\bar{\sigma}_J^1 = \frac{g_J}{3} \bar{\sigma}^1 \tag{15}$$

The GDR gamma strength function therefore becomes

$$f_{GDR}(E_{\gamma}) = 3 \cdot 29 \; 10^{-6} \frac{NZ}{A} \cdot S(E1) \cdot \frac{\Gamma_G E_{\gamma}^2}{(E_{\gamma}^2 - E_G^2)^2 + \Gamma_G^2 E_{\gamma}^2} \quad (\text{MeV})^{-3} \tag{16}$$

neglecting the contribution from exchange forces (x=0) in eq.(14)). The parameter S(E1) represents the fraction of the dipole sum rule needed in order to reproduce the data.

Several analyses of high energy gamma ray spectra along these lines have now been done [12-19], and it is, for heavy systems, possible to reproduce quite accurately the experimental spectra from approximately E_{γ}=5-6 MeV up to E_{γ}>20 MeV, thereby spanning over more than 5 decades in intensity. This success emphasizes the largely statistical nature of the observed GDR decay, at least in heavier systems.

The non-statistical nature of the (p,γ) reaction is well known (see for example Snower [25]). A recent investigation [26] of the ^{3}He+^{25}Mg reaction at E_{beam} (^{3}He) > 20 MeV indicates that nonstatistical components may play an inportant role in lighter systems.

3.2 The High-Energy Gamma Decay from newly formed Compound States

By neglecting all the summations over angular momentum quantum numbers in eqs.(7) and (8) and assuming a simple exponential dependence of the level density on excitation energy, the probability for emitting dipole radiation in competition with neutrons from a given level may be expressed:

$$\frac{dP(E_\gamma)}{dE_\gamma} \propto \frac{NZ}{A} \frac{E_\gamma^3}{T^2} \cdot \exp(\frac{S_n - E_\gamma}{T} - \frac{\hbar^2 I \Delta I}{JT}) \cdot f(\bar{E}_{GDR}, \Gamma_{GDR}, \beta, \gamma, E^*, I..)$$

$$(17)$$

Here T is the nuclear temperature ($\propto (E^*/a)^{1/2}$), S_n the (neutron) separation energy and J the nuclear moment of inertia. The factor T^{-2} appears from an analytical form of the neutron transmission coefficients [27]. The gamma ray strength function f represents the effect of the GDR on the dipole decay and can be described by a Lorentzian function of average energy $\bar{E}_{GDR}$ and width Γ_{GDR} (see section 3.1). Nevertheless the structure of this strength function will in general be more complex than a single Lorentzian and is expected to depend on quantities such as the nuclear deformation and shape parameters (β, γ). Furthermore it may be sensitive to changes in T and I. It is the purpose of the experimental work on excited state GDR's to study the structure of this strength function.

A consequence of eq.(17) is that from level density arguments alone, the gamma ray strength (for $E_\gamma > S_n$) is expected to increase strongly with excitation energy (temperature). This implies that the gamma rays with transition energies in the GDR region are predominantly emitted from the very highest region of E^* populated in the reaction, and therefore constitute a tool to study this region.

In this section, the contribution to the GDR from the γ decay of states produced immediately after compound formation (i.e. in competition with the first emitted particle) is discussed [12]. The selection of the first step of the decay chain offers the possibility of studying the resonant states in a rather narrow interval of E^*. In addition, it permits a stringent test of the statistical nature of the decay.

The method consists of creating the difference of gamma ray spectra obtained from the decay of compound systems differing by

the average energy (binding + kinetic) removed by the particle. In
the region of rare earth nuclei around A=165 the evaporated particles are almost exclusively neutrons, and the energy match can be
performed quite well, although the difference in rotational energy
at the highest I introduces some spread if both systems are not
populated up to the same maximum angular momentum.

In the study reported here, this procedure was carried out on
the ^{166}Er* and ^{165}Er* systems formed by the ^{16}O + ^{150}Nd (E_{beam}=84
MeV) and ^{17}O + ^{148}Nd (E_{beam}=72 MeV) reactions (see table 1). The
spectra of the gamma rays emitted in the decay of ^{166}Er* and ^{165}Er*
are displayed in fig.7a. Before subtracting the spectra from each
other they have been normalized to the same number of counts per
channel at 6-7 MeV. The philosophy underlying this procedure is,
that the major part of the γ rays in this energy range originates
from a region of E* below the particle binding energy which is
common to the final step of the decay of both systems. In addition,
this part of the spectrum is free from the possible influence of a
change in the discriminator thresholds in the two experiments. The
resulting difference spectrum is displayed in fig.7b. Under the
above assumptions, the measured γ ray yield from 10 to 20 MeV
accounts for 24±5% of the yield of the total ^{166}Er* decay in the
same E_γ range. This number, which is practically independent of
the detector response, may be compared to the predictions obtained
by using the statistical decay code CASCADE [22] in which the
energy dependence of the E1 matrix elements is given by a single
Lorentzian function. Calculations for ^{166}Er* and ^{165}Er* are shown
in fig.7c. The input parameters for the ^{166}Er* calculation have
been adjusted to reproduce the shape of the measured spectrum from
7 to 20 MeV. For the GDR an average energy $E_G{}^C$ = 14.5 MeV and
width $\Gamma_G{}^C$ = 9.5 MeV have been used together with a level density
parameter value of a=7.5. This average energy is in good agreement
with the ground state systematics although the width is larger than
typical ground state widths. For ^{165}Er* the same parameters have
been used except for the decrease of E* and I necessary to match
the experimental conditions. Again the difference spectrum (fig.
7d) was obtained by normalizing the calculated spectra at 6-7 MeV.
In fact, this normalization agrees within 2% with the normalization factor obtained from the calculated cross sections. By integrating this spectrum from 10 to 20 MeV it is found that the region
of highest E* accounts for ≃29% of the total ^{166}Er* decay yield in
good agreement with experiment.

In fig.8 the contribution from the various decay steps to the
calculated ^{166}Er* spectrum is displayed. The calculated yields of
the first 5 steps in the range 10 to 20 MeV amount to 28.6%, 26.2%,
23.9%, 17.5%, and 3.5% of the total, respectively. It is noted
that the shape of the spectrum corresponding to the first decay
step (gamma decay in competition with the first particle) agrees
well with the shapes of both the calculated and measured difference
spectra.

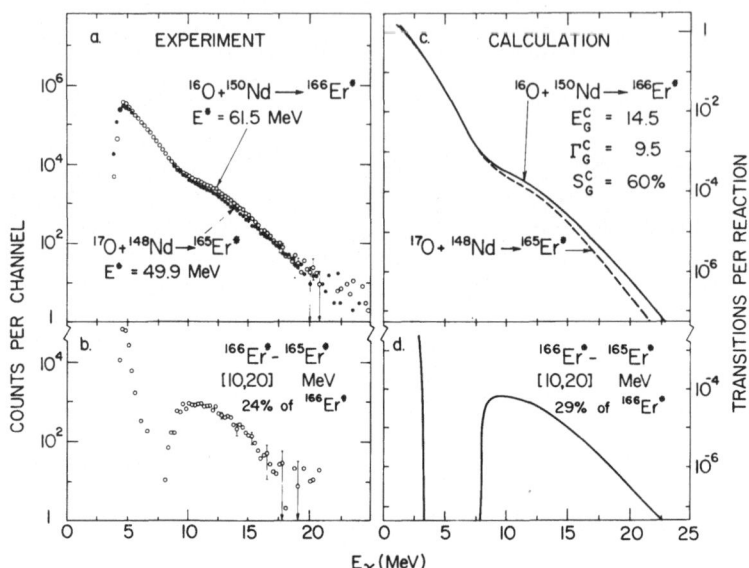

Fig.7. Comparison of experimental spectra for ^{166}Er* and ^{165}Er*
(a) to spectra calculated with the statistical model code
CASCADE (c). The spectra obtained by creating the difference of the ^{166}Er* and ^{165}Er* spectra are shown in (b)
and (d). The excess of counts at $E_\gamma \approx 5$ MeV in the measured
^{166}Er* spectrum, as compared to the ^{165}Er* spectrum, is
due to slightly different settings of the discriminator
levels in the two experiments. The percentage of the total
γ decay associated with the highest excitation energy in
the range $10 \leq E_\gamma \leq 20$ MeV is indicated.

At $E_\gamma \geq 16$ MeV the yield difference between the measured ^{166}Er*
and ^{165}Er* spectra seems smaller than predicted by the calculations. Above this energy, however, the statistics of the present
spectra is not sufficient to determine whether this is due to a
genuine change in the structure of the resonance with increasing
nuclear temperature and angular momentum.

These results therefore indicate, that the observed GDR decay
occurs predominantly from an equilibrated system, and that it can
be analyzed with some confidence within the framework of the statistical theory for nuclear decay.

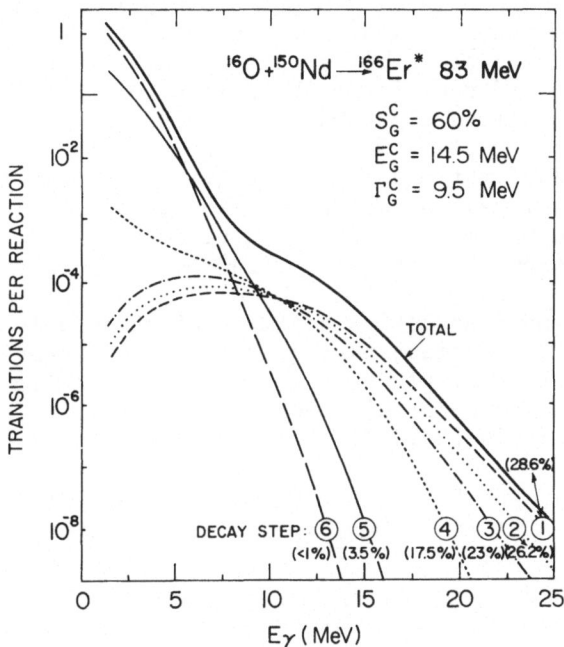

Fig.8. Same CASCADE calculation for ^{166}Er* as in fig.7c, showing
the contribution to the total spectrum from the first 6
decay steps. The yield ratio of each step to the total
spectrum for $10 \leq E_\gamma \leq 20$ MeV is listed in parentheses. The
GDR parameters used in the calculations are indicated.

4. THE NUCLEAR SHAPE AT HIGH EXCITATION ENERGY

The shapes of nuclei arise from the basic correlations which
exist between the nucleons. When the nucleus is subjected to
extreme conditions such as high rotation and internal excitation
energy these correlations tend to break down, ultimately resulting
in changes of the shape. By studying the dynamics of the nuclear
shape as a function of the temperature degree of freedom, it may
therefore be possible to probe in detail the microscopic structure
of hot nuclei.

4.1 Information from the Spectrum

It is well known [28] from photoabsorption measurements in deformed nuclei that the shape of the strength function for GDR's built on the ground state reflects the shape of the nucleus. Indeed, in deformed nuclei the GDR splits into components corresponding to vibrations along the 3 principal nuclear axes. In axially symmetric nuclei two of these components are degenerate in frequency. To first order, the splitting between these components is proportional to the deformation ($\Delta E_{GDR}/\overline{E}_{GDR} \propto \delta$). In rotating nuclei the GDR may be further divided into as many as 5 components, seen from the laboratory frame of reference.

A simple and instructive estimate of the magnitude of these splittings can be obtained by considering the eigenvalues of a Hamiltonian using a single-particle harmonic oscillator potential plus a two-body isovector dipole-dipole interaction, which rotates about the 1-axis [29,7,30]. The triaxial Hamiltonian is then:

$$H = \frac{1}{2} \sum_{i=1}^{3} \left\{ \sum_{n=1}^{A} \left(\frac{P_i^2}{M} + M\omega_i^2 x_i^2 \right)_n + \chi_i \left[\sum_{n=1}^{A} (\tau_3 x_i)_n \right]^2 \right\} - \omega L_1$$

(18)

Here the index $i=(1,2,3)=(x,y,z)$ labels the nuclear axes in the intrinsic frame. The oscillator frequencies ω_i are related to the deformation parameters β and γ by

$$\omega_i = \omega_o \exp\left(-\sqrt{\frac{5}{4\pi}} \beta \cos\left(\gamma - \frac{2\pi}{3} i\right)\right)$$

(19)

where $\hbar\omega_o = 41A^{-1/3}$ [MeV]. The coupling constant is given by $\chi_i = 3 \ A/NZ \ M\omega_i^2$. A schematic representation of the nuclear shapes corresponding to different γ values is given in fig.9. The eigenvalues of eq.(18) are E_1

$$E_1 = 2\hbar_1$$

(20)

$$E_{2,3} = 2\hbar \left(\frac{\omega^2}{4} + \frac{1}{2}(\omega_2^2 + \omega_3^2) \pm \frac{1}{2}\sqrt{(\omega_2^2 - \omega_3^2)^2 + 2(\omega_2^2 + \omega_3^2)\omega^2} \right)^{1/2}$$

(21)

where ω denotes the rotational frequency.

In fig.10 the eigenvalues at $\hbar\omega=0$, are plotted in units of $2\hbar\omega_o$ (roughly the average resonance energy of the non-deformed system) as a function of β, and for different γ values. It is seen that for axially symmetric nuclei ($\gamma=0$, prolate and $\gamma=+60^o$,

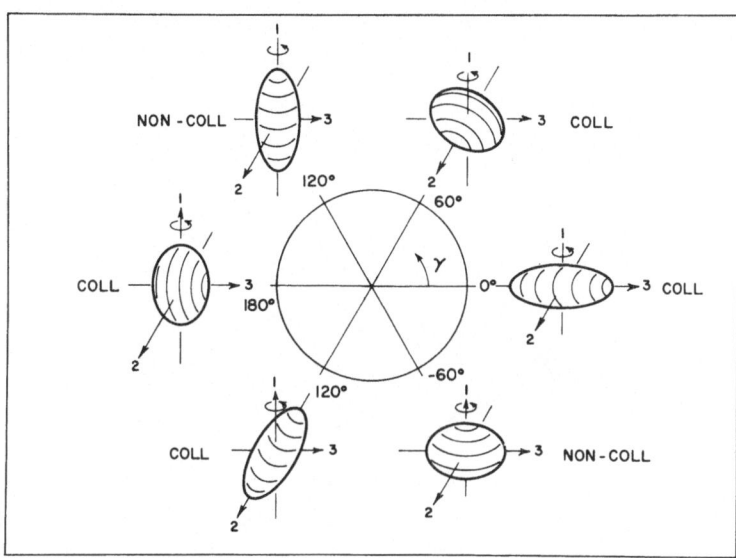

Fig.9. Schematic illustration of the nuclear shapes corresponding to various values of the triaxiality parameter γ
(see eq.(19). The nucleus is assumed to rotate about the
1-axis. If this axis is perpendicular to the symmetry
axis, the rotation is collective.

oblate) two of the frequencies are degenerate. When the system is
rotating, this degeneracy is lifted (fig.11). Seen from the laboratory system the $K=\pm1$ components, corresponding to vibrations
perpendicular to the 1 axis, are (for rotations $\perp$ to the symmetry
axis) further split according to $E_{2,3}\pm\omega$ (right hand side of fig.
11). It is noted that the absolute magnitude of the rotational
splitting in deformed nuclei is small for realistic rotational
frequencies. As an example, the largest splitting for an A=160
nucleus is 0.38 E_R^{SPH} at I=50, which is to be compared to the
static splitting at I=0 of 0.28 E_R^{SPH}.

An interesting result is obtained for a nucleus with β=0 for
which eq.(21) reduces to $E_{2,3}=2\hbar\omega_0(1\pm\omega/2\omega_0)$. Whereas there is no
splitting at I=0, the maximum splitting at I=70(for A=100) is 30%
of the average resonance energy, although 3 of the components now
are degenerate at $2\hbar\omega_0$ (fig.12). A similar linear dependence of
$E_{2,3}$ on ω is obtained for axially symmetric nuclei rotating about
the symmetry axis (non-collective rotation) for which $\omega_2=\omega_3$
(fig.13).

From this simple model it therefore seems that the effect of
rotational splitting will be difficult to observe experimentally,
when the intrinsic width of the individual components (Γ>4 MeV)
is taken into account.

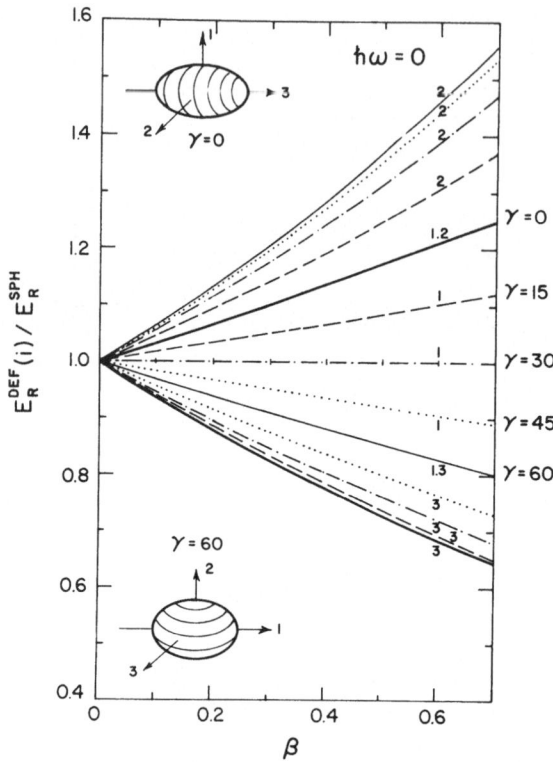

Fig.10. Eigenvalues (at ħω=0) of the Hamiltonian of eq.(18)
plotted as a function of the deformation parameter β for
γ values between γ=0 (prolate) and γ=60 (oblate). The
eigenvalues are given in units of the corresponding
eigenvalues for a spherical nucleus.

Nevertheless, the magnitude of the splitting due to deformation is such that it should be apparent in the experimental spectra, also at finite temperature, although the individual peaks
may be substantially broadened due to a statistical distribution
of shapes and orientations of the spin vector [31]. For axially
symmetric nuclei the observed spectrum will then most likely
appear as two major bumps with the dominant fraction of the total
strength in the lower or higher peak, according to whether the
shape is oblate or prolate. In particular, it should be possible
to obtain an estimate of the nuclear shape from experiment.

Recently, the dependence of the strength function on spin
and temperature has been investigated using a realistic potential
and in self-consistent calculations [32,33,34,35]. It is found
that the inclusion of pairing affects the strength function. The
pairing, however, is expected to have disappeared completely at

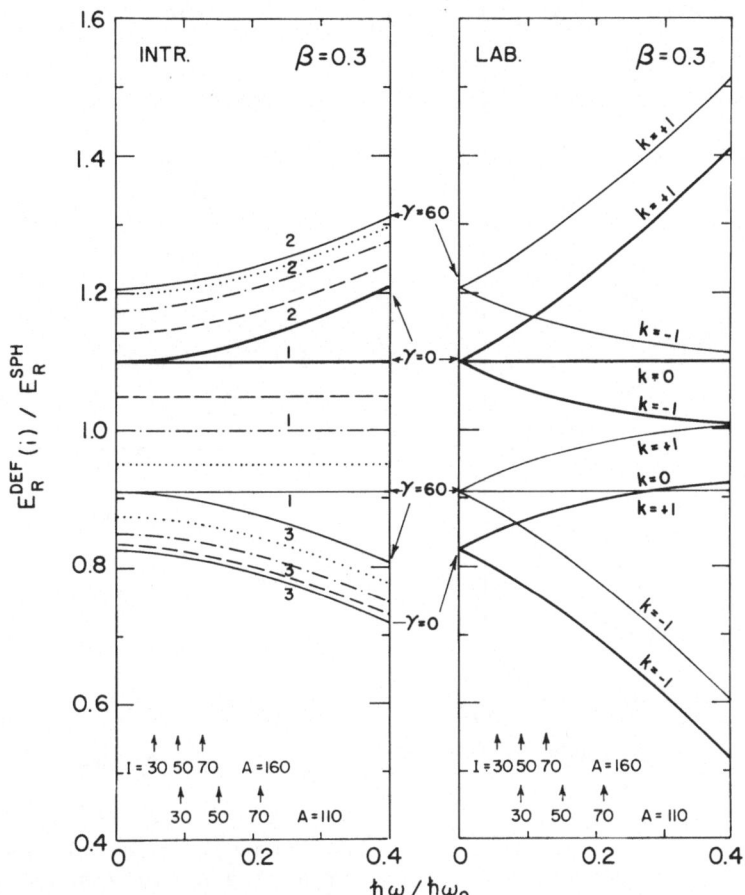

Fig.11. Eigenvalues of eq.(18) as a function of the rotational
frequency (in units of the oscillator frequency $\hbar\omega_o$).
The eigenvalues are calculated for $\beta=0.3$ and $\gamma = 0$, 15,
30, 45, 60 and are given in units of the non-rotating
spherical value. Seen from the laboratory system the
eigenvalues corresponding to vibrations perpendicular
to the axis of rotation (1-axis) are further split into
two components (right hand side of the figure). Also
indicated are the rotational frequencies corresponding
to realistic spin values for A=110 and A=160 nuclei.

the temperatures relevant for the reactions discussed here (T>1
MeV).

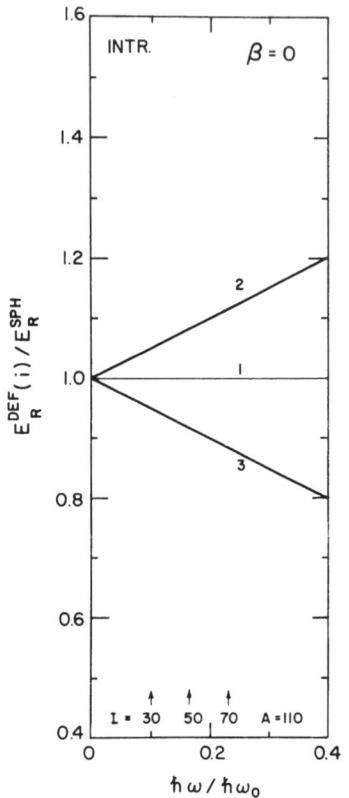 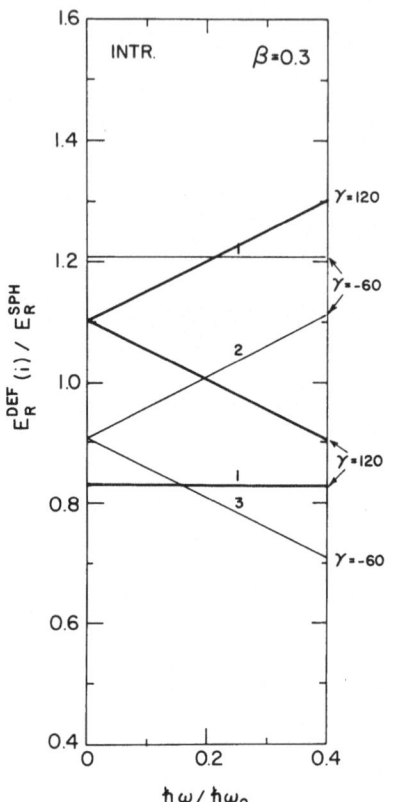

Fig.12. Same as fig.11 for β=0

Fig.13. Same as fig.11 for β=0,3, but assuming non-collective rotation of a prolate shape (γ=120°) and an oblate shape (γ=-60°).

4.2 Information from the Angular Distribution

It is apparent from the discussion in the previous section that it is possible, by measuring the spectrum in the GDR region, to discriminate between different nuclear shapes. It is not possible by this method, however, to discriminate between different types of rotation (collective or non-collective). This may in principle be accomplished by measuring the angular distribution of the emitted high energy gamma rays, since the K=0 and K=±1 components of the GDR have different angular dependences.

It is interesting to note that the so far most direct proof of the dipole nature of the GDR-bump has not been obtained by measuring angular distributions. Instead, in a recent measurement by E. Garman et al. [26], the isovector nature of the bump has been demonstrated by comparing the gamma ray spectra from compound systems formed with total isospin T=0 and T≠0, respectively. Since a GDR transition must connect states differing by ΔT=1, the decay of T=0 systems is greatly inhibited, and can only proceed to higher-lying T=1 analog states.

4.2.1 Dipole angular distribution with respect to the spin direction. Consider a transition from a state $|i\rangle$, of definite spin J_i with projection M_i onto some appropriately chosen axis of quantisation, to a state $|f\rangle$ of definite J_f and M_f. According to de Shalit and Feshbach [36] the intensity emitted into an element of solid angle $d\omega$ is given by

$$\frac{d\omega}{d\Omega} (J_i, M_i, \pi_i \to J_f, M_f, \pi_f, \bar{k}) = \frac{k(2j+1)^2}{64\pi^3 \hbar c^2} \cdot \begin{pmatrix} J_f & j & J_i \\ -M_j & m & M_i \end{pmatrix}$$

$$\cdot |\langle J_f M_f || M_j^{(\sigma')} || J_i \pi_i \rangle|^2 \cdot [[d_{m,1}^{(j)}(\theta)]^2 + [d_{m,-1}^{(j)}(\theta)]^2]$$

(22)

Here $m \equiv M_f - M_i$, $j \equiv J_i - J_f$ and k is the photon wave number. The angular information is contained in the last bracket, which depends on the angle θ between the emitted photon and the chosen axis of quantisation. This expression is obtained by summing over the polarisation of the gamma rays and therefore contains the well known result that an angular distribution measurement cannot distinguish between electric and magnetic radiation.

In treating the present problem, it is convenient to choose the direction of the total angular momentum as the axis of quantisation. If we furthermore assume that there is no loss of alignment connected with the transition (semiclassical approximation, valid for I>>1), we have m=ΔI. For dipole transitions which have j=1, m=-1,0,+1.

It now remains to relate the orientation of the total angular momentum I to the axis of vibration under consideration. This is done via the angle β which takes the value β=0 for a vibration parallel to I (this corresponds to a non-stretched transition, since the angular momentum associated with the vibrational mode then is perpendicular to I) and the value β=90° for a vibration perpendicular to I (stretched transition).

The total angular distribution function may then be written as a weighted sum of terms corresponding to different m values. For dipole radiation

$$\omega_1(\theta, \beta) = \sum_m |a_m^{(1)}(\beta)|^2 \{|d_{m,1}^{(1)}(\theta)|^2 + |d_{dm,-1}^{(1)}(\theta)|^2\}$$

(23)

The probability amplitudes $a_m^{(1)}(\beta)$ are in general the rotational D-function or, since we are only interested in the polar angle β, the reduced functions

$$d^{\ell}_{o,m}(\beta) = (-1)^m \sqrt{\frac{4\pi}{2\ell+1}} \; Y^{\ell}_{\mu}(\beta) \tag{24}$$

Upon insertion of the appropriate values and excution of the summation we obtain

$$\omega_1(\theta,\beta) = \cos^2(\beta)\sin^2(\theta) + \frac{1}{2}\sin^2(\beta)(1+\cos^2\theta) \tag{25}$$

which may be written in a more compact form by introducing the Legendre polynomials $P_2(\cos(\theta)) \equiv 1/2(3\cos^2(\theta)-1)$ yielding

$$\omega_1(\theta,\beta) = \frac{2}{3}[1-P_2(\cos(\beta)) \cdot P_2(\cos(\theta))] \tag{26}$$

Thus, for stretched transitions

$$\omega_1^{\Delta I=\pm 1}(\theta) = \frac{2}{3}\,[1+\frac{1}{2}P_2(\cos(\theta))] \tag{27}$$

and for non-stretched transitions

$$\omega_1^{\Delta I=0}(\theta) = \frac{2}{3}\,[1-P_2(\cos(\theta))] \tag{28}$$

4.2.2 Dipole angular distributions with respect to the beam direction. Measurements of the gamma ray angular distribution with respect to the spin direction, may in principle be done in large detector arrays such as the Heidelberg "Crystal Ball" or the Oak Ridge "Spin Spectrometer". A recent experiment to this effect is discussed by D. Schwalm in a lecture to this school.

A more commonly employed method is to measure angular distributions with respect to the beam direction. The necessary transformation of eq.(26) to the laboratory frame of reference, may be done by introducing the polar angle ν of the emitted gamma ray with respect to the direction of the incident beam, and the azimuthal angle α of the direction of I, noting that in HI reactions I lies in a plane perpendicular to the beam direction.

Upon integration over all azimuthal angles we obtain the angular distribution of the gamma rays with respect to the beam direction

$$\omega(\nu,\beta) = \frac{2}{3} [1+\frac{1}{2}P_2(\cos(\beta)) \cdot P_2(\cos(\nu))] \tag{29}$$

It is noted that the angular dependence of this function is opposite the one calculated with respect to the spin direction. It is also attenuated by a factor 2, reflecting the loss of information resulting from the integration over all possible orientations of I.

   As an example, the GDR strength function has been calculated for A=160 at I=40 and for various shapes within the framework of the rotating harmonic oscillator model described in ref. [30], (fig.12). The influence of the temperature has only been included in the form of the exponential level density factor $\exp(\pm I/JT)$ = $\exp(\pm\omega/T)$ which affects the strengths of the K=±1 components. The displayed functions have been calculated in the laboratory frame of reference at 0° and 90° with respect to the beam direction.

   4.2.3 Attenuation of angular distributions. The angular distributions described by eqs.(26) and (29) will in general be attenuated due to a statistical distribution of the orientation of the spin vector with respect to the axis of vibration under consideration. Statistical distributions of the projection K of the total angular momentum onto the nuclear symmetry axis have been measured for fission and may be described by [3]

$$\rho(K,I) = \rho(K=0,I) \cdot \exp(-K^2/2K_o^2) \tag{30}$$

with

$$K_o^2 = T \cdot J_{diff} \tag{31}$$

and

$$J_{diff} = (1/J_3-1/J_\perp)^{-1} \simeq J_{sph}(1- \frac{2\delta}{3})(1+ \frac{\delta}{3})\sigma^{-1} \tag{32}$$

Where J_3 and $J_\perp$ are the moments of inertia parallel and perpendicular to the symmetry axis respectively. δ is the nuclear deformation parameter.

   Assuming $K=I\cdot\cos(\beta)$, eq.(30) describes a statistical distribution of β around 90°

$$\rho(\beta,I) = \rho(\beta=90^o,I) \cdot \exp(-I^2\cos^2(\beta)/2K_o^2) \tag{33}$$

which may be folded into eq.(26).

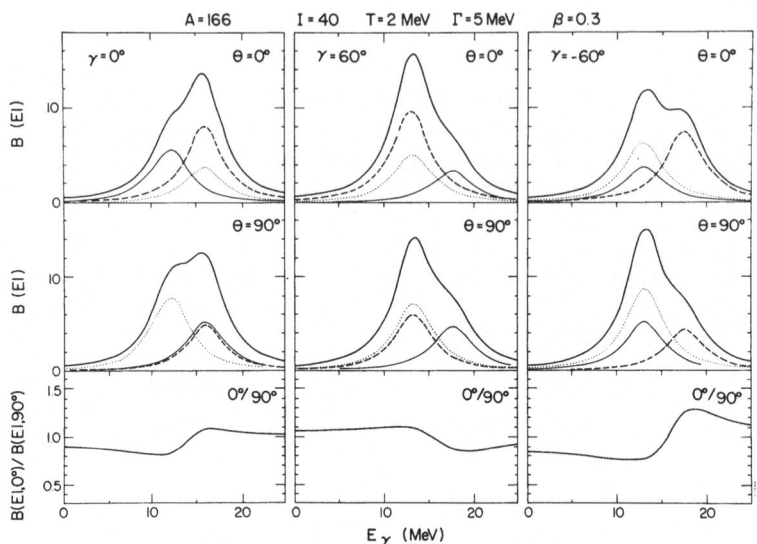

Fig.14. Calculated GDR strength function using the rotating
harmonic oscillator model described in ref.[30]. The
strength function is decomposed into the contribution
from the K=0 (dashed curve), and K=±1 (dotted and full-
drawn curves) components, which have been folded with
a Lorentzian function with a 5 MeV width. The function
is shown as it would appear at 0° and 90° with respect
to the beam direction in the laboratory system. Also
shown are the corresponding 0° to 90° anisotropies.
ERRATA: in the spectrum corresponding to γ=0° and
θ=90°, the dotted and full drawn lines should be exchanged.

For large values of the parameter $A=I^2/(2TJ_{diff})$ the integral can
be worked out explicitly yielding

$$\omega(\theta) \propto [1 + (\frac{1}{2} - \frac{3}{4A}) \cdot P_2(\cos(\theta))] \qquad (34)$$

which should be compared to eq.(27).

Similarly, with respect to the beam direction

$$\omega(\nu) \propto [1-(\frac{1}{4} - \frac{3}{8A}) \; P_2(\cos(\nu))] \tag{35}$$

It is seen that this attenuation can be quite important for smaller A values (a reduction from 1.5 to 1.1 in the 90° to 0° anisotropy for A=160, I=20, δ=0.3 and T=1.9 MeV). This emphasizes the need to select light nuclei, low temperatures and high spins for such studies.

4.3 The Shapes of ^{108}Sn* and ^{166}Er* at high E*.

It was seen in section 4.1 that the structure of the GDR at high excitation energy should be sensitive to the nuclear shape, as is the case for GDR's built on the ground state. In this section the spectra [13] from the decay of the systems ^{108}Sn* and ^{166}Er* are discussed. These nuclei were chosen since they are known to be spherical and prolate deformed, respectively, in the ground state. The details of the reactions are listed in table 1.

Spectra from the decay of ^{108}Sn* produced at E*=61 MeV with a ^{16}O beam and at E*=51 MeV with ^{16}O and ^{12}C beams are displayed in fig.15. Also shown are spectra calculated with the statistical model code CASCADE, in which the energy dependence of the electric dipole matrix elements is given by a Lorentzian function. The calculations, which represent best fits to the data, have been obtained by varying the centroid E_G^C, width Γ_G^C and strength S_G^C of the Lorentzian. The strength is taken to be the fraction of the classical E1 sumrule needed to reproduce the data. For all calculations discussed in this section a level density parameter a=A/7.5 was used. Due to the large statistical fluctuations at high E_γ, the experimental spectra cannot be unfolded. Furthermore, it is difficult to measure the detector response at these high transition energies. The procedure adopted has therefore been to fold the calculated spectra with a response function calculated with the EGS (Electron-Gamma-Shower) computer program (which was discussed in section 2) before comparing them to the data. The energy axis in fig.15 therefore denotes the energy recorded in the detector, which is lower than the true gamma ray energy due to the effect of the detector response.

The spectra in fig.15 are well described by statistical calculations using a single Lorentzian function with E_G^C = 15.5 MeV and S_G^C = 65%(E1) class. Due to the lack of an accurate absolute gamma ray normalization the total dipole sumrule strength is poorly determined. However, the resonance parameters E_G^C and Γ_G^C are quite insensitive even to substantial variations of S_G^C. The estimated uncertainties are ±0.5 MeV in E_G^C and ±0.7 MeV in Γ_G^C. In the case of the 84 MeV ^{16}O reaction, an increase of Γ_G^C from 6.0 to 6.5 MeV produces a better fit.

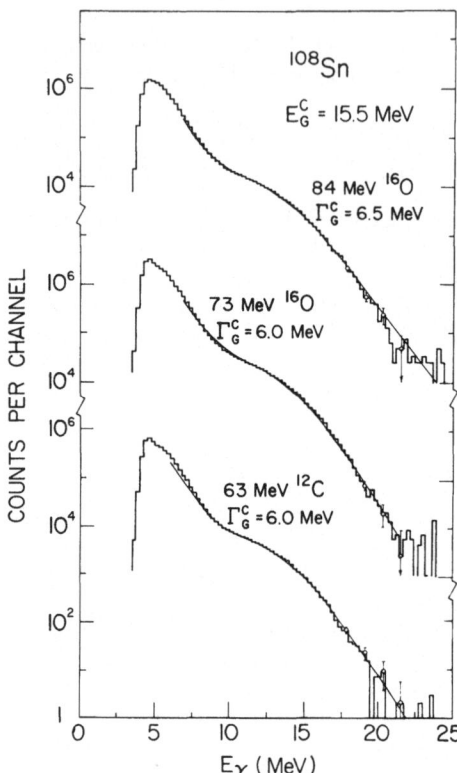

Fig.15. Measured gamma ray spectra from the decay of ^{108}Sn*. The abscissa represents the energy registered in the detectors. The data have been fitted with statistical model calculations done with the code CASCADE and folded with a calculated detector response function. The used GDR parameters are indicated in the figure and are explained in the text.

These widths are in excess of typical T=0 MeV values ($\Gamma \approx 4.9$ MeV for ^{120}Sn [2]). As previously mentioned, broadenings due to Coriolis contributions and effects related to the statistical distribution of shapes and spin orientations are investigated in ref.[31]. The authors find a 1-2 MeV increase in Γ_{GDR}, as compared to the ground state width for the temperature and spin region discussed here.

In fig.16a, the measured gamma ray spectrum from the ^{16}O + ^{150}Nd $\rightarrow$ ^{166}Er* reaction at 84 MeV is compared to the spectrum of the decay of ^{108}Sn*, formed using the same beam energy. The ^{166}Er* spectrum has also been fit with a statistical model calculation. In this case a reduced resonance energy (E_G^C=14.1 MeV), and a substantially larger width (Γ_G^C=9.5 MeV) would be needed in order to describe the spectral shape. However, an accurate fit to the data is obtained only if two Lorentzian functions are used. In fig.16a the width of each Lorentzian has been fixed to Γ_G^C=5.2 MeV. The energies and relative strengths of the two components are E_G^C(1)= 12.6 MeV, S_G^C(1)=60% and E_G^C(2)=16.4 MeV, S_G^C(2)=40%, respectively. The quality of these fits may be judged with more sensitivity in fig.16b. Here the GDR region has been emphasized in both experimental and calculated spectra by multiplying them by the statistical level density factor $\exp(E_\gamma/T_G)$, taking into account the different average temperatures associated with the two systems. It is clearly seen that the shape of the GDR in ^{166}Er* reflects a composite structure. The energy splitting deduced by using two GDR components of Lorentzian shape implies a nuclear deformation $\delta = \Delta E_G^C/E_G^C \approx 0.27 \pm 0.07$.

The deduced relative strengths of the ^{166}Er* GDR components ($\approx$60% in the lower peak) indicate a predominantly oblate shape although a weak admixture (<5%) from the isoscalar giant quadrupole resonance (GDR), expected to be centered at an energy approximately that of the lower GDR component, may be present. This pattern indicates a change of shape from prolate to oblate of ^{166}Er* with increasing I and/or E*.

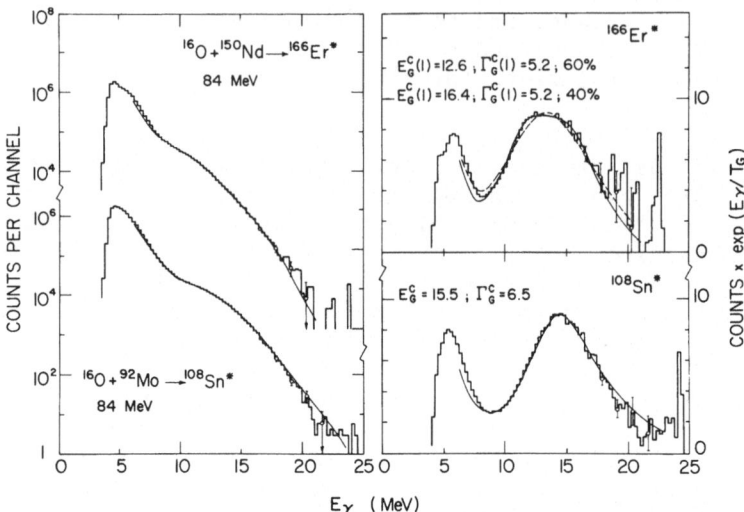

Fig.16. (a) Measured gamma ray spectra from the decay of ^{166}Er*
        (top) at E*=61.5 MeV and ^{108}Sn* (bottom) at E*=61.2 MeV,
        plotted as a function of the detected gamma ray energy.
        The ^{166}Er* spectrum has been fitted with a statistical
        model calculation using an electric dipole strength func-
        tion given by a double Lorentzian shape with the para-
        meter values indicated in (b). (b) Same spectra multi-
        plied by $\exp(E_\gamma/T_G)$, using $T_G(Sn)=1.6$ MeV and $T_G(Er)=1.4$
        MeV, and normalized to approximately the same height at
        E_γ=14-15 MeV. The ordinate is in arbitrary units. The
        quality of the fits and the different structures of the
        GDR's in the two systems is hereby emphasized.

5. THE HIGHEST TRANSITION ENERGIES

5.1 The Giant Quadrupole Resonance

The rapid and drastic improvement of the experimental techniques which we have witnessed over the last 3 years, warrants a
search for the next multipole of giant resonances built on excited
states in nuclei, namely the Giant Quadrupole Resonance (GQR).
While the isoscalar dipole mode is a spurious mode corresponding

to a tranlation of the nucleus as a whole, two quadrupole modes
exist.

The average excitation energy of the isoscalar mode has been
predicted by Mottelson [37], from microscopic considerations, to
be $E_{GQR}^{T=0} \simeq 58\ A^{-1/3}$ [MeV]. The energy of the isovector mode may be
calculated from the liquid drop model: $E_{GQR}^{T=1} \simeq 127\ A^{-1/3}$ [MeV].
While the isoscalar mode lies only slightly below the GDR in
energy, the typical excitation energy for the isovector mode is
above 20 MeV, i.e. well above $\bar{E}_{GDR}$.

The strength of the GQR relative to that of the GDR may be
estimated [3] from the decay rates for electric radiation of multipolarity λ

$$T(E\lambda;I_1 \rightarrow I_2) = \frac{8\pi(\lambda+1)}{\lambda[(2\lambda+1)!!]^2} \cdot \frac{1}{\hbar} \cdot q^{2\lambda+1} \cdot B(E\lambda;I_1 \rightarrow I_2) \qquad (36)$$

with $q = 6.1\ 10^{-3}A^{1/3}E_\gamma$ [MeV]. The reduced transition matrix elements $B(E\lambda)$ may in the first rough approximation be estimated
from the single particle moments or more accurately by relating
them via eq.(3) to the sumrules for dipole (eq.(4)), and quadrupole
multipole operators.

By assuming that $\langle r^2 \rangle = \frac{3}{5}(1.2)^2 A^{2/3}$ the classical quadrupole
sumrule may be written in closed form

$$S(E2)_{class} = \frac{15}{2\pi} \frac{\hbar^2}{2M}\ Z\ e^2(1.2)^2 A^{2/3} \qquad (37)$$

Insertion of the appropriate values and units yields a ratio (for
$A \approx 100$):

$$[\Gamma(E1) \cdot E_\gamma^{-3}]\ /\ [\Gamma(E2) \cdot E_\gamma^{-5}] \simeq 10^{-4}$$

Although the energy factor in eq.(36) will favour quadrupole radiation at the very high transition energies, the expected admixture
of isoscalar GQR modes into the GDR bump should be small ($=<2-3\%$).
At higher E_γ however, on the high energy tail of the GDR, the
isovector GQR should contribute significantly to the spectrum.
As for the GDR, the GQR is expected to split into various components in deformed nuclei [38]. In axially symmetric nuclei 3 components are expected corresponding to $K=\pm2$, $K=\pm1$ and $K=0$. For
oblate nuclei the $|K|=2$ modes will lie lowest in energy and $K=0$
highest. In rotating nuclei the degeneracy of the $K=\pm2$ and the

K=±1 modes is lifted. The angular distribution of the various K components can be worked out similarly to the deviations in section 4, yielding the angular distribution function for quadrupole radiation with respect to the spin direction

$$\omega_2(\theta,\beta) = \frac{2}{5}[1+\frac{5}{7}P_2(\cos(\beta))\cdot P_2(\cos(\theta)) - \frac{12}{7}P_4(\cos(\beta))\cdot P_4(\cos(\theta))]$$

(38)

5.2 The Isovector Giant Quadrupole Resonance in ^{111}Sn*

Spectra [14] of ^{111}Sn* at E*≈66 MeV and E*≈99 MeV are displayed in fig.17. It is noted that the spectrum corresponding to the reaction of highest E* extends to E_γ≈35 MeV, i.e. covering almost twice the transition energy range so far measured in this type of study. This improvement is due to the refinement of the experimental techniques, which has reduced the cosmic ray background by a factor of ≈10^3 as compared to the earliest measurements (fig.7), but also to the influence of the higher temperature. The latter effect may be estimated from the exponential factor appearing in eq.(17) which increasingly favours the emission of very high energy gamma rays when the temperature is raised. This transition energy range, apart from ensuring complete statistical definition of the GDR region, is also wide enough to encompass the transition energy region where gamma rays from the decay of isovector giant quadrupole resonance (GQR) modes are expected.

It is noted that in the spectrum of highest E_γ there is a change of slope around E_γ≈20 MeV. Attempts to fit both spectra with statistical model calculations using a gamma emission probability in competition with particles given, in analogy with eq. (17), by

$$\frac{\Gamma_\gamma}{\Gamma_p} \propto [K(E1)\cdot S(E1)\cdot E_\gamma^3\cdot f_{GDR}+K(E2)\cdot S(E2)\cdot E_\gamma^5\cdot f_{GQR}]\cdot\exp[\frac{S_p-E_\gamma}{T}]\cdot\frac{1}{T^2}$$

(39)

show that consistent fits for both energies are only obtained by including a GQR component at E_γ≈27 MeV with full EWSR strength [14].

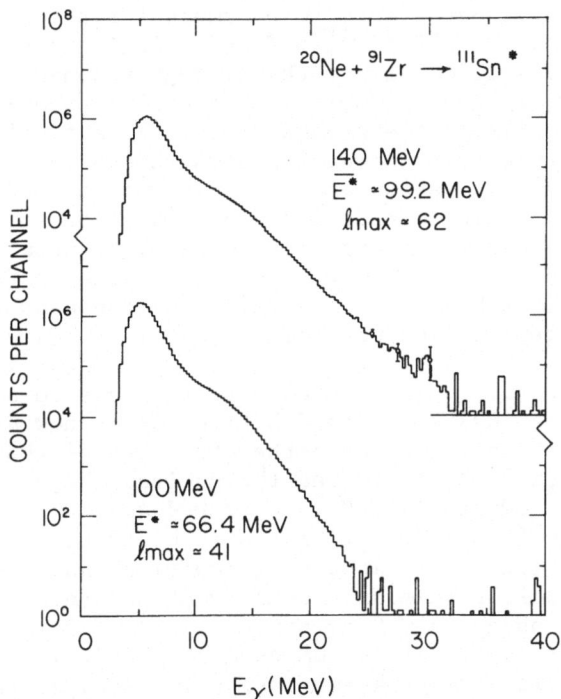

Fig.17. Measured gamma ray spectra from the decay of ^{111}Sn*,
       formed with a ^{20}Ne beam at E*=99.2 MeV (top) and
       E*=66.4 MeV (bottom).

6. SUMMARY AND CONCLUSION

The gamma rays from the deexcitation of GDR modes are mainly emitted from the region of highest excitation energy populated in fusion reactions. Consequently, the GDR strength function which is measured contains mainly information about the structure of the compound nucleus and of the first few evaporation products. This selectivity can be exploited to study the influence of the spin and temperature degrees of freedom on states far above the yrast line.

The extracted characteristic parameters for the GDR's built on excited states in ^{166}Er*, ^{108}Sn* and ^{111}Sn* are summarized in table 2. For all studied cases, the average resonance energies, measured at $T \neq 0$, agree very well with the corresponding values for resonances built on the ground state. In particular, no change of E_{GDR} is observed with increasing temperature in the Sn isotopes, for which a large excitation energy range was covered. This indicates that the coupling constant of the vibration is not significantly affected by the temperature increase. It should be noted that this conclusion is in contradiction to data on Er isotopes presented at this school by D. Schwalm. The cause of this discrepancy is at present not understood.

Although the inclusion of temperature tends to smooth the fine detail of the GDR strength function, it is now established that a clear correlation between the structure of the observed GDR's built on excited states and the nuclear shape exists. A splitting of the GDR in ^{166}Er* is observed which can be related to a deformation $\beta=0.27\pm0.07$. This in in clear contrast to ^{108}Sn* where no such splitting is apparent, implying that the spherical shape of this nucleus is mostly conserved. The relative intensities of the two observed GDR components in ^{166}Er* indicate a dominant proportion of oblate shapes at high E*. This is in contrast to the prolate shape known at T=0 for this nucleus and suggests a change of deformation with increasing E* and I. This means that changes in deformation may be followed, if differential measurements as a function of E* and I are done. It may then also be possible to study in detail the transition between the low E* shell-structure dominated region and the region of higher E* where the contribution from the shell structure energy to the free energy has vanished. In particular future improvement of angular distribution measurements should provide an important tool for such studies.

The systematics of the GDR width as a function of excitation energy is exhibited in fig.18 for the studied Sn isotopes. It was seen in section 4 that the structure of the GDR in ^{108}Sn* was well described by a single Lorentzian function with the same average energy as for resonances built on the ground state. The

Table 2. Average resonance energies and widths for the GDR's built on excited states. The parameters have been obtained by fitting the experimental data with statistic model calculations as discussed in sections 3 and 4. In ^{166}Er* the GDR is split into two components whose energies and widths are listed in addition to the parameters obtained by fitting the resonance with a single component.

Compound	Beam	E(MeV)	$\bar{E}_{GDR}$(MeV)	Γ_{GDR}(MeV)
^{166}Er*	^{16}O	61	14.1	9.5-10
			(12.6;16.4)	(5.2;5.2)
^{108}Sn*	^{16}O	61	15.5	6.5
	^{16}O	51	15.5	6.0
	^{12}C	51	15.5	6.0
^{111}Sn*	^{20}Ne	99	15.5	7.5
	^{22}Ne	66	15.5	11.0

width, however, was found to be 1-2 MeV larger than for measurements at T=0 MeV. There may be several possible explanations for this effect: i) a change in the nuclear deformation towards an oblate shape when I and E* increase; ii) a classical broadening effect due to the coexistence of many shapes at high T and to the wobbling of the nucleus around the direction of the spin vector, and iii) an increase in the spreading width $\Gamma\downarrow$ (which is a measure of the damping of the vibration). Including the measurements on ^{111}Sn* shows that a strong correlation between Γ_{GDR} and E* exists. The present experiments cannot distinguish between the 3 possibilities listed above, although it seems unlikely that possibility 2 alone could account for a width increase of the observed magnitude. It might, however, explain the widths at E*≃50-60 MeV.

An example of $\Gamma\downarrow$ calculations at T=0 in spherical nuclei is given in ref.[39]. Generally speaking an increase in the damping width of the vibration is expected with increasing temperature, due to the possibility of creating more complex excitations (2p2h,....). At higher temperatures recent microscopic calculations of Γ_{GDR} by I. Gallardo et al. [40] are now available for the Sn isotopes (contribution to this school). These calculations include the effect of a changing and fluctuating deformation as a function of T and I. The calculations can essentially reproduce the observed widths at E*≃60 MeV (Γ_{GDR}(calc)=7 MeV) and at E*=100 MeV

($\Gamma_{GDR}(calc)=10.5$ MeV). The main reason for this width increase is, according to these calculations which, do not include the coupling to complex excitations, due to a change of the nuclear shape from spherical to oblate (with β=0.25) driven by rotation.

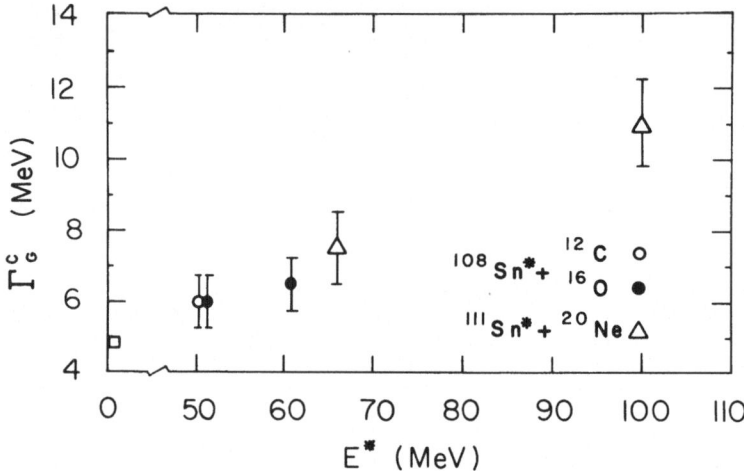

Fig.18. Evolution of the GDR width Γ_G^C in the studied Sn
        isotopes with excitation energy. The width has been
        obtained by fitting the experimental spectra with
        statistical model calculations, as described in the
        text.

If the gamma rays in the GDR region mainly are emitted in the first stages of the deexcitation of the compound nucleus, this selectivity is strongly enhanced for large transition energies (see fig.8). Indeed, for transition energies above ≈25 MeV, the gamma rays will almost exclusively be emitted from the compound nucleus. The study of the GQR may therefore constitute a unique tool to investigate highly excited nuclear states in a narrow E* interval. The experiments on [111]Sn* discussed here are the first reaching to such high gamma ray energies. A statistically significant excess of counts above the GDR contribution is seen, which most likely is related to the deexcitation of GQR modes built directly on states in the compound nucleus.

While the field of excited state GDR's is still in its infancy, the remarkable progress which has been achieved by the groups working in this field over the last few years, already shows that detailed nuclear structure information may be obtained far from the yrast line, presenting nuclear spectroscopy with a new degree of freedom to explore.

REFERENCES

[1]   G. G. Baldwin and G. S. Klaiber, Phys. Rev. 71:3 (1947).
[2]   B. L. Berman, Nucl. Data Tables 15:321 (1975).
[3]   A. Bohr and B. Mottelson, Nuclear Structure, vols.1 and 2, Benjamin,Reading, Mass. (1975).
[4]   M. Goldhaber and E. Teller, Phys. Rev. 74:1046 (1948).
[5]   H. Steinwedel and J. H. D. Jensen, Z. Naturf. 5a:413 (1950).
[6]   J. P. Elliott and B. H. Flowers, Proc. Roy. Soc. A242:57 (1957).
[7]   D. M. Brink, Ph.D. Thesis, University of Oxford (1955)
[8]   G. E. Brown and M. Bolsterli, Phys. Rev. Lett. 3:472 (1959).
[9]   P. P. Singh, R. E. Segel, L. Meyer-Schutzmeister, S. S. Hanna and R. G. Allas, Nucl. Phys. 65:577 (1965).
[10] J. O. Newton, B. Herskind, R.M. Diamond, E. L. Dines, J. E. Draper, K. H. Lindenberger, C. Schuck, S. Shih and F. S. Stephens, Phys. Rev. Lett. 46:1383 (1981).
[11] J. J. Gaardhøje, XX Int. Winter Meeting on Nuclear Physics, Bormio, Italy (1982).
[12] J. J. Gaardhøje, O. Andersen, R. M. Diamond, C. Ellegaard, L. Grodzins, B. Herskind, Z. Sujkowski and P. M. Walker, Phys. Lett. 139B:273 (1984).
[13] J. J. Gaardhøje, C. Ellegaard, B. Herskind and S. G. Steadman, Phys. Rev.Lett. 53:148 (1984).
[14] J. J. Gaardhøje, C. Ellegaard, B. Herskind, M. A. Deleplanque, R. M. Diamond, E. L. Dines, A. Machiavelli and F. S. Stephens (to be published in Phys. Rev. Lett.).
[15] J. J. Gaardhøje, XXII Int. Winter Meeting on Nuclear Physics, Bormio, Italy (1984), vol.2.
[16] J. E. Draper, J. O. Newton, L. G. Sobotka, K. H. Lindenberger, G. J. Wozniak, L. G. Moretto, F. S. Stephens, R. M. Diamond and R. J. McDonald, Phys. Rev. Lett. 49:434 (1982)
[17] W. Hennerici, V. Metag, H. J. Hennrich, R. Repnov, W. Wahl, D. Habs, K. Helmer, U.V. Helmolt, H. W. Heyng, B. Kolb, D. Pelte, D. Schwalm, R. S. Simon and R. Albrecht, Nucl. Phys. A396:329 (1983).
[18] A. M. Sandorfi, J. Barrette, M. T. Collins, D. H. Hoffmann, A. J. Kreiner, D. Branford, S. G. Steadman and J. Wiggins, Phys. Lett. 130B:19 (1983).
[19] E. F. Garman, K. A. Snover, S. H. Chew, S. K. B. Hesmondhalgh and W. N. Catford (to be published).

[20] R. L. Ford and W. R. Nelson, SLAC report no. 210 (1978).
[21] T. D. Thomas, Nucl. Phys. 53:577 (1964).
[22] F. Pühlhofer, Nucl. Phys. A280:267 (1977).
[23] P. Axel, Proc. of Int. Symp. on Nuclear Structure, Dubna
      1968. IAEA Vienna (1968) p.299.
[24] E. Hayward, Nucl.Structure and Electromagn.Interactions, Scot-
      tish Summer School 1964. Oliver and Boyd Press (1965) p.141
[25] K. Snover, Proc. of Masurian Summer School on Nuclear Phy-
      sics, Mikolajki, (1982) Poland.
[26] E. Garman: private communication (to be published).
[27] J. E. Lynn, The Theory of Neutron Resonance Reactions,
      Clarendon Press, Oxford (1968).
[28] E. G. Fuller and E. Hayward, Nucl. Phys. 30:   (1962)
[29] J. G. Valatin, Proc. Royal Society (London) A238:132 (1956).
[30] K. Neergaard, Phys. Lett. 110B:7 (1982).
[31] T. Døssing and K. Neergaard, Nordic Meeting on Nuclear
      Physics (Fuglsø 1982), and to be published.
[32] J. L. Egido, H. J. Mang and P. Ring, Nucl. Phys. A339:390
      (1980).
[33] J. L. Egido and P. Ring, Nucl. Phys. A338:19 (1982) and
      Phys. Rev. C25:3339 (1982).
[34] M. Faber, J. L. Egido and P. Ring, Phys. Lett. 127B:5
      (1983).
[35] P. Ring, L. M. Robledo, J. L. Egido and M. Faber (to be
      published).
[36] A. de Shalit and H. Feshbach, Theoretical Nuclear Physics,
      vol.1, John Wiley & Sons (1974).
[37] B. Mottelson, Proc. Int. Conf. on Nuclear Structure, King-
      ston. Ed. by D. A. Bromley and E. W. Vogt, University of
      Toronto Press, p.525.
[38] S. Åberg, private communication and Z. Szymanski, private
      communication.
[39] P. F. Bortignon and R. A. Broglia, Nucl. Phys. A371:405
      (1981).
[40] I. Gallardo, R. Broglia, M. Diebel, private communication
      (to be published).

QUANTUM TUNNELING MULTIDIMENSIONAL SYSTEMS

D.M. Brink

Department of Theoretical Physics
1 Keble Road
Oxford OX1 3NP

INTRODUCTION

The description of quantum tunneling in systems with many degrees of freedom is a fundamental problem in many areas of theories,[1],[2] tunneling of magnetic flux in solids,[3] sub-barrier fusion in nuclei,[4],[5],[6] fission,[7] and transfer of nucleons between nuclei,[8],[9],[10]. The purpose of these lectures is to discuss various approaches to the problems of tunneling in several dimensions and to give some examples of applications. Four basic methods will be considered. They are perturbation theory, coupled channels, WKB and path integrals with approximately one lecture devoted to each approach. The first lecture will begin with a review of the WKB method in one dimension.

TUNNELING BY WKB IN ONE DIMENSION

The penetration of a particle through a potential barrier can be calculated approximately by using the WKB method. A typical barrier is shown in Fig. 1.

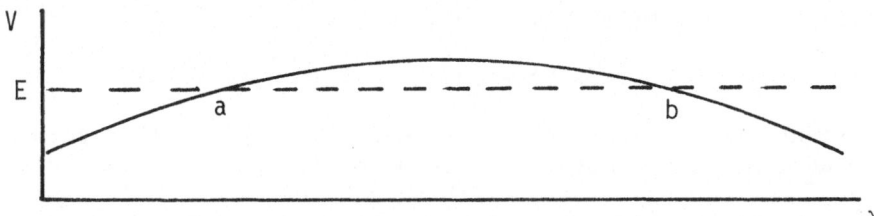

Fig. 1. A potential barrier with classical turning points at x = a and x = b.

The classical turning points are determined by the condition
$V(x) = E$.   The incident wave function to the left of the barrier
and the transmitted wave to the right are

$$\psi(x) = k^{-\frac{1}{2}} \, [\exp(i\omega(x,a)) + R \, \exp(-i\omega(x,a))]; \; x < a \qquad (1)$$

$$= k^{-\frac{1}{2}} \, T \, \exp(i\omega(x,b)) \qquad\qquad ; \; x > b$$

where $k(x) = [(2m/h^2)(E - V(x1)]^{\frac{1}{2}}$, $\omega(x,x') = \int_{x'}^{x} k(x)dx.$ (2)

Underneath the barrier when $E < V(x)$ the wave function is a super-
position of exponentially decreasing and increasing components

$$\psi(x) = \gamma^{-\frac{1}{2}}[B_- \, \exp(-Q(x,a) + B_+ \, \exp(Q(x,a))] \qquad (3)$$

when $a < x < b$ with

$$\gamma(x) = [(2m/\hbar^2)(V(x) - E)]^{\frac{1}{2}} , \; Q(x,x') = \int_{x'}^{x} \gamma(x) \, dx. \qquad (4)$$

WKB connection formulae can be found by using an Airy function
connection formula,[11] or by analytic continuation,[12].   The
connection formula relating an oscillating solution for $x < a$ to a
decaying solution for $x > a$ is

$$\psi(x) \sim A \, k^{-\frac{1}{2}} \, \sin(\omega(a,x) + \tfrac{1}{4}\pi) \; ; \; x < a$$
$$\qquad\qquad\qquad\qquad\qquad\qquad\qquad\qquad (5)$$
$$\psi(x) \sim \tfrac{1}{2}A\gamma^{-\frac{1}{2}} \, \exp(-Q(x,a)) \qquad ; \; x > a$$

while near the turning point b

$$\psi(x) \sim A'k^{-\frac{1}{2}} \, \sin(\omega(x,b) + \tfrac{1}{4}\pi) \; ; \; x > b$$
$$\qquad\qquad\qquad\qquad\qquad\qquad\qquad\qquad (6)$$
$$\psi(x) \sim \tfrac{1}{2}A'\gamma^{-\frac{1}{2}} \, \exp(-Q(b,x)) \qquad ; \; x < b$$

Equation (6) relates an exponentially small solution for $x < b$ to
an oscillating solution for $x > b$.   There is another connection
formula which relates an exponentially large solution for $x < b$ to
an oscillating solution for $x > b$.   It is

$$\psi(x) \sim D\gamma^{-\frac{1}{2}} \, \exp(Q(b,x)) \qquad ; \; x < b$$
$$\qquad\qquad\qquad\qquad\qquad\qquad\qquad\qquad (7)$$
$$\sim k^{-\frac{1}{2}} \, [D \cos(\omega(x,b) + \tfrac{1}{4}\pi) + E \sin(\omega(x,b) + \tfrac{1}{4}\pi)]$$

The constant E in the connection formula (7) is not well determined
because adding a component of the solution (6) to (7) hardly changes
the wave function for $x < b$, but changes E for $x > b$.   The WKB
formula for the barrier penetration coefficient is obtained by
using the connection formulae (5) at a and (7) is b.   At $x = a$
we obtain

$$R = \exp(-i\pi/2) \quad ; \quad B_- = \exp(-i\pi/4) \tag{8}$$

At $x = b$ we put $D = T$ and $E = iT$ in order to get an outgoing wave. Then the connection formula gives

$$T = B_- \exp(-Q(b,a) + i\pi/4) = \exp(-Q(b,a)) \tag{9}$$

An interesting point about the derivation is that there is no formula for B_+. This component adjusts to a value which gives the outgoing wave boundary condition for $x > b$. Equation (9) gives a barrier penetration coefficient

$$P = |T|^2 = \exp(-2Q(b,a)) \tag{10}$$

The WKB formula (10) is valid only if the incident energy is well below the top of the barrier in the sense that $Q(b,a)$ is large compared with unit. Kemble,[13] has given a more accurate formula which is valid both below and above the barrier top

$$P = |T|^2 = (1 + \exp 2Q)^{-1} \quad , \quad |R|^2 = (1 + \exp(-2Q))^{-1} \tag{11}$$

In the case of a parabolic barrier

$$V(x) = V_B - \tfrac{1}{2} m \omega^2 x^2 \tag{12}$$

$$Q = \pi (E - V_B)/\hbar\omega \tag{13}$$

A PERTURBATION FORMULA

In this section we will derive a perturbation formula for a multidimensional barrier penetration problem. Consider a single particle moving in a potential field $V(r) = V_1(r) + V_2(r)$ and suppose that the two particles do not overlap much so that the particle has to tunnel through a potential barrier to reach the final state if it is initially in a state ψ_i in the potential V_1 and finally in a state ψ_f in the potential V_2. The perturbation approach gives a Fermi Golden Rule formula for the transition probability ω per unit time for making the transition

$$\omega = (2\pi/h)| < \psi_f | V_1 | \psi_i > |^2 \rho_f (E) \tag{14}$$

In (14) the initial and final states are eigenstates for the particle moving in the potentials V_1 and V_2 respectively

$$(T + V_1)\psi_i = E_i \psi_i \quad , \quad (T + V_2)\psi_f = E_f \psi_f \tag{15}$$

The initial and final states must have almost the same energy $(E_i \simeq E_f)$ and the final potential $V_2(r)$ has to be extensive enough for there to be an almost continuous distribution of final states. The density of final states is $\rho_f(E)$.

The derivation of (14) follows the usual route.   First we obtain a transition probability to a definite final state, then we sum over all final states with $E_f \simeq E_i$.   When making the first step we allow the potentials $V_1(\vec{r})$ and $V_2(r)$ to be time-dependent and replace (15) by the equivalent time-dependent equations

$$i\hbar\, \delta\psi_i/\delta t \;=\; (T + V_1)\psi_i \tag{16}$$

$$i\hbar\, \delta\psi_f/\delta t \;=\; (T + V_2)\psi_f \tag{17}$$

The exact wave function satisfies the equation

$$i\hbar\;\; \partial\psi/\delta t \;=\; (T + V_1 + V_2)\psi \tag{18}$$

because the particle is influenced by both potentials. Eq. (18) has to be solved with the initial condition

$$\psi(r,t) \;=\; \psi_i(r,t) \quad ; \quad t = t_o \tag{19}$$

The amplitude for the particle to be in the state ψ_f at time t is the overlap

$$A_{fi}(t,t_o) \;=\; \langle\psi_f(r,t)\,|\psi(r,t)\rangle \tag{20}$$

Using (16) and (18) we obtain

$$i\hbar\;(\delta/\delta t)\langle\psi_f|\psi\rangle \;=\; \langle\psi_f|\, T + V_1 + V_2|\psi\rangle - \langle\psi_f\,|\, T + V_2\,|\psi\rangle$$

$$=\; \langle\psi_f|\, V_1|\psi\rangle \tag{21}$$

$$=\; \langle\psi_f|\, V_1|\psi_i\rangle \tag{22}$$

Equation (21) is exact while (22) is obtained by making the perturbation approximation $\psi \simeq \psi_i$.   Integrating both sides of (22)

$$A_{fi}(t_1,t_o) \;=\; (1/i\hbar) \int_{t_o}^{t_1} \langle\psi_f\,|\,V_1\,|\,\psi_i\rangle \; dt \tag{23}$$

if $A_{fi}(t_o,t_o) = 0$.   Equation (23) has the usual structure of a time-dependent perturbation formula.   The step from (23) to (14) follows the usual path in the derivation of a Fermi Golden Rule formula.

The perturbation formula (14) has been used by Macdonald,[14] and Tobocman,[15].   It is given in a somewhat more general form in the second French edition of Messiah,[16] p.856.

The exact formula for the penetration factor of a square potential barrier of height V_o and thickness b is

$$P = \frac{4k^2\gamma^2}{4k^2\gamma^2 \cosh^2(\gamma b) + (k^2-\gamma^2)^2 \sinh^2(\gamma b)}$$

where

$$k = (2mE/\hbar^2)^{\frac{1}{2}} \qquad ; \qquad \gamma = (2m(V_o-E)/\hbar^2)^{\frac{1}{2}}$$

and E is the incident energy.   If $\gamma b \gg 1$ then

$$\cosh^2(\gamma b) \simeq \sinh^2(\gamma b) \simeq \exp(2\gamma b)/4$$

and $\quad P = \dfrac{16k^2\gamma^2\exp(-2\gamma b)}{(k^2 + \gamma^2)^2}$ \hfill (24)

It is an interesting exercise to derive (24) from the perturbation formula (14).

The WKB Penetrability from Perturbation Theory

The WKB penetrability formula (10) can be derived from the perturbation equation (14).   The derivation cannot be made by simply substituting WKB wave functions into (14) because these diverge at the classical turning points.   The first step is to write the matrix element of V_1 in another way

$$\langle\psi_f | V_1 | \psi_i\rangle \simeq (\hbar^2/2m) \int_\Sigma ds. \ [(\psi_f^*\nabla\psi_i) - (\psi_i \nabla \psi_f^*)] \qquad (25)$$

We use (15) to derive equation (25) and suppose that $E_f = E_i = E$. We assume that the two potentials V_1 and V_2 do not overlap so that it is possible to draw a surface Σ between the potentials in order that $V_1(r) = 0$ for all points r in the region R_2 to the right of Σ while $V_2(r) = 0$ for all r in the region R_1 to the left of Σ. In one dimension the surface Σ is a point and it is always possible to satisfy the above condition by choosing $V = 0$ at the top of the barrier between the two potentials.

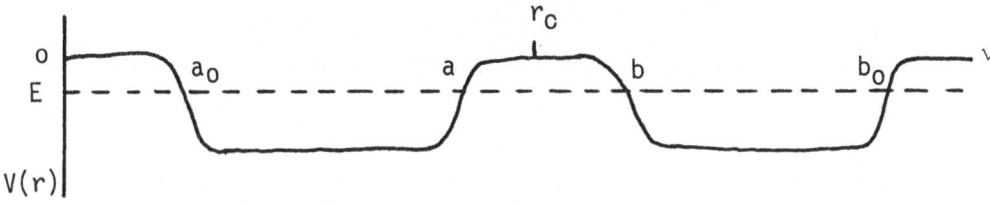

Fig. 2.   Tunneling between the two potentials.

To prove (25) we write

$$\langle \psi_f | V_1 | \psi_i \rangle = \int_{R_1} \psi_f^* V_1 \psi_i \, d\underset{\sim}{r}$$

$$= \int_{R_1} \psi_f^* (E + (\hbar^2/2m)\nabla^2) \psi_i \, d\underset{\sim}{r}$$

$$= (\hbar^2/2m) \int_\Sigma ds. \, [\psi_f^* \nabla \psi_i - \psi_i \nabla \psi_f^*]$$

$$+ \int_{R_1} (V_2 \psi_f)^* \psi_i \, d\underset{\sim}{r} \tag{26}$$

The first step holds because $V_1(r) = 0$ if r is in the R_2. The second step replaces V_1 by $E - T$ using (15). Then there is an integration by parts. Finally in the second integral in (26) is zero because $V_2(r) = 0$ for r in R_1. There is an analogous result which holds for time-dependent potentials

$$\int_{-\infty}^{\infty} dt \, \langle \psi_f | V_1 | \psi_i \rangle = \frac{\hbar^2}{2m} \int dt \int_{-\infty}^{\infty} ds. \, [\psi_f^* \nabla \psi_i - \psi_i \nabla \psi_f^*] \tag{27}$$

It can be proved in a similar way to (25) but now ψ_i and ψ_f satisfy (16) and (17).

In one dimension the surface reduces to a single point r_o between the two potentials. Then (25) can be evaluated using WKB wave functions (5) and (6).

$$\psi_i = \tfrac{1}{2} C_1 \gamma^{-\frac{1}{2}} \exp\left(-\int_a^{r_o} \gamma(r) dr\right), \quad \psi_f = \tfrac{1}{2} C_2 \gamma^{-\frac{1}{2}} \exp\left(\int_b^{r_o} \gamma(r) dr\right) \tag{28}$$

and the transition probability becomes

$$\omega = (2\pi/h)(\hbar^2/4m)^2 C_1^2 C_2^2 \exp(-2Q) \, dn/dE \tag{29}$$

This equation can be simplified by using WKB normalization conditions for C_1 and C_2,[12]

$$\int \psi_1^2 \, dr = 1 = \tfrac{1}{2} C_1^2 \int_{a_o}^{a} dr/k(r) = (C_1^2 \, \hbar/2m) \int_{a_o}^{a} dr/\dot{r} = C_1^2 \hbar T_1/4m$$

where T_1 is the period of the classical motion in V_1. Hence

$$C_1^2 = (4m/\hbar T_1) \quad ; \quad C_2^2 = (4m/\hbar T_2) \tag{30}$$

The density of states is calculated from Bohr's quantization condition $\int k(x) dx = n\pi$

$$\pi dn/dE = \int_b^{b_o} dk/dE \, dr = (1/h) \int_b^{b_o} dr/\dot{r} = T_2/2\hbar \tag{31}$$

Using (3 and (31) the perturbation formula (29) simplifies to the
WKB result,

$$\omega = (1/T_1) \exp (-2Q) \tag{32}$$

THE COUPLED CHANNELS METHOD

Now we consider the effects of intrinsic degrees of freedom on
the quantum tunneling of a collective variable by the coupled
channels method.   We consider a hamiltonian of the form

$$H = - (\hbar^2/2m)(d^2/dr^2) + V_o(r) + h(r,\xi) \tag{33}$$

$$h(r,\xi) = H_o(\xi) + V(r,\xi) \tag{34}$$

Here the tunneling through a barrier in $V_o(r)$ is modified by coupling
to the intrinsic degree of freedom ξ; $H_o(\xi)$ is the hamiltonian for
the intrinsic degrees of freedom and $V(r,\ )$ represents the coupling.
The Schrödinger equation can be written as a set of coupled
equations by expanding the wave function

$$\Psi(r,\xi) = \Sigma_\sigma \phi_\alpha (r) \chi_\alpha(\xi) \tag{35}$$

where $\chi_\alpha(\xi)$ is a complete basis of intrinsic states.   The coupled
equations are

$$(T + V_o(r) - E) \phi_\alpha(r) + \Sigma_\beta h_{\alpha\beta} (r) \phi_\beta(r) = 0 \tag{36}$$

where $h_{\alpha\beta} = \langle\chi_\alpha|h|\chi_\beta\rangle$ and $T = -(h^2/2m)(d^2/dr^2)$ is the kinetic energy
operator associated with r.

Two Level Model

In general coupled equations like (36) can only be solved
numerically but there are some simple special cases which can be
solved by analytical methods.   One is a two level model considered
by Dasso et al.,[17]

$$(T + V_o(r) - E) \phi_1 = F(r)\phi_2$$
$$(T + V_o(r) - E) \phi_2 = F(r)\phi_1 \tag{37}$$

It is clear that these equations can be decoupled by introducing

$$\phi_\pm = \phi_1 \pm \phi_2 \tag{38}$$

to give uncoupled equations for $\phi_\pm(r)$

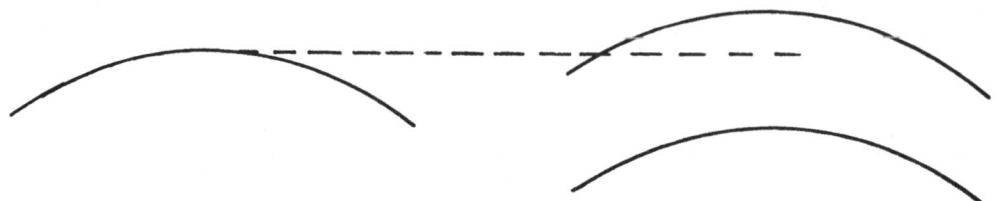

Fig. 3.   Barrier split in two by coupling

$$(T + V_o(r) \pm F(r) - E) \; \phi_{\pm}(R) = 0 \tag{39}$$

By satisfying the appropriate boundary conditions one finds that
the penetration coefficient P_1 for a wave incident on the target
in its ground state χ_1 is

$$P_1 = q_+ \, P_+ + q_- \, P_- \tag{40}$$

where $P_\pm$ are the penetration coefficients for the barriers
$V_o(r) \pm F(r)$ and $q_\pm$ are the probabilities of being in the uncoupled
states at the barrier.   In this example $q_+ = q_- = \frac{1}{2}$.

      This result is illustrated by the following figures,[17].   In
the absence of coupling the incident flux encounters the barrier
to the left.   When the coupling is turned on the barrier splits in
two.   Half of the flux encounters the lower barrier and half the
higher barrier.   The effect on the transmission coefficient as a
function of energy is illustrated in Fig. 4.   The left hand side·
of Fig. 4 shows $P(E)$ without coupling and the right hand side shows
the effect of coupling.   We see immediately that the effect of
coupling is to enhance the transmission coefficient below the
barrier and to reduce it above the barrier.

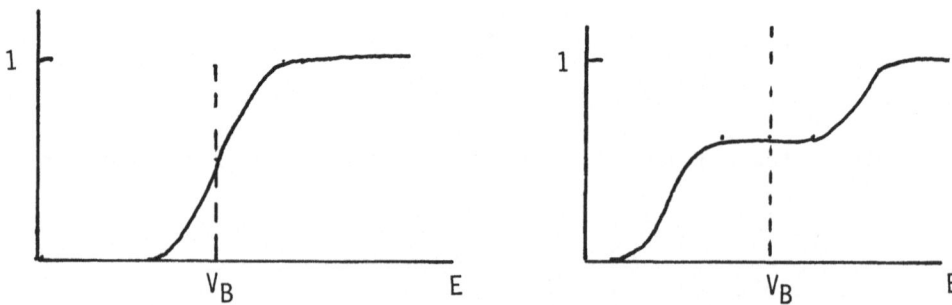

Fig. 4.   Effect of coupling on penetration coefficient.

Constant Coupling Model

Another exactly solveable model is obtained by assuming that the coupling matrix elements $h_{\alpha\beta}$ in (36) are constant, that is independent of position. Then the equations can be uncoupled by diagonalizing the coupling.[18]

$$\xi_i(r) = \Sigma_\alpha U_\alpha \phi_\alpha(r)$$

$$\Sigma_{\alpha\beta} U_{i\alpha} h_{\alpha\beta} U_{\beta j}^{-1} = \lambda_i \delta_{ij} \qquad (41)$$

The uncoupled equations which result from this transformation are

$$(T + V_0(r) + \lambda_i - E) \xi_i(r) = 0 \qquad (42)$$

and the penetration coefficient in the channel α is

$$P_\alpha(E) = \Sigma_i |U_{i\alpha}|^2 P_0(E - \lambda_i) \qquad (43)$$

where $P_0(E)$ is the penetration coefficient with no coupling. The interpretation of (43) is that transmission occurs through certain eigenchannels of transition states labelled by i. The transition state i has an effective barrier height $V_B + \lambda_i$. The quantity $q_i = |U_{i\alpha}|^2$ is the probability of arriving at the barrier in the transition state i and $P(E - \lambda_i)$ is the transmission probability in that state.

There is an interesting special case of a two level problem[18] with a coupling matrix

$$h_{\alpha\beta} = \begin{pmatrix} O & F \\ F & -Q \end{pmatrix}$$

In this example the entrance channel is coupled to a reaction channel with a finite Q-value. Diagonalizing the coupling matrix gives

$$\lambda_\pm = (- Q \pm (Q^2 + 4F^2)^{\frac{1}{2}})/2 \qquad (44)$$

The probabilities $q_\pm$ for the two eigenchannels are

$$q = F^2/ (F^2 + \lambda_\pm^2) \qquad (45)$$

The eigenchannel with the larger value of λ^2 has the smaller probability.

The consequences of this coupling depend dramatically on the sign of Q. When $Q < 0$ there is a small change in penetrability for energies above the unperturbed barrier and the effect is hardly

noticeable.   When $Q > 0$ there is a large enhancement for energies
below the barrier and the penetrability can be changed by an order
of magnitude.

Angular Momentum Coupling

This case has been discussed by Lindsay and Rowley,[19] in the
context of sub-barrier fusion reactions.   Consider the case where
a 0^+ ground state of the target is coupled to a 2^+ excited state.
The simplest description is obtained by choosing the quantization
axis along the line joining the centers of the two nuclei.   The
intrinsic states χ_α can be specified by the quantum numbers (I,K)
where I is the total angular momentum component along the
quantization axis.   Intrinsic states with different K are coupled
by a Coriolis interaction and for sub-barrier fusion it is probably
a reasonable approximation to neglect this coupling.   Then K is a
conserved quantum number.   As the initial state has $K = 0$ only the
excited states with $K = 0$ need be considered.   For states with total
angular momentum J we have a set of coupled equations

$$(T + V_0(r) + \frac{\hbar^2}{2m} \frac{J(J+1)}{r^2} + \varepsilon_i - E)\phi_\alpha^J(r) + \Sigma_\beta h_{\alpha\beta}^J(r)\phi_\beta^J(r) = 0$$

$$(46)$$

All the states α have $K = 0$.   If the excitation energies are
neglected and if

$$h_{\alpha\beta}^J(r) = M_{\alpha\beta}^J F(r)$$

then the equations (46) can be uncoupled as before and

$$(T + V_0(r) + \frac{\hbar^2}{2m} \frac{J(J+1)}{r^2} + \lambda_i F(r) - E)\xi^J(r) = 0 \qquad (47)$$

The fusion cross-section calculated from (47) has the same structure
as (43)

$$\sigma_{fus} = (\pi/k^2)\Sigma_J(2J+1) q_i^J P_i^J(E) \qquad (48)$$

where $P_i^J(e)$ is the penetration factor for the eigenchannel i and
q_i^J is the probability of arriving at the barrier in that eigenchannel.

WKB IN SEVERAL DIMENSIONS

We discuss the WKB method in three dimensions,[16].   The extension to n dimensions is essentially the same.   We seek a solution
of the Schrödinger equation of the form

$$\psi(\underset{\sim}{r}) = A(\underset{\sim}{r}) \exp(iW(\underset{\sim}{r})/\hbar) \tag{49}$$

This wave function is substituted into the Schrödinger equation and terms of order h^2 are neglected. Collecting terms independent of $\hbar$ gives the Hamilton-Jacobi equation

$$(\nabla W)^2/2m + V(\underset{\sim}{r}) = E \tag{50}$$

and terms of order $\hbar$ give the continuity equation

$$\mathrm{div}(A^2 \nabla W) = 0 \tag{51}$$

Although the Hamilton-Jacobi equation can be written down in any number of dimensions it can be solved by elementary means only in one dimension or in problems that can be reduced by separation of variables to one dimension.

There is a close connection between the solutions of the Hamilton-Jacobi equation and the corresponding classical equations of motion. Any solution of (50) determines a family of classical trajectories with energy E. These are orthogonal to the surfaces $W(\underset{\sim}{r})$ = const. and the momentum of the trajectory at any point is

$$\underset{\sim}{p} = \partial W/\partial \underset{\sim}{r} \tag{52}$$

There is a definite relation between the boundary condition satisfied by $W(\underset{\sim}{r})$ and the initial conditions of the associated family of trajectories. These are straightforward for a scattering problem because the trajectories have a fixed initial momentum. They are more difficult for a bound state problem. The connection problem at a boundary between a classically allowed region ($E > V(\underset{\sim}{r})$) and a forbidden region ($E < V(\underset{\sim}{r})$) is also much more difficult in several dimensions than in one dimension.

PENETRATION INTO POTENTIAL BARRIERS IN SEVERAL DIMENSIONS

The problem of tunneling into multidimensional barriers was studied by Kapur and Peierls,[20] using a WKB approach. They formulated it in the following way: A particle with energy E moves in a potential field $V(\underset{\sim}{r})$. We wish to know the probability of finding it at a point $\underset{\sim}{r}$ in the classically forbidden region $V(\underset{\sim}{r}) < E$.

In the neighbourhood of $\underset{\sim}{r}$ the wave function is assumed to have the WKB form

$$\psi(\underset{\sim}{r}) = A(\underset{\sim}{r}) \exp(-S(\underset{\sim}{r})/\hbar) \tag{53}$$

The action $S(\underset{\sim}{r})$ satisfies an equation

$$(\nabla S)^2 = 2m(V(\underset{\sim}{r}) - E) \tag{54}$$

which is identical with the Hamilton-Jacobi equation except for a
sign.   The solution W(r) of the Hamilton-Jacobi equation (50) in
an accessible region is normally real.   The corresponding function
S(r) in (53) is in general complex.   This is because of the boundary
conditions which S must satisfy on the surface Σ with V(r) = E, which
separates the accessible and inaccessible regions.   When a particle
in the allowed region is reflected at Σ then ∇S has a real component
at right angles to the surface and an imaginary one parallel to it.
The real part of S is of particular interest because it determines
the modulus of the wave function $\psi(r)$.

Let

$$S = S_1 + iS_2$$

with S_1 and S_2 real.   Then (54) can be written as

$$(\nabla S_1)^2 = 2m \ (V(r) - E) + (\nabla S_2)^2 \tag{55}$$

$$> 2m \ (V(r) - E)$$

It is evident that the equality holds only if S is real.   Kapur
and Peierls are interested in the quantities defined by the line
integral

$$S_1(r) - S_1(r_o) = \int_{r_o}^{r_1} | \nabla S_1 | \ ds \tag{56}$$

where r_o lies on the line of steepest gradient of S_1 through r and
on the boundary surface Σ.   They argue that

$$|\psi(r)|^2 \sim |\psi(r_o)|^2 \exp[-2 \ (S_1(r) - S_1(r_o))] \tag{57}$$

so that the magnitude of the wave function ψ at r is governed mainly
by the difference (56).   It is difficult to calculate this
difference exactly because S is not known but using (55) they prove
the inequality

$$S_1(r) - S_1(r_o) > \min \int_{r_o'}^{r} [2m \ (V(r) - E)]^{\frac{1}{2}} ds \tag{58}$$

The path of integration which yields the minimum of the right hand
side of (58) is determined from the variational principle

$$\delta \int_c [2m(V(x) - E)]^{\frac{1}{2}} ds \tag{59}$$

where the upper limit r is fixed and the lower limit r_o' is allowed
to vary on the surface.

Banks, Bender and Wu,[21,22] and Gervais and Sakita,[23] developed the
ideas of Kapur and Peierls.   The basic argument of ref.[21] was that
barrier penetration occurs mostly in small tubes in configuration
space around certain classical solutions so that the WKB approx-

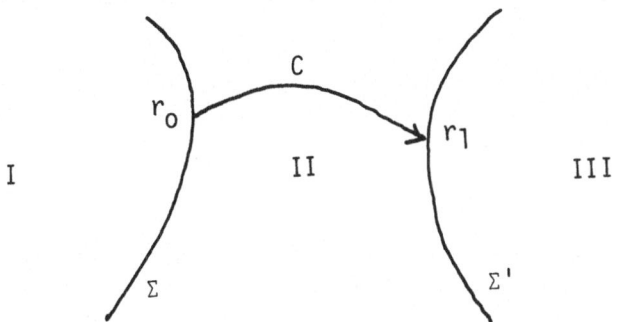

Fig. 5.   Illustration of the favoured tunneling path.

imation is essentially one dimensional.   This is illustrated in
Fig. 5.   The particle tunnels from region I to region III through
the classically forbidden region II.   The surfaces   and   are
the boundaries of the accessible regions I and III.   The favoured
tunneling path C is determined by a variational principle like (59).
Gervais and Sakita,[23] use the idea of ref.[21] and introduce a
collective coordinate associated with the path C so that the
dominant effect is contained in this degree of freedom.   They
construct a multi-dimensional wave function by studying small
fluctuations around the path C.   In this way they are able to
discuss the connection problem at $r_{\sim 0}$.

MODIFIED WKB SOLUTION

The generalization of the WKB method to several dimensions
presents many problems.   The equations are simple but it is almost
impossible to satisfy boundary conditions and find connection
formulae.   Progress can be made in cases where it is possible to
reduce the problem to the one dimensional case,[21,23].

In this section we present a modification,[24] of the method of
Gervais and Sakita,[23].   It is less general but is better adapted
for many physical applications.   We use the Hamiltonian (33)
(replacing r by x) in which the collective variable x is coupled
to intrinsic degrees of freedom ξ and look for a solution of the
Schrödinger equation of the form

$$\Psi(x,\xi) = \phi(x,\xi) \ [dW/dx]^{-\frac{1}{2}} \ \exp(i \ W(x)/\hbar) \tag{60}$$

where $\varepsilon = \pm 1$ in a classically allowed region and $\varepsilon = \pm i$ in a
classically forbidden region.   The general multidimensional WKB
method would allow W to be a function of x and ξ.   We simplify the

method by putting all the ξ dependence in the pre-exponential factor ϕ.

We substitute the trial wave function (60) into the Schrödinger equation and assume that $W(x)$ satisfies a one-dimensional Hamilton-Jacobi equation

$$(\varepsilon^2/2m)(dW/dx)^2 + V_o(r) = E \tag{61}$$

This is the first WKB equation. If ϕ = const. then (60) is a WKB wave function in the potential $V_o(x)$. The second equation is

$$\frac{i\varepsilon\hbar}{m}\frac{dW}{dx}\frac{\partial\phi}{\partial x} - h(x,\xi)\phi = -\frac{\hbar^2}{2m}\frac{dW}{dx}^{\frac{1}{2}}\frac{d^2}{dx^2}\phi\frac{dW}{dx}^{-\frac{1}{2}} \simeq 0 \tag{62}$$

In the last step the term on the right hand side of (62) is neglected. This is analogous to the approximation made in the usual WKB method.

The classically allowed region corresponds to $E > V_o(x)$ and the forbidden region of $E < V_o(x)$. In each case the solution of (61) is

$$W(x) = \int [2m \mid E - V_o(x) \mid]^{\frac{1}{2}} dx \tag{63}$$

In order to simplify equation (62) we follow Gervais and Sakita,[23] and replace x by a new variable τ defined by

$$dx/d\tau = (dw/dx)/m = [(2/m) \mid E - V_o(x) \mid]^{\frac{1}{2}} \tag{64}$$

so that

$$\frac{\partial}{\partial\tau} = \frac{1}{m}\frac{dW}{dx}\frac{\partial}{\partial x}$$

Then equation (62) can be written as

$$i\varepsilon\hbar\partial\phi/\partial\tau \simeq h(x(\tau),\xi)\phi \tag{65}$$

Equation (65) resembles a time-dependent Schrödinger equation for ϕ. The ratio τ/ε plays the role of a time. It is real in a classically allowed region and pure imaginary in a forbidden region.

Solving (65) and substituting into (60) gives the basic WKB-like solutions. The remaining step is to study connection formulae and to construct a complete solution. The connection problem is very similar to the case where there is no coupling with the additional constraint that the ξ dependence is continuous at $x = a$ and $r = b$ in Fig. 1. We consider the case of a wave

coming from the left.   In the notation of eqs. (1) and (2) the
wave functions in the three regions are:

Region I, x < a

$$\psi(x,\xi) = k^{-\frac{1}{2}}[\phi_i(\xi,\tau) \exp(i\omega(x,a) + \phi_r(\xi,\tau) \exp(-i\omega(x,a)-i\pi/2)]$$

(66)

Region II, a < x < b

$$\psi(x,\xi) = \gamma^{-\frac{1}{2}}\phi_-(\xi,\tau) \exp(-Q(x,a) - i\pi/4)$$

(67)

Region III, x > b

$$\psi(x,\xi) = k^{-\frac{1}{2}}\phi_t(\xi,\tau) \exp(-i\omega(x,b)) \exp(-Q)$$

(68)

The functions ϕ_i, ϕ_r and ϕ_t describing the propagation of the
internal state in the incident reflected and transmitted waves
satisfy equation (65) which is like a time-dependent Schrödinger
equation.   In particular the normalization

$$\int |\phi(\xi,\tau)|^2 d\xi$$

remains constant as τ changes for each of these functions.   The
situation inside the barrier is different.   There eq. (65) reads

$$\hbar \, \delta\phi_-/\delta\tau = - h(x(\tau),\xi) \, \phi_-$$

(69)

and the normalization changes with τ.   Let us denote the normalization constant by

$$K(\tau) = \int d\xi \, |\phi_1(\xi,\tau)|^2$$

(70)

Then the penetration coefficient

$$P = K_b \exp(-2Q)$$

(71)

is modified by the factor K_b which is the normalization (70) when
the particle emerges from the barrier b.

In general K_b can be anything.   There is a simple result
when $V_o(x)$ is chosen to be the adiabatic barrier.   By this we mean
that the interaction is divided between V_O and $h(x,\xi)$ so that the
lowest eigenvalue of $h(x,\xi)$ is zero for all x.   Then $h(x,\xi)$ is a
positive operator and it is easy to show that $K_b < 1$ so that

$$P < \exp(-2Q)$$

(72)

In this case the coupling reduces the penetrability.

THE PATH INTEGRAL METHOD

Feynman's original path integral approach to quantum mechanics was given in Lagrangian form,[25]. The propagator in coordinate space from an initial point x_0 to a final point x_1 after time interval T is expressed as a functional integral

$$K(x_1,x_0,T) = \int D[x(t)] \, \exp(i/\hbar)S[x(t)] \tag{73}$$

where S is the action functional

$$S[x(t)] = \int_0^T L(x,\dot{x}) \, dt = \int_0^T (\tfrac{1}{2}m\dot{x}^2 - V(x)) \, dt \tag{74}$$

The integral in (73) is over all points x(t) satisfying boundary conditions

$$x(0) = x_0 \quad , \quad x(T) = x_1 \tag{75}$$

The semi-classical approximation for K is obtained by evaluating the path integral (73) by the stationary phase approximation. The basic assumption is that the most important contributions to the path integral come from the neighbourhood of paths for which the action integral is stationary

$$\delta S[x(t)] = 0$$

These paths satisfy classical equations of motion and the boundary condtions (75) at t = 0 and t = T. If only one stationary path exists the semi-classical formula for the propagator is,[26]

$$K(x_1,x_0,T) \sim [\frac{i}{2\pi\hbar} \cdot \frac{\delta^2 S}{\delta x_0 \delta x_1}]^{\frac{1}{2}} \exp \, [(i/\hbar) \, S(x_1,x_0,T) - i\pi\nu/2] \tag{76}$$

Here $S(x_1,x_0,T)$ is the action integral (74) calculated along the classical path. The integer ν is a Morse or Maslov index and we do not need to discuss it here. It is easy to show that

$$\delta S/\delta x_1 = p_1 \quad , \quad \delta S/\delta x_0 = -p_0 \quad , \quad \delta S/\delta T = -E \tag{77}$$

where p_0 and p_1 are the momenta at t = 0 and t = T. The pre-exponential factor in (76) can be written in a number of different ways because

$$\frac{\delta^2 S}{\delta x_o \delta x_1} = -\frac{\delta p_o}{\delta x_1} = \frac{\delta p_1}{\delta x_o} \tag{78}$$

If there are several paths for which the action S is stationary then (76) contains a sum of terms, one for each stationary path.

For applications to barrier penetration we need the propagator G(E) in the energy representation. It is the Fourier transform of $K(x_1, x_o, T)$

$$G(x_1, x_o, E) = (1/i\hbar) \int_o^\infty dT \, K(x_1, x_o, T) \, \exp(iEt/h) \tag{79}$$

The semi-classical formula for G is obtained by evaluating the time integral in (79) by the stationary phase approximation. Using the semi-classical formula for K we have

$$G(x_1, x_o, E) \sim \frac{1}{ih} \int_o^\infty dT \, [\frac{1}{2\pi i\hbar} \frac{\delta p_o}{\delta x_1}]^{\frac{1}{2}} \exp[(i/h)(S(x_1, x_o, T)] + ET)$$

$$\tag{80}$$

In evaluating the integral by the stationary phase approximation the saddle point T_s is given by

$$\delta S/\delta T = -E$$

The pre-exponential factor is assumed to be slowly varying and is evaluated at the saddle point T_s. The saddle point condition (80) has the following meaning: Take the classical path with energy E which starts at x_o and ends at x_1. The time taken to move from x_o to x_1 is T_s. The saddle point evaluation of (80) gives

$$G(x_1, x_o, E) = \frac{1}{i\hbar} (\dot{x}_o \dot{x}_1)^{-\frac{1}{2}} \exp[(i/h) W (x_1, x_o, E) - i\chi] \tag{81}$$

$$W(x_1, x_o, E) = \int_{x_o}^{x_1} p(x) dx$$

$$p(x) = [2m (E - V(x))]^{\frac{1}{2}} \tag{82}$$

The phase χ in (81) comes from the Maslov index. If there are several saddle points in the time integration the total G would be a sum of contributions like (81). Note that the integral W in the phase of (81) has the same form as the WKB integrals in (1).

The propagators K and G in Feynman's theory are related to matrix elements of certain operators in the coordinate representation

$$K(x_1,x_0,T) = \langle x_1 | \exp(-iHT/\hbar) | x_0 \rangle$$

$$G(x_1,x_0,E) = \langle x_1 | (E + i\varepsilon - H)^{-1} | x_0 \rangle$$

This makes a link with conventional quantum mechanics. Using standard methods

$$G(x_1,x_0,E) = B \, \psi_+(x_1)\psi_-(x_0) \qquad ; \quad x_1 > x_0$$

$$= B \, \psi_-(x_1)\psi_+(x_0) \qquad ; \quad x_1 < x_0 \qquad (83)$$

$$B = (2m/\hbar^2) \, (\psi_- \psi_+' - \psi_+ \psi_-')$$

where $\psi_\pm(x)$ are solutions of the Schrödinger equation satisfying outgoing wave boundary conditions as $x \to \pm\infty$. Using these relations it is possible to find the general form of G for a tunneling problem. Let x_0 and x_1 be on opposite sides of a barrier like the one shown in Fig. 1. Using (83) we get

$$G(x_1,x_0,E) = (1/i\hbar)(\dot{x}_0 \dot{x}_1)^{-\frac{1}{2}} \, T \exp[(i/h)(W(x_1,b) + W(a,x_0))]$$

$$(84)$$

The derivation of (84) uses WKB wave functions near x_0 and x_1 but no assumption is made about the wave function in the barrier region. Equation (84) can be taken as a definition of the transmission amplitude T. In the next section $G(x_1,x_0,T)$ is calculated by the path integral method and T can be extracted by comparing with (84).

A PATH INTEGRAL FOR TUNNELING

Consider a potential barrier like the one shown in Fig. 1. Let x_0 and x_1 be on opposite sides of the barrier and suppose that the incident energy is higher than the top of the barrier. The semi-classical propagator is given by (81) and the stationary time for propagation from x_0 to x_1 is

$$T_s = \int_{x_0}^{x_1} (m/p(x)\,dx \qquad (85)$$

When the incident energy is below the top of the barrier the transition from x_0 to x_1 is classically forbidden. The integral (80) has no stationary point for real T because there is no real classical path connecting x and x. There are, however, saddle points in the complex T-plane which correspond to complex times

$$T_s = T_0 - (2n + 1)i\tau \qquad (86)$$

where n is an integer. T_O is the classical time to propagate from x_O to a, then from b to x_1 and $(2n + 1)i\tau$ is an imaginary time for propagating under the barrier from a to b.   If the incident energy is well below the top of the barrier ($E \ll V_B$) the main contributions to the integral (8) comes from a steepest descent path starting at $T = 0$ and passing over the n = 0 saddle in the lower half of the complex T-plane.   The saddle point contributions is still given by (81) but now

$$W(x_1,x_o,E) = W(x_1,b) + W(a,x_o) + i\hbar Q(b,a) \qquad (87)$$

where Q(b,a) is the WKB penetration integral.   Comparing (81) with (84) gives the transmission amplitude

$$T = \exp(-Q(b,a)) \qquad (88)$$

which is exactly the WKB result (eq. (9)).

    If the incident energy is near the barrier top the result (88) is no longer valid because the saddles become very flat.   Various authors (Freed,[27], Coleman,[28]) have suggested that a better result can be obtained by summing amplitudes of multiple reflections under the barrier.   This is equivalent to summing contributions from the line of saddles (86).   Such a procedure is unreliable because the saddles become very wide and the approximation of summing contributions of independent saddles is no longer reliable.   Brink and Smilansky,[29], have shown that the contribution of all the saddles can be summed by using a uniform approximation.   The method yields the Kemble,[13] formula (11).

    Now we consider the coupled channels problem with the Hamiltonian (33)

$$H = -(\hbar^2/2m)d^2/dx^2 + V_o(x) + h(x,\xi)$$

$$h(x,\xi) = H_o(\xi) + \mathbf{v}(x,\xi)$$

Pechukas[30] has given a path integral formula for the propagator $K_{\beta\alpha}$ (x_1,x_o,T) for the system to move from an initial position x_o to a final position x_1 at t = T, while the internal state changes from α to β.   The internal states α and β are eigenstates of $H_o(\xi)$. The propagator is

$$K_{\beta\alpha} (x_1,x_o,T) = \int D[x(t)]T_{\beta\alpha} [x(t)] \exp(i/\hbar) S_o[x(t)]) \qquad (89)$$

In (89) $S_o[x(t)]$ is the action for the relative motion calculated for the potential $V_o(x)$.   The factor $T_{\beta\alpha}[x(t)]$ is an amplitude for a transition between the intrinsic states α and β as the relative coordinate passes along x(t).

$$T_{\beta\alpha}[x(t)] = < B \mid U [x(t), r, o] \mid \alpha> \tag{90}$$

In (90) $U[x(t), t, t']$ is the time development operator for the internal structure when the relative motion is given by $x(t)$. It satisfies

$$ih \, \delta U/\delta t = h(x(t),\xi)U \tag{91}$$

with $U[x(t), t, t'] = 1$ when $t = t'$. The integral (89) is over all paths satisfying the boundary conditions (75). We evaluate (89) using the stationary phase approximation. The stationary path is determined by the condition $\delta S_0[x(t)] = 0$ and is a classical path in the potential $V_0(x)$ with boundary conditions (75). The factor $T_{\beta\alpha}[x(t)]$ is assumed to be a slowly varying functional of the path. It is evaluated for the stationary path and taken outside the integral to give

$$K_{\beta\alpha} (x_1,x_0,T) \simeq T_{\beta\alpha} [x(t)] K_0(x_1,x_0,T) \tag{92}$$

where K_0 is the semi-classical propagator (76) for the potential $V_0(x)$. The next step is to evaluate the energy protector. Going through a similar argument gives

$$G_{\beta\alpha} (x_1,x_0,T) \simeq T_{\beta\alpha} [x(E,t)] G_0(x_1,x_0,E) \tag{93}$$

Well above the barrier ($E \gg VB$) (93) gives the propagator $G_{\beta\alpha}$ as the product of the propagator G without coupling multiplied by the transition amplitude $T_{\beta\alpha}$ for the internal motion calculated for the classical path $x(E,t)$ with energy E in the potential $V_0(x)$. When the incident energy is well below the barrier top ($E \ll V_B$) the semi-classical propagator is still given by equation (93) but there are some important differences in interpretation. The first is that $G_0(x_1,x_0,E)$ contains the barrier penetration factor (88). The second difference is that the transition amplitude is not unitary. This is becasue the stationary time

$$T_s = T_0 - i\tau$$

is complex. Here T_0 is the real time for propagation from x up to the nearside of the barrier and from the farside of the barrier to x, while $-i\tau$ is the imaginary time for propagation through the barrier. In fact, the form (93) of the propagator is exactly equivalent to the WKB form for the wave function in eqs. (66) – (68).

The non-unitary character of $T_{\beta\alpha}$ under the barrier can be seen most clearly if the intrinsic Hamiltonian $h(x,\xi)$ including effects of coupling is independent of x in the barrier region

$$h(x,\xi) \;=\; h(x_B,\xi) \;=\; h_B(\xi) \tag{94}$$

Let ϕ_i be the eigenstates of h_B

$$h_B\phi_i(\xi) \;=\; \varepsilon_i\phi_i(\xi)$$

and ε_i the corresponding eigenvalues.   We choose the ground state energy of the internal Hamiltonian $H_0(\xi)$ to be zero.   Hence the lowest eigenvalue ε_0 of h_B is a measure of the energy shift due to the interaction.   The sign and magnitude of the shift depend on the nature of the intrinsic variables and of the interaction.   The ϕ_i are the transition states or eigenchannels referred to in an earlier section.   The coupled equations (91) for $T_{\beta\alpha}$ can be solved in this case and

$$T_{\beta\alpha}\left[x(E,t)\right] \;=\; \Sigma_i \; B_{\beta i} \; \exp(-\varepsilon_i\tau/h)A_{i\alpha} \tag{95}$$

Here $A_{i\alpha}$ is a transition amplitude from the initial intrinsic state α to the transition state i as the projectile comes up to the barrier.   The exponential in (95) describes propagation through the barrier and $B_{\beta i}$ is a transition amplitude from the state i to the final state β after passage through the barrier.   Both the matrices $A_{i\alpha}$ and $B_{\beta i}$ are unitary.   Non-unitary effects due to barrier penetration are contained in the exponential factor in (95). Because $B_{\beta i}$ is unitary

$$\Sigma_\beta \mid T_{\beta\alpha}\mid^2 \;=\; \Sigma_i \mid A_{i\alpha}\mid^2 \exp(-2\varepsilon_i\tau/h) \tag{96}$$

and the total penetration probability is

$$P = \exp(-2Q) \; \Sigma_i \; \mid A_{i\alpha}\mid^2 \; \exp(-2\varepsilon_i\tau/h) \tag{97}$$

For a parabolic barrier like equation (12)

$$\tau = \pi/\omega \tag{98}$$

Using eq. (13) for $Q(E)$ and (98) we have

$$Q(E) + \varepsilon_i\tau/h \;=\; Q(E-\varepsilon_i)$$

so that equation (97) can also be written as

$$\begin{aligned}P(E) &= \Sigma_i \mid A_{i\alpha}\mid^2 \exp(-2Q(E-\varepsilon_i)) \\ &= \Sigma_i \mid A_{i\alpha}\mid^2 P_0(E-\varepsilon_i)\end{aligned} \tag{99}$$

where P_0 is the simple penetration factor (10).

Thus the total penetration factor is a sum of contributions from each eigenchannel weighted by probability $|A_{i\alpha}|^2$ of reaching the barrier in that eigenchannel.

Equation (97) cannot be used if the incident energy is near the barrier top. This is because the stationary time for passage from x_0 to x_1 has a logarithmic divergence when $E \rightarrow V_B$. When E is near V_B the velocity is very slow near the barrier and the projectile takes a long time to cross it.

In the case of constant coupling (95) near the barrier T can be written in a form which is similar to (95)

$$T_{\beta\alpha} [x(E,t)] = \Sigma_i \, b_{\beta i} \, \exp(-\varepsilon_i T/h) a_{i\alpha} \qquad (100)$$

and $b_{\beta i}$ and $a_{i\alpha}$ have the same physical significance as in (95). They differ from $B_{\beta i}$ and $A_{i\alpha}$ only by a phase. Now all the singular behaviour of T as $E \rightarrow V_B$ is concentrated in the exponential factor in (100) and the $b_{\beta i}$ and $a_{i\alpha}$ are slowly varying functions of E. Equation (93) is replaced by,[28].

$$G_{\beta\alpha} (x_1,x_0,E) = \Sigma_i \, b_{\beta i} \, a_{i\alpha} \, G_0(x_1,x_0,E - \varepsilon_i) \qquad (101)$$

If G_0 is evaluated by the uniform approximation of Ref.[27] then equation (101) gives a penetration factor of the same form as (99)

$$P(E) = \Sigma_i \, |a_{i\alpha}|^2 \, P_0 \, (E - \varepsilon_i) \qquad (102)$$

except that now P_0 is given by the Kemble formula (11) which holds both above and below the barrier top.

Equations (101) and (102) can be derived without making the constant coupling assumption (95). It is only necessary to calculate $b_{\beta i}$ and $a_{i\alpha}$ in an interaction representation with respect to $h(x_\beta,\xi)$ in order to isolate the singular dependence as $E \rightarrow V_B$.

COMPARISON OF THE METHODS

Three essentially different methods for studying quantum tunneling in multidimensional systems have been discussed in these lectures.

The perturbation method is equivalent to the WKB approximation in one dimension, but it is somewhat apart in several dimensions. It would be interesting to study the connection with the other methods but as presented here it is suited to a different kind of problem.

The coupled channels method is completely general if the equations are solved exactly. Normally this is possible only by using numerical methods, but there are some special cases where the equations decouple and can be solved simply yielding results like (40) or (43). These special cases give a lot of physical insight into the problem of barrier penetration with channel coupling.

The WKB method can be developed either by using wave functions or by path integral methods. For energies $E \ll V_B$ both methods give identical results if suitable approximations are made. The path integral method is easier to use because it avoids the WKB matching problem. It can also be extended to give a useful formula (102) which holds both near and below the barrier. There is a very strong similarity between the WKB formula (99) and (102) and the coupled channels formula (43). The WKB formula is more general in the sense that it does not require constant coupling.

NUCLEON TRANSFER BETWEEN HEAVY IONS

In this section we discuss nucleon transfer between nuclei in a heavy ion collision as an illustration of the perturbation method. We consider the case of neutron transfer. The neutron is initially in a state ψ_i bound in a single nucleon potential V_1 which represents the shell model potential V_1 of the first nucleus. It is transferred into a single particle state ψ_f in the final Potential V_2 of the second nucleus. The potential V_1 moves past the potential V_2 during the transfer and the relative motion is described by an orbit $\underline{s}(t)$. It is not necessary to specify the details of the orbit at this stage and at first we assume that they are general functions of the neutron coordinate r and the time t. They should be separated by a large distance as $t \to t \pm \infty$.

The initial state ψ_i satisfies equation (16) and the final state ψ_f equation (17). The perturbation formula for the transfer amplitude is obtained by combining equations (23) and (27),

$$A_{fi} = \frac{h}{2mi} \int_{-\infty}^{\infty} dt \int_{\Sigma} d\underline{s} \cdot (\psi_f^* \nabla \psi_i - \psi_i \nabla \psi_f^*) \qquad (100)$$

The surface integral in (103) is over a surface Σ which lies between the potentials V_1 and V_2 at all times. Equation (103) has some nice features. It is symmetric between the initial and final states. It involves the wavefunctions ψ_i and ψ_f only on the surface Σ as Σ is assumed to lie outside V_1 and V_2 the wavefunctions ψ_i and ψ_f can be replaced by their asymptotic forms. It is interesting to note that the integrand in (103) resembles a quantal probability current.

We would like to evaluate the amplitude (103) when the orbit or motion s(t) is a Rutherford orbit. If the scattering angle is small then the Rutherford orbit can be replaced by a constant velocity orbit tangential to it at the point of closest approach. This is a reasonable approximation for high energy scattering, because the transfer takes place near the point of closest approach and because the acceleration would be small. It would not be a good approximation for low energy, large angle scattering. The transfer amplitude depends only on the relative velocity so we can assume that V_2 is at rest and V_1 has a velocity v. The distance of closest approach is d. It is convenient to take the z-axis Parallel to the direction of relative motion and the x axis in the plane of the orbit in the direction from V_2 to V_1. The equation of the orbit relative to the centre of V_2 is

$$\underset{\sim}{s}(t) = \underset{\sim}{d} + \underset{\sim}{v}t \tag{104}$$

where v is in the z-direction and d is in the x-direction.

If $\phi_i(\underset{\sim}{r})$ and $\phi_f(\underset{\sim}{r})$ are bound state wave functions in the static potentials V_1 and V_2 with energies ε_i and ε_f then

$$\psi_i(\underset{\sim}{r},t) = \phi_i(\underset{\sim}{r}-\underset{\sim}{s}(t))\exp\{(i/h)(m\underset{\sim}{v}.\underset{\sim}{r}-(\varepsilon_i + \tfrac{1}{2}mv^2)t)\} \tag{105}$$

$$\psi_f(\underset{\sim}{r},t) = \phi_f(\underset{\sim}{r})\exp(-i\varepsilon_f\, t/h) \tag{106}$$

The wave function ψ_i is obtained from ϕ_i by a galilean transformation. If the surface Σ in equation (103) is taken to be $x = d_2$ then the integral is over the variables y, z and t. By making the charge of variables

$$(z,t) \to (z,\, z_1 = z - vt)$$

the integrals over z and t can be evaluated with the result

$$A_{fi} = \frac{i\hbar}{2mv} \int dy \{\tilde{\phi}_f^*(d_1,y,k_z)\frac{\partial}{\partial d_2} \tilde{\phi}_i(d-d_1,y,k_1)$$

$$- \tilde{\phi}_i(d-d_2,y,k_1) \frac{\partial}{\partial d_2} \tilde{\phi}_f^*(d_2,y,k_z)\} \tag{107}$$

where

$$\tilde{\phi}(x,y,k) = \int_{-\infty}^{\infty} \exp(-ikz)\phi(x,y,z)dz \tag{108}$$

$$k_i = -(Q + \tfrac{1}{2}mv^2)/\hbar v, \quad k_f = -(Q - \tfrac{1}{2}mv^2)/\hbar v \tag{109}$$

and $Q = \varepsilon_i - \varepsilon_f$ is the Q-value of the reaction.

The asymptotic forms of ϕ_i and ϕ_f which hold outside the range of the potentials V_1 and V_2 are proportional to Hankel functions

$$\chi_\ell(\gamma r) = -i^\ell h_\ell^{(1)}(i\gamma r)$$

$$\phi_a(\underset{\sim}{r}) = C_\alpha \gamma_\alpha \chi_{\ell\alpha}(\gamma_\alpha r) Y_{\ell m}(\sigma,\phi)$$

$$\chi_\ell(\gamma r) \sim \exp(-\gamma r)/\gamma r \qquad (110)$$

when r is very large. In equation (110) C_i and C_f are normalization constants and γ_i and γ_f are related to the bound state energies by

$$\varepsilon_\alpha = -(h^2/2m)\gamma_\alpha^2 \qquad (111)$$

If we use the form (110) in (107) the Fourier transform (108) can be evaluated analytically (Lo Monaco and Brink[31])

$$\tilde{\phi}_{\ell m}(x,y,k) = 2\, C_\ell\, Y_{\ell m}(\beta,\phi) K_m(\eta\rho)$$

where

$$x = \rho \cos\phi \quad , \quad y = \rho \sin\phi$$

$$\rho = r \cos\theta \quad , \quad z = r \sin\theta$$

and

$$\cos\beta = ik/\gamma \quad , \quad \sin\beta = \eta/\gamma \qquad (112)$$

$$\eta = (\gamma^2 + k^2)^{\frac{1}{2}} \qquad (113)$$

The function K_m is a Bessel function. Next the $\tilde{\phi}$ can be substituted into (107) and the y integral evaluated to give[31].

$$A(\ell_f m_f, \ell_i m_i) = -4\pi i\,(\hbar/mv)\,(-1)^{m_1}\, C_2^* C_1\, K_{m_1-m_2}(\eta d)$$

$$\times Y_{\ell_2 m_2}^*(\beta_2,0) Y_{\ell_1 m_1}(\beta_1,0) \qquad (114)$$

The complex angles β_i and β_f are given by (112) with k_α from (109) and γ_α from (111), while η is given by

$$\eta^2 = k_i{}^2 + \gamma_i{}^2 = k_f{}^2 + \gamma_f{}^2 \tag{115}$$

By using the expressions for k_i and k_f from (109) η can also be written in the form

$$\eta^2 = -(2m/h^2)\bar{\varepsilon} \tag{116}$$

$$\bar{\varepsilon} = \tfrac{1}{2}(\varepsilon_i + \varepsilon_f) - \tfrac{1}{4}[Q^2/(\tfrac{1}{2}mv^2) + \tfrac{1}{2}mv^2] \tag{117}$$

Here $\bar{\varepsilon}$ is a kind of average bound state energy which depends on ε_i, ε_f, Q and the relative velocity v. The reation is kinematically favoured if η is as small as possible. For a fixed ε_i this means

$$\varepsilon_f = \varepsilon_i + \tfrac{1}{2}mv^2$$

Equations (109) for k_i and k_f are classical relations. To see this consider classical conditions for smooth transfer of a nucleon from V_1 to V_2 (fig. 6).

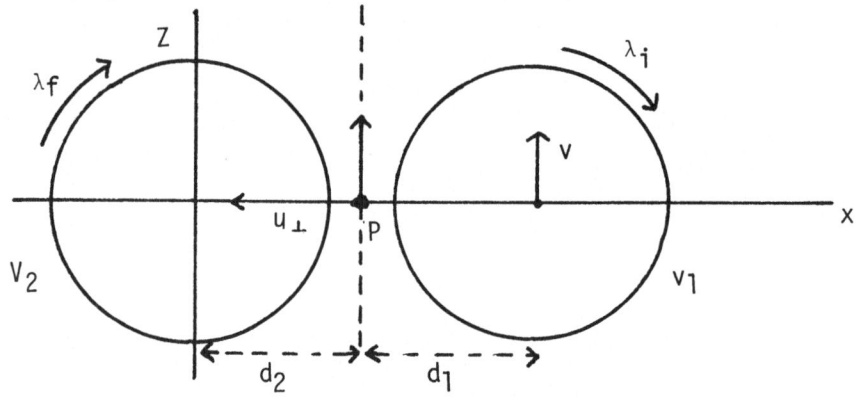

Fig. 6.   Diagram for transfer kinematics.

The transfered neutron at P has velocity component v_i in the z direction relative to V_1 and v_f relative to V_2. Smooth transfer demands that

$$v_i + v = v_f \tag{118}$$

If the velocity component of the neutron perpendicular to the z - direction is $u_\perp$ and if $V_1 = V_2 = 0$ at P then -

$$\varepsilon_i = \tfrac{1}{2}m(v_i{}^2 + u_\perp{}^2), \quad \varepsilon_f = \tfrac{1}{2}m(v_f{}^2 + u_\perp{}^2) \tag{119}$$

Subtracting gives

$$Q = \varepsilon_i - \varepsilon_f = \tfrac{1}{2}m(v_i^2 - v_f^2) \qquad (120)$$

Now we solve equations (118) and (120) and obtain

$$v_i = -(Q + \tfrac{1}{2}mv^2)/mv, \qquad v_f = -(Q - \tfrac{1}{2}mv^2)/mv \qquad (121)$$

These relations are the same as (109) because $k_i = mv_i/\hbar$ and similarly for k_f. The perpendicular component of momentum is

$$p_\perp = mv_\perp = i\hbar\eta \qquad (122)$$

and is pure imaginary. This is because ε_i and ε_f are negative for bound states and (119) can be satisfied only if $u_\perp$ is imaginary.

Finally, the angular momentum components in the y-direction of the transfered nucleon in the initial and final states are

$$\lambda_i = d_1 k_1 \quad , \quad \lambda_f = -d_2 k_f \qquad (123)$$

Using (128) and (120)

$$\lambda_i/d_1 + \lambda_f/d_2 = -mv/\hbar$$

$$\lambda_i - \lambda_f = -Qd/\hbar v - (\tfrac{1}{2}mv/\hbar)(d_1 - d_2)$$

These are equivalent to the kinematical conditions of ref. 32. The sign differences are due to a different coordinate system.

SUB-BARRIER FUSION

Now we apply the coupled channels method to study sub-barrier fusion between heavy ions. Recently this problem has been studied by Esbenson[5], Dasso et al.[6], Lindsay and Rowley[19], Jacobs and Smilansky[33], Lee and Takigawa[34] and many others.

The quantal formula for a fusion cross-section is

$$\sigma_f = (\pi/k^2)\Sigma_\ell (2\ell + 1)P_\ell \qquad (124)$$

where the transmission coefficient P_ℓ is the probability of fusion for an incident partial wave ℓ. If we appropriate the Coulomb barrier by an inverted parabola then the transmission coefficient is given by the formula (11).

$$P_\ell = \frac{1}{1 + \exp((V_{B\ell} - E)/\varepsilon)} \tag{125}$$

where

$$\varepsilon = \hbar w_B/2\pi \quad , \quad w_B^2 = \frac{1}{\mu}(\partial^2 V/\partial r^2)_{r_B}$$

$$V_{B\ell} = V(r_B) + h^2 \ell(\ell+1)/(2\mu r_B^2)$$

and $V(r)$ is the Coulomb plus nuclear potential. The quantity $V_{B\ell}$ is the barrier height in the partial wave ℓ and ε is a measure of the energy interval over which barrier penetration is important. Equation (125) contains some simplifications because the barrier radius r_B and w_B are assumed to be independent of ℓ. With these assumptions the sum (124) can be evaluated to quite a good approximation by replacing it by an integral to give a closed formula for σ_f due to Wong[35].

$$\sigma_f = \pi r_B^2 (\varepsilon/E) \ell n[1 + \exp((E - V_B)/\varepsilon)] \tag{126}$$

where $V_B = V(r_B)$. For most systems ε lies in the range $0.5 - 1$ MeV. The number of partial waves ℓ_B which contribute significantly to the sum (124) for sub-barrier fusion can be estimated from

$$\hbar^2 \ell_B(\ell_B + 1) \simeq \varepsilon \, 2\mu \, r_B^2$$

where μ is the reduced mass of the two nuclei involved. The number ℓ_B varies from 2 for light nuclei to more than 20 for heavier systems.

When the incident energy is well above the top of the Coulomb barrier then equation (126) simplifies to give

$$\sigma_f \approx \pi r_B^2 (1 - V_B/E) \tag{127}$$

For sub-barrier energies

$$\sigma_f \approx \pi r_B^2 (\varepsilon/E) \exp((E - V_B)/\varepsilon) \tag{128}$$

and the fusion cross-section decreases exponentially with $(E - V_B)$.

Recent experiments (cf. ref.[4]) show that the simple barrier penetration model of Wong (126) cannot account for the energy variation of the fusion cross section for sub-barrier energies.

Typically equation (126) underestimates σ_f by several orders of magnitude if $E \ll V_B$. The authors of ref. 4,5,6,19,33,39 all point out that the effects of coupling of intrinsic degrees of freedom to the tunneling coordinate are important for understanding the experimental results. The general conclusion reached by these authors is that the heavy ion fusion cross section is given by the approximate relation

$$\sigma_f \simeq \Sigma_i \ q_i \ \sigma_w(E - \varepsilon_i) \tag{129}$$

where $\Sigma q_i = 1$ and $\sigma_w (E - \varepsilon_i)$ are fusion cross sections calculated from a simple barrier penetration model like the Wong formula (128). The interpretation of (129) is that fusion occurs through certain eigen channels or transition states labelled by i. The transition state i has an energy ε_i and the effective barrier height in that channel is $V_B + \varepsilon_i$. The quantity q_i is the probability of arriving at the barrier in the transition state i and $\sigma_w(E - \varepsilon_i)$ is the fusion probability in that state. It is clear that the simple models of Dasso et al.[6] will lead to formulae like (129). The path integral method leads to a similar result (99). The work of Lindsay and Rowley[19] shows that proper treatment of angular momentum coupling does not make any significant change in the structure of (129).

When the incident energy is well above the top of the barrier then $\sigma_w(E)$ is a slowly varying function of E and (129) gives

$$\sigma_f(E) \simeq \sigma_w(E - \bar{\varepsilon})$$

where $\bar{\varepsilon} = \Sigma_i \ q_i \ \varepsilon_i$ is an average of the transition state energies. For energies $E \ll V_B$ the term in (129) with the smallest energy ε_o gives the dominant contribution and

$$\sigma_f(E) \simeq q_o \ \sigma_w(E - \varepsilon_o)$$

Thus the energy width of the transition region is approximately $\bar{\varepsilon} - \varepsilon_o$. Deformations enhance the sub-barrier fusion cross-section. Then the barrier height depends on deformation and

$$\bar{\varepsilon} - \varepsilon_o \simeq \tfrac{1}{2}[V_B(max) - V_B(min)]$$

The existence of a neutron transfer channel with a negative Q would give $\bar{\varepsilon} - \varepsilon_o \simeq |Q|$ and would produce an enhancement of the sub-barrier cross section.

REFERENCES

1. C.G. Callan and S. Coleman, Phys. Rev. D16: 1762 (1977).
2. T. Banks, C.M. Bender and T.T. Wu, Phys. Rev. D8: 3346; 3366
    (1973).
3. A.O. Caldeira and A.J. Leggett, Phys. Rev. Lett. 46: 211 (1981).
4. M. Beckerman, M. Saloma, A. Sperduto, H. Enge, J. Ball, A. Di
    Rienzo, S. Gazes, J.D. Molitoris and Mas Nai-feng, Phys. Rev.
    Lett. 45: 1472 (1981).
5. H. Esbensen, Nucl. Phys. A352: 147 (1981).
6. C.H. Dasso, S. Landowne and A. Winther, Nucl. Phys. A405: 381:
    A407: 221 (1983).
7. D.M. Brink, M.C. Nemes and D. Vautherin, Ann. Phys. 147: 171
    (1983).
8. S. Landowne, C.H. Dasso, B.S. Nilsson, R.A. Broglia and A.
    Winther, Nucl. Phys. A259: 99 (1976).
9. H. Hasan and D.M. Brink, J. Phys. G4: 1573 (1978).
10. G. Pollarolo, R.A. Broglia and A. Winther, Nucl. Phys. A361:
    307 (1981).
11. L.D. Landau and E.M. Lifshitz, "Quantum Mechanics", Pergamon
    Press: London (1958).
12. A.B. Migdal, "Qualitative Methods in Quantum Theory", Benjamin
    Inc. Reading Mass. (1977).
13. E.C. Kemble, Phys. Rev. 48: 549 (1935).
        Macdonald, Nucl. Phys. 54: 393 (1964).
        Tobocman, Phys. Rev. C17: 2205 (1978).
16. A. Messiah, "Méchanique Quantique", Dunod, Paris (1959).
17. C.H. Dasso, S. Landowne and A. Winther, Nucl. Phys. A405: 381,
    (1983).
18. C.H. Dasso, S. Landowne and A. Winther, Nucl. Phys. A407: 221,
    (1983).
19. R. Lindsay and N. Rowley, J. Phys. G10: 805 (1984).
20. P.L. Kapur and R. Peierls, Proc. Roy. Soc., A163: 606 (1937).
21. T. Banks, C.M. Bender and T.T. Wu, Phys. Rev. D8:3346, 3366,
    (1973).
22. C.M. Bender and T.T. Wu, Phys. Rev. D7: 1620 (1973).
23. J.L. Gervais and B. Sakita, Phys. Rev. D16: 3507 (1977).
24. D.M. Brink, M.C. Nemes and D. Vautherin, Ann. Phys. 147: 171
    (1983).
25. R.P. Feynman and A.R. Hibbs, "Quantum Mechanics and Path
    Integrals", McGraw Hill, New York (1965).
26. S. Levit and U. Smilansky, Ann. Phys. 103: 198, 108: 165 (1977).
        K. Möhring, S. Levit and U. Smilansky, Ann. Phys. 127:
        198 (1980).
27. K.F. Freed, J. Chem. Phys., 56: 692 (1972).
28. S. Coleman, Phys. Rev. D15: 2929 (1977).
29. D.M. Brink and U. Smilansky, Nucl. Phys. A405: 301 (1983).
30. P. Pechukas, Phys. Rev. 181: 166 (1969).
31. L. Lo Monaco and D.M. Brink, to be published.

32. D.M. Brink, Phys. Lett. 40B: 37 (1972), H. Hasan and D.M.
     Brink, J. Phys. G4: 1573 (1978).
33. P.M. Jacobs and U. Smilansky, Phys. Lett. 127B: 313 (1983).
34. S.Y. Lee and N. Takigawa, Phys. Rev. C28: 1123 (1983).

GRAZING COLLISIONS IN LOW ENERGY HEAVY ION REACTIONS

Aage Winther

The Niels Bohr Institute
DK-2100   Copenhagen Ø
Denmark

The present series of lectures is based on joint work with staff and visitors in Copenhagen. I should especially mention Ricardo Broglia, Carlos Dasso, Steven Landowne, Giovanni Pollarolo, Henning Esbensen, Andrea Vitturi and Jose Manuel Quesada. Much of what I am going to say has already been published. The material which is not published will be included in a forthcoming Volume II of a monography of which volume I has already appeared (Broglia and Winther 1981).

I   SEMICLASSICAL THEORY OF REACTIONS

We shall consider the collisions between a projectile nucleus a and a target A. The entrance channel where both nuclei are in their ground states will be denoted by α . In the reaction particles may be exchanged between the nuclei and such reaction channels are denoted β , γ etc. and the nuclei appearing as the recoil and ejectile as B, F etc. and b, f etc. respectively. Inelastic channels are denoted by primes e.g. α'.

We consider mainly low energy collisions of energy 5-10 MeV per nucleon.

a.   The Coupled Equations

Because of the short wavelength of relative motion (typically ~ 0.05 fm) we may use a classical descrip-

tion for this degree of freedom. We thus assume that
the relative center of mass coordinate $\vec{r}$ is a definite function of time which satisfies the Newtonian
equation of motion

$$m_{aA} \ddot{\vec{r}} = - \vec{\nabla} U , \tag{1}$$

where m_{aA} is the reduced mass and U the interaction
energy. In channel α it is given by

$$U_{aA} = \frac{Z_a Z_A e^2}{r} + U_{aA}^N(r) , \tag{2}$$

where U_{aA}^N is the socalled ion-ion potential; the
attractive nuclear interaction between the two nuclear
surfaces. It is noticed that the potential (2) is different in different channels especially if proton
transfer has taken place. In fact the trajectory will
also be influenced by energy and angular momentum loss
to intrinsic motion in the two nuclei. These effects
will be neglected here, as they become small for grazing collisions with very heavy nuclei.

The intrinsic motion of the colliding nuclei is
now a timedependent problem. The Schrödinger equation
for the intrinsic motion of the two nuclei is thus

$$i \hbar \dot{\psi}(t) = H(t) \psi(t) . \tag{3}$$

The Hamiltonian

$$H(t) = H_a + H_A + V_{aA} - U_{aA}(r) \tag{4}$$

contains besides the intrinsic Hamiltonians of the colliding nuclei H_a and H_A the interaction V_{aA} between the nucleons in a and in A. The Hamiltonian
may also be written in the form

$$H(t) = H_b + H_B + V_{bB} - U_{bB}(r) \tag{5}$$

which is appropriate when acting on eigenstates of
channel β .

We try to solve (3) by the ansatz

$$\psi(t) = \sum_{\beta} a_{\beta}(t) \, \psi_{\beta}(t) \, e^{-\frac{i}{\hbar} E_{\beta} t}, \tag{6}$$

where the channel wavefunctions $\psi_{\beta}(t)$ are

$$\psi_{\beta}(t) = \psi_i^b(\xi_b) \, \psi_{\jmath}^B(\xi_B) \, e^{i\delta_{\beta}}. \tag{7}$$

The wavefunctions ψ_i^b and $\psi_{\jmath}^B$ are eigenstates of the intrinsic Hamiltonians H_b and H_B respectively i.e.

$$\begin{aligned} H_b \, \psi_i^b &= E_i^b \, \psi_i^b \\ H_B \, \psi_{\jmath}^B &= E_{\jmath}^B \, \psi_{\jmath}^B \end{aligned} \tag{8}$$

and

$$E_{\beta} = E_i^b + E_{\jmath}^B. \tag{9}$$

The phase factor is a Galilean transformation which follows the relative motion and ensures that the wavefunctions are given in the same system of reference i.e.

$$\delta_{\beta} = \frac{1}{\hbar} \left[m_{bB} \, \dot{\vec{r}} \, (\vec{r}_{bB} - \vec{r}(t)) - \int_0^t (U(r(t)) - \frac{1}{2} m_{bB} (\dot{\vec{r}}(t))^2) dt \right] \tag{10}$$

Inserting the ansatz (6) in (3) we find that the unknown coefficients a_{β} must satisfy the equations

$$i\hbar \sum_{\gamma} \dot{a}_{\gamma} \psi_{\gamma} e^{-\frac{i}{\hbar} E_{\gamma} t} = \sum_{\gamma} (V_{\gamma} - U) \psi_{\gamma} e^{-\frac{i}{\hbar} E_{\gamma} t} a_{\gamma}. \tag{11}$$

In trying to isolate $\dot{a}_{\beta}$ it is tempting to multiply by ψ_{β}, but we meet here the difficulty that the states ψ_{β} and ψ_{γ} are orthogonal only if γ is an excited state (β') of the nuclei b and B. If β and γ belong to different partitions of the nucleons the overlap

$$\langle \psi_{\beta} | \psi_{\gamma} \rangle = g_{\beta \gamma}(t) \rightarrow \delta_{\beta \gamma} \quad \text{for} \quad t \rightarrow \pm \infty \tag{12}$$

is nonvanishing, when the two nuclei (around t = 0)
have overlapping densities. We may still isolate the
timederivative by introducing the (adjoint) state

$$\omega_\rho = \sum_\gamma g_{\gamma\rho}^{-1} \psi_\gamma \to \psi_\rho \quad \text{for } t \to \pm\infty, \quad (13)$$

where g^{-1} is the reciprocal overlap matrix. Utilizing that

$$\langle \omega_\rho | \psi_\gamma \rangle = \delta_{\rho\gamma} \quad (14)$$

we find (cf. Broglia and Winther 1972)

$$i\hbar \dot{a}_\rho = \sum_\gamma \langle \omega_\rho | V_\gamma - U | \psi_\gamma \rangle e^{\pm\frac{i}{\hbar}(E_\rho - E_\gamma)t} a_\gamma. \quad (15)$$

These are the semiclassical coupled equations for reactions. They should be solved with the initial condition that

$$a_\rho(-\infty) = \delta_{\rho\alpha} \quad (16)$$

and the final amplitudes $a_\gamma(+\infty)$ describe the nuclei
after the reaction.

     The unpleasant feature of non-orthogonality also
appears in a quantal formulation. It is possible to
avoid it by using an ansatz where the channel wavefunctions ψ_ρ are the eigenstates of the two nuclei at a
fixed center of mass distance. Such description would
be appropriate if the relative motion was very slow,
and is often used in atomic collisions. It is however
inconvenient by the fact that the coupling between the
states does not vanish outside the interaction region.

     Another peculiarity of the coupled equations (15)
is that they are asymmetric, and therefore do not conserve the quantity $\sum_\rho |a_\rho(t)|^2$ at intermediate times.
If one substitutes the amplitudes a_ρ by the adjoint
amplitudes

$$\bar{a}_\rho(t) = \sum_\delta g_{\rho\delta} \, a_\delta \, e^{\frac{i}{\hbar}(E_\rho - E_\delta)t} \rightarrow a_\rho(t) \text{ for } t \rightarrow \pm\infty, \quad (17)$$

which signify the amplitudes on the states ω_ρ, one finds the equation

$$i\hbar \dot{\bar{a}}_\rho = \sum_\delta \langle \psi_\rho | V_\rho - U | \omega_\delta \rangle \, e^{\frac{i}{\hbar}(E_\rho - E_\delta)t} \, \bar{a}_\delta. \quad (18)$$

These equations constitute the coupled equations in the socalled post form, since the interaction is the one belonging to the partition in the final state. The equations (15) are analogously called the prior form of the coupled equations.

It is noted that the quantity $\sum_\delta \text{Re}(\bar{a}_\delta^* (t) \cdot a_\delta(t))$ is conserved and equal to unity if the equations are solved with the initial condition (16).

b. Perturbation Theory

The coupled equations are of course most easily solved if the interaction is weak, i.e. more precisely if the action integrals

$$\chi_{\rho\alpha} = \frac{1}{\hbar} \int_{-\infty}^{\infty} \langle \omega_\rho | V_\alpha - U | \psi_\alpha \rangle \, dt \ll 1 \quad (19)$$

measured in units of $\hbar$ are small. However, because of the nonorthogonality i.e.

$$g_{\rho\alpha} = \delta_{\rho\alpha} + \varepsilon_{\rho\alpha}, \quad (20)$$

we also have to assume that the deviation from orthogonality is small i.e.

$$\varepsilon_{\rho\alpha} \ll 1. \quad (21)$$

If both conditions (19) and (21) are fulfilled, we may to first order in ε write

$$g_{\rho\alpha}^{-1} = \delta_{\rho\alpha} - \varepsilon_{\rho\alpha} \qquad (22)$$

i.e.

$$\omega_\beta = \psi_\rho - \langle \psi_\alpha | \psi_\rho \rangle \psi_\alpha . \qquad (23)$$

To first order in the action we thus find

$$i\hbar \, \dot{a}_\rho = \left[\langle \psi_\beta | V_\alpha - u | \psi_\alpha \rangle \right. $$
$$\left. - \langle \psi_\beta | \psi_\alpha \rangle \langle \psi_\alpha | V_\alpha - u | \psi_\alpha \rangle \right] e^{\frac{i}{\hbar}(E_\beta - E_\alpha)t} \qquad (24)$$
$$= \langle \psi_\rho | V_\alpha - \langle V_\alpha \rangle_\alpha | \psi_\alpha \rangle e^{\frac{i}{\hbar}(E_\beta - E_\alpha)t}$$

where $\langle V_\alpha \rangle_\alpha = \langle \psi_\alpha | V_\alpha | \psi_\alpha \rangle$ is the expectation value of V_α in channel α. We thus find

$$a_\rho(t) = \frac{1}{i\hbar} \int_{-\infty}^{t} \langle \psi_\rho | V_\alpha - \langle V_\alpha \rangle | \psi_\alpha \rangle e^{\frac{i}{\hbar}(E_\beta - E_\alpha)t} dt . \qquad (25)$$

It is noticed that the second term $\langle V_\alpha \rangle$ is important for transfer reactions since the states ψ_ρ and ψ_α are then non-orthogonal.

For two-nucleon transfer it is necessary to solve the coupled equations to second order. One should then expand the overlap also to second order in ε according to

$$g^{-1} = 1 - \varepsilon + \varepsilon^2 \qquad (26)$$

The resulting amplitude may then be written

$$a_\rho^{(2)}(\infty) = a_\rho^{(1)}(\infty) + a_\rho^{(No)} + a_\rho^{(suc)}, \qquad (27)$$

where $a^{\omega}(\infty)$ is given by (25). This term describes the simultaneous transfer of the two nucleons as a cluster.

The second term

$$a_{\rho}^{(No)} = \frac{-1}{i\hbar} \sum_{\gamma} \int_{-\infty}^{\infty} \langle \psi_{\rho} | \psi_{\gamma} \rangle \langle \psi_{\gamma} | V_{\alpha} - \langle V_{\alpha} \rangle | \psi_{\alpha} \rangle e^{\frac{i}{\hbar}(E_{\rho}-E_{\alpha})t} dt \quad (28)$$

is the second order non-orthogonality correction while

$$a_{\rho}^{(suc)} = \frac{1}{(i\hbar)^2} \int_{-\infty}^{\infty} dt \sum_{\gamma} \langle \psi_{\rho} | V_{\gamma} - \langle V_{\gamma} \rangle | \psi_{\gamma} \rangle e^{\frac{i}{\hbar}(E_{\rho}-E_{\gamma})t}$$
$$\int_{-\infty}^{t} dt' \langle \psi_{\gamma} | V_{\alpha} - \langle V_{\alpha} \rangle | \psi_{\alpha} \rangle e^{\frac{i}{\hbar}(E_{\gamma}-E_{\alpha})t'} \quad (29)$$

is the term describing the successive (timeordered) transfer of two nucleons, γ being a channel where one nucleon has been transferred.

It is easy to prove that in the limit where the two nucleons are moving independently the quantity a^{NO} is equal to $a^{(1)}$ but with the opposite sign. In this case only the successive transfer is possible. In the opposite limit of a strongly coupled pair a^{NO} would cancel the third term and only the simultaneous transfer would survive.

c.   Coupled Channel with Weak Transfer

For most nuclear collisions with heavy target and projectile the action integral (strength parameter) (cf. (19)) is larger than unity for the excitation of the low lying rotational or vibrational state mostly because of the Coulomb interaction (Coulomb excitation). In this case the rotational (or vibrational) states should be included through the coupled equations while the transfer may often still be treated by first order perturbation theory.
The coupled equations thus take the form

$$i\hbar \dot{a}_{\alpha'} = \sum_{\alpha} \langle \psi_{\alpha'} | V_{\alpha} - \langle V_{\alpha} \rangle | \psi_{\alpha} \rangle e^{\frac{i}{\hbar}(E_{\alpha'}-E_{\alpha})t} a_{\alpha} \quad (30a)$$

$$i \hbar \, a_{\rho} - \sum_{\rho} \langle \psi_{\rho'} | V_{\rho} - \langle V_{\rho} \rangle | \psi_{\rho} \rangle \, e^{\frac{i}{\hbar}(E_{\rho'} - E_{\rho})t} \, a_{\rho}$$

$$= \sum_{\alpha} \langle \psi_{\rho'} | [V_{\alpha}, P] | \psi_{\alpha} \rangle \, e^{\frac{i}{\hbar}(E_{\rho'} - E_{\alpha})t} \, a_{\alpha} \,, \tag{30b}$$

where α, α' and β, β' indicate states of the two nuclei before and after the transfer, respectively. The operator P is the projection operator

$$P = \sum_{\alpha'} | \psi_{\alpha'} \rangle \langle \psi_{\alpha'} | \tag{31}$$

on the states α' that were included in the coupled equations. It arises from the non-orthogonality correction since to the order we consider

$$\omega_{\rho} = \psi_{\rho} - \sum_{\alpha'} \langle \psi_{\rho} | \psi_{\alpha'} \rangle \, \psi_{\alpha'} \,. \tag{32}$$

If we include all states in the rotational (or vibrational) band built on the ground state of, say, the target nucleus the operator (31) is equal to

$$P = \delta(\vartheta - \vartheta') | \psi_i^a(\zeta_{\alpha}) \, \varphi_j^A(\zeta_A, \vartheta) \rangle \langle \psi_i^a(\zeta_{\alpha}') \varphi_j^A(\zeta_A', \vartheta') | \tag{33}$$

where $\varphi_i^A(\zeta_A, \vartheta)$ is the intrinsic wavefunction for a definite value ϑ of the collective coordinates (orientation angle or vibrational amplitude).

The matrix element appearing in (30b) is therefore

$$\langle \psi_{\rho'} | [V_{\alpha}, P] | \psi_{\alpha} \rangle$$

$$= \langle \psi^b(\zeta_b) \, \psi^B(\zeta_B, \vartheta) \, e^{i \delta_{\rho}} | V_{\alpha}(\vartheta) - \langle V_{\alpha} \rangle_{\vartheta} | \psi^a(\zeta_{\alpha}) \psi^A(\zeta_A, \vartheta) \, e^{i \delta_{\alpha}} \rangle \tag{34}$$

where the average potential $\langle V_{\alpha} \rangle_{\vartheta}$ in the entrance channel is to be evaluated for a definite value of the collective coordinate i.e.

$$\langle V_\alpha \rangle_\vartheta = \langle \psi^a \varphi^A | V_\alpha | \psi^a \varphi^A \rangle_\vartheta . \qquad (35)$$

d.   The Ion-Ion Potential

An important ingredient in all heavy ion collisions is the expectation value of the interaction between the two ions i.e.

$$U_\alpha = \langle V_\alpha \rangle$$

$$= \langle \psi_0^a \psi_0^A | V_\alpha | \psi_0^a \psi_0^A \rangle \qquad (36)$$

$$= \frac{Z_a Z_A e^2}{r} + U_{aA}^N(r)$$

where Z_a and Z_A are the charge numbers of projectile and target nuclei. The nuclear ion-ion potential is

$$U_{aA}^N(r) = \int d^3x \, d^3x' \, \rho_a(\vec{x}') \rho_A(\vec{x}) V_{12}(\vec{x}-\vec{x}'+\vec{r}) \qquad (37)$$

where V_{12} is the effective nucleon-nucleon interaction while ρ_a and ρ_A are the particle densities in the two nuclei. Because of the antisymmetrization that should be made also between nucleons in a and A there is an exchange correction to (37) which approximately may be included in the effective interaction as a repulsive δ-function potential.

For grazing collisions we are only interested in the potential U^N in the situation where the distance r between the centers of mass of the two nuclei is larger than the sum of the mean radii R_a and R_A. In this region the potential is essentially of exponential form with a decay length, a, characteristic for the exponential slope of the densities a ≈ 0.6 fm. If one uses empirically determined densities and effective interactions (cf. Aküyz and Winther 1981) one may derive the following parametrization

$$U_{aA}^{N}(r) = - \frac{R_a R_A}{R_a + R_A} \frac{16 \pi \gamma a}{1 + exp\left(\frac{r - R_a - R_A}{a}\right)} \tag{38a}$$

with

$$R_i = (1.20 A_i^{1/3} - 0.09) \text{ fm}$$
$$a^{-1} = 1.17(1+0.53(A_a^{-1/3}+A_A^{-1/3})) \text{fm}^{-1} \tag{38b}$$

and

$$\gamma = 1 \text{ MeV/fm}^2.$$

The expression (37) is only applicable for $r \gtrsim R_a + R_A$. At the distance $R_a + R_A$ it has the maximum slope

$$\left(\frac{\partial U^N}{\partial r}\right)_{max} = 4\pi\gamma \frac{R_a R_A}{R_a + R_A}. \tag{39}$$

Blocki, Randrup and Swiatecki (1977) have given a general argument for the functional form of the short range proximity interaction between two curved surfaces. If we denote by e(s) the interaction energy per unit area of two almost parallel surfaces at a distance s we can calculate the total interaction energy between two curved surfaces as

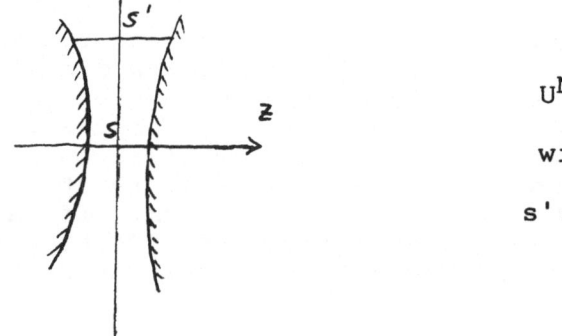

$$U^N = \int dx \, dy \, e(s')$$

with

$$s' = s + \frac{1}{2}\left(\kappa_{\shortparallel} x^2 + \kappa_{\perp} y^2\right)$$

where we used a parabolic aproximation for the distance s' in terms of the minimum distance and the maximum and minimum "rates of curvature" $\kappa_{\shortparallel}$ and $\kappa_{\perp}$. Changing variables to $x' = \sqrt{\kappa_{\shortparallel}}\, x$ and $y' = \sqrt{\kappa_{\perp}}\, y$ one may integrate over the azimuthal angle around z to find

$$\mathcal{U}^{N}(s) = \frac{2\pi}{\sqrt{\kappa_{\shortparallel}\kappa_{\perp}}} \int_{s}^{\infty} e(z)\, dz \quad . \tag{40}$$

Since for two spheres

$$\kappa_{\shortparallel} = \kappa_{\perp} = \frac{1}{R_a} + \frac{1}{R_A} = \frac{R_a + R_A}{R_a R_A} \tag{41}$$

we recognize the reduced radius factor in front of (37). On the other hand we may use the result (40) to generalize (37) to describe the interaction between deformed nuclei by the substitution

$$\frac{R_a R_A}{R_a + R_A} \;\to\; \frac{1}{\sqrt{\kappa_{\shortparallel}\kappa_{\perp}}} \quad , \tag{42}$$

We may also use the proximity form (40) to calculate the slope for s = 0. One finds

$$\left(\frac{\partial \mathcal{U}}{\partial r}\right)_{s=0} = -2\pi\, \frac{R_a R_A}{R_a + R_A}\, e(0)$$

As is seen from the case of two planar surfaces that are brought into contact (s = 0) the energy e(o) is equal to -2γ where γ is the surface tension. This very general argument for the maximum attraction was the reason for the appearance of the surface tension in the expression (37) for the ion-ion potential.

II   ELASTIC SCATTERING

a.   Classical Cross Section

The trajectory of relative motion is determined by the ion-ion potential from the energy and impact parameter ϱ since for grazing collisions we neglect the change in the trajectory due to the transfer of mass, charge, energy and angular momentum. We may thus solve the coupled equations with the initial condition

$$a_\gamma(-\infty) = \delta_{\gamma\alpha} , \tag{1}$$

where α is the entrance channel. The probability P that the nuclei after the collision are in the channel β is then

$$P_\beta = |a_\beta(+\infty)|^2 \tag{2}$$

and the cross section for the reaction is

$$d\sigma_\beta = P_\beta \, d\sigma_{geom} . \tag{3}$$

The geometric cross section is given by

$$d\sigma_{geom} = \varrho \, d\varrho \, d\varphi$$

$$= \frac{\varrho \, d\varrho}{\sin\vartheta \, d\vartheta} \, d\Omega \tag{4}$$

where $\vartheta = \vartheta(\varrho)$ is the deflection angle as a function of ϱ.

In any heavy ion collision with bombarding energy above the Coulomb barrier one cannot reproduce the angular distribution without seriously considering the solution of the coupled equations. Small impact parameters will lead to a strong overlap of the nuclei where perturbation theory is not applicable.

It is however a great simplification that there are usually many reaction channels that become im-

portant at about the same distance of closest approach
of the two ions. The situation is illustrated in Fig.
1 where the probability of staying in the entrance
channel P_α is indicated as a function of the impact
parameter ρ . Also indicated are the transition probabilities p_n of the various reactions depopulating
this channel. They all start to become important at
the grazing impact parameter ρ_g where correspondingly the probability P_α quickly drops to zero. For
most heavy ion collisions there will be a strong depo-
pulation already for $\rho > \rho_g$ due to Coulomb excitation. We shall for simplicity neglect this for the
moment.

We may then estimate the probability P_α by the
expression

$$P_\alpha = \prod_n (1 - p_n) \tag{5}$$

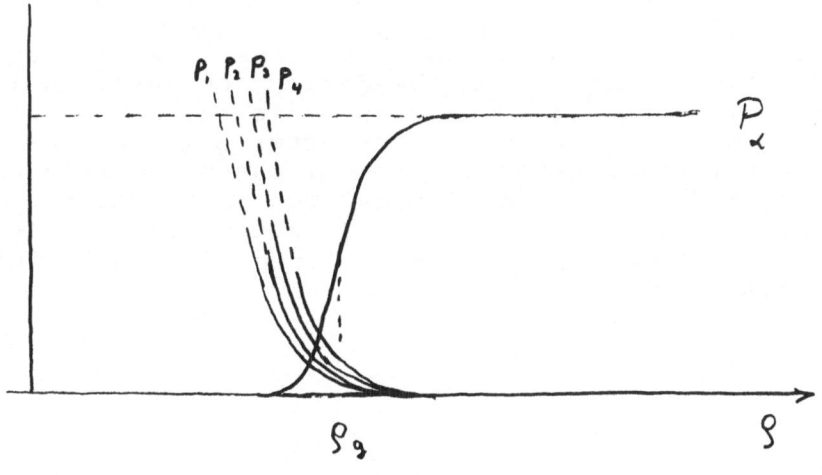

Fig. 1

We have here assumed that the transitions n which are
mainly single particle stripping and pick-up reactions
to various states in target and projectile are independent of each other. Since there may be typically 50
such transitions of similar magnitude we may safely assume that all p_n are small numbers. The error that
we make by this assumption for smaller impact parameters is uninteresting for elastic scattering since P_α
is then very small indeed. This observation leads to
the great simplification that p_n can be calculated by
first order perturbation theory and leads furthermore
to the following approximation

$$P_0 = e^{-\sum_n P_n} \tag{6}$$

It should be noted that since p_i is the transition probabilities $0 \to i$ the probability P_i that the
reaction ends in the channel i is given by

$$P_i = P_i \prod_{n \neq i} (1 - P_n)$$

$$\approx P_i \, e^{-\sum_n P_n} \tag{7}$$

where we may, with a small error only, include n = i in
the summation.

In order to calculate the cross section for
elastic scattering (neglecting Coulomb excitation) we
should evaluate (3) with $P_\alpha = P_0$. The geometrical
cross section for a typical case is illustrated in Fig.
2(b). It is calculated from the deflection function in
Fig. 2(a) which shows the characteristic rainbow phenomenon where for a given impact parameter ρ_R the deflection function achieves a maximum, the so-called
rainbow angle ϑ_R. The Coulomb deflection function
which applies for large values of ρ would lead to the
dotted curve. If the maximum in the effective potential
for the radial motion

$$U_{eff}(r) = \frac{Z_a Z_A e^2}{r} + U^N(r) + \frac{L^2}{2 m_0 r^2} \tag{8}$$

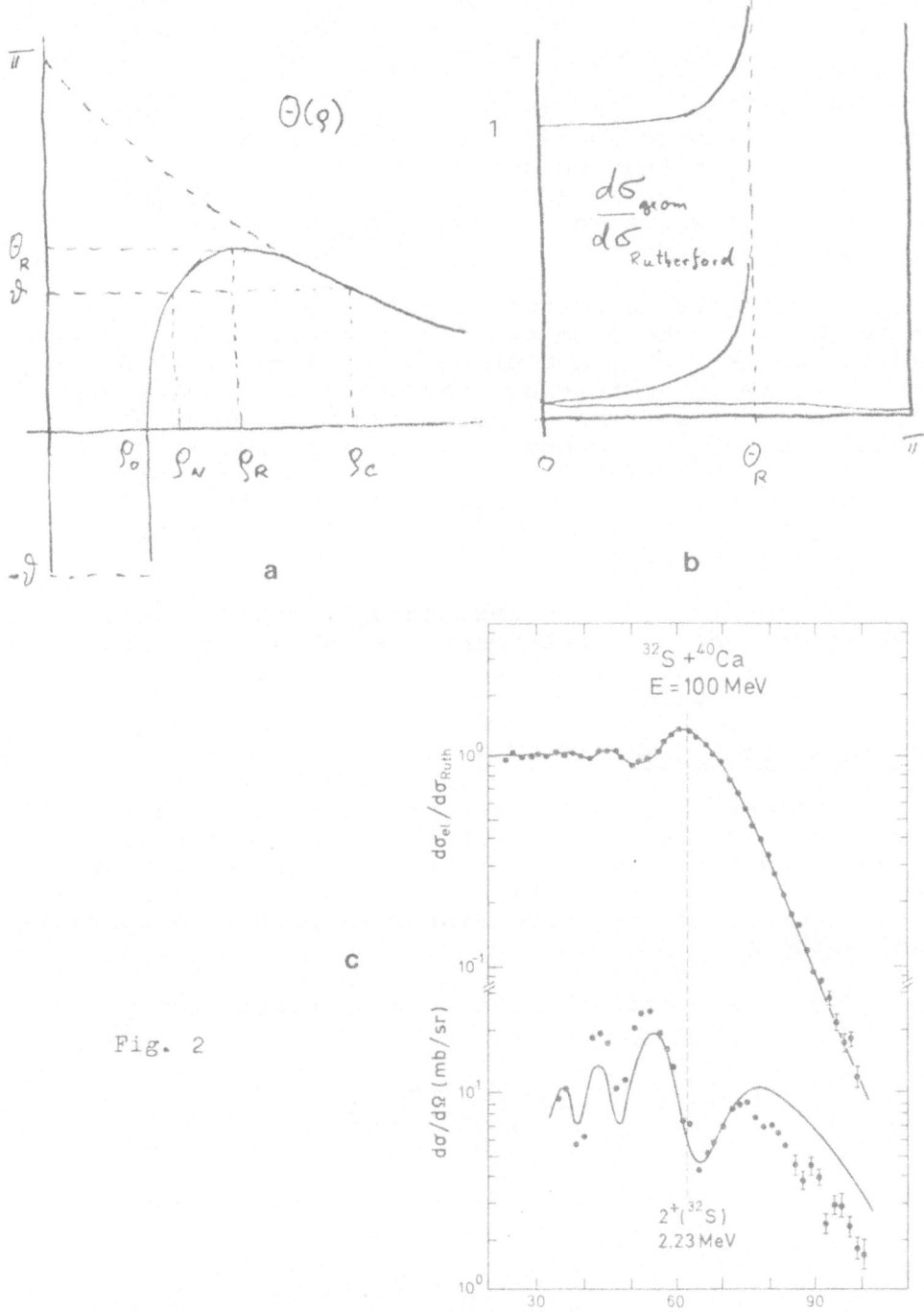

Fig. 2

with orbital angular momentum

$$L = m_o v \, \rho \tag{9}$$

becomes equal to the bombarding energy in the center of mass system the deflection function approaches $-\infty$. This orbiting situation occurs for $\rho = \rho_o$. For smaller impact parameters the deflection function is uninteresting since we may usually assume that $P_\alpha \, (\rho < \rho_o) = 0$.

We note that a scattered particle detected at an angle ϑ may come from two different trajectories with deflection $\Theta = \vartheta$, a Coulomb scattering with $\rho > \rho_R$ and a "nuclear" scattering with $\rho < \rho_R$. Furthermore it may also arise from a scattering with $\Theta = -\vartheta$ behind the target nucleus, and in fact also with $\Theta = -2\pi + \vartheta$, $\Theta = -2\pi - \vartheta$ etc. The corresponding contributions to the cross section are indicated. For $\vartheta > \Theta_R$ only far side collisions ($\Theta < 0$) contribute to the cross section.

The resulting cross section only bears a qualitative resemblance to experimental elastic scattering data (cf. Fig. 2(c)), and there is clearly need for a better description.

b)   Classical Limit of DWBA

In order to obtain a better description we shall, for later use, study the slightly more general problem of deriving the classical limit of the quantal scattering amplitude for a reaction in first order perturbation theory, the so-called distorted wave Born approximation (DWBA).

The cross section is in this approximation given by

$$\left(\frac{d\sigma}{d\Omega}\right)_\beta = \frac{k_\beta}{k_\alpha} \frac{m_\beta m_\alpha}{(2\pi\hbar^2)^2} |T_{\beta\alpha}|^2, \tag{10}$$

where k_α and k_β are the wavenumbers in entrance and exit channels while m_α and m_β are the reduced masses. The T-matrix is

$$T_{\beta\alpha} = \langle \vec{k}_\beta \beta | V_{aA} - U_{aA} | \alpha \, \vec{k}_\alpha \rangle, \tag{11}$$

where $|\vec{K}_\alpha >$ describes the relative motion in the entrance channel in the potential U_{aA} which at large distances behaves like a plane wave with wavenumber $\vec{k}_\alpha$. For inelastic processes the matrix element (11) can be separated into an integral over the nuclear degrees of freedom and an integral over the relative center of mass coordinate $\vec{r}$ of the two ions by introducing the formfactor

$$f_{\beta\alpha}(\vec{r}) = < \beta |V_{aA} - U_{aA}|\alpha >, \qquad (12)$$

which is only a function of $\vec{r}$. This is not possible for transfer reactions where the center of mass coordinate is different in entrance $(\vec{r}_\alpha)$ and exit $(\vec{r}_\beta)$ channel. If we denote $(\vec{r}_\alpha + \vec{r}_\beta)/2 = \vec{r}$ we have

$$\vec{r}_\alpha = \vec{r} + \vec{\xi}/2$$
$$\vec{r}_\beta = \vec{r} - \vec{\xi}/2, \qquad (13)$$

where $\vec{\xi} = \vec{r}_\alpha - \vec{r}_\beta$. The distorted waves in entrance and exit channel can be written

$$|\chi^{(+)}(\vec{k}_\alpha, \vec{r}_\alpha)> = e^{i\vec{\xi}\cdot\vec{P}_\alpha/(2\hbar)} |\chi^{(+)}(\vec{k}_\alpha, \vec{r})>$$
$$|\chi^{(-)}(\vec{k}_\beta, \vec{r}_\beta)> = e^{-i\vec{\xi}\cdot\vec{P}_\beta/(2\hbar)} |\chi^{(-)}(\vec{k}_\beta, \vec{r})>, \qquad (14)$$

where the momentum operators $\vec{P}_\alpha$ and $\vec{P}_\beta$ in the semiclassical limit can be substituted by the local momenta $\vec{P}_\alpha(\vec{r})$ and $\vec{P}_\beta(\vec{r})$.

We may now define the formfactor for transfer in analogy to (12) by the expression

$$f_{\beta\alpha}(\vec{K}, \vec{r}) = <\psi^b \psi^B| e^{i\vec{K}\cdot\vec{\xi}} (V_{aA} - U_{aA})|\psi^a \psi^A>, \qquad (15)$$

where $\vec{K} = (\vec{p_\alpha} + \vec{p_\beta})/2\hbar$ is the average wavenumber.

   In order to evaluate the   T-matrix (11) one expands the distorted waves (14) in partial waves

$$|\chi(\vec{k_\alpha}, \vec{r})\rangle = \frac{4\pi}{k_\alpha r} \sum_{\ell_\alpha m_\alpha} i^{\ell_\alpha} e^{i\beta_{\ell_\alpha}} \chi_{\ell_\alpha}(r) \, Y^*_{\ell_\alpha m_\alpha}(\hat{k_\alpha}) \, Y_{\ell_\alpha m_\alpha}(\hat{r}), \quad (16)$$

where β_{ℓ_α} is the phaseshift and $\chi_{\ell_\alpha}(r)$ the radial wavefunction. It is the regular solution of the radial equation in channel α i.e.

$$\left(\frac{d^2}{dr^2} + \left(k_\alpha(r)\right)^2 \right) \chi_{\ell_\alpha}(r) \;=\; 0 \tag{17}$$

where

$$\left(k_\alpha(r)\right)^2 \;=\; \frac{2 m_{aA}}{\hbar^2} \left(E(\alpha) - U_{aA}(r) - i \, W(r) - \frac{\ell_\alpha(\ell_\alpha+1)\hbar^2}{2 m_{aA} r^2} \right), \tag{18}$$

At large distances $k_\alpha(r)$ $\hat{k_\alpha} = k_\alpha$ $\hat{k_\alpha} = \vec{k_\alpha}$. We introduced here besides the ion-ion potential (I.37) also an imaginary part i $W(r)$ which takes into account the absorption from the elastic channel.

   The nuclear matrix element can be expanded in terms of spherical tensors by the following decomposition

$$f_{\beta\alpha}(\vec{K}, \vec{r}) \;=\; \sum_{JJ'\lambda} \langle I_A M_A J M | I_B M_B \rangle \langle I_b M_b J'M' | I_a M_a \rangle \\ \langle \lambda m J M | J'M' \rangle \quad f^{JJ'}_{\lambda m}(K, r) \tag{19}$$

where we specified the spin quantum numbers I and M of the nuclei a,A,b and B in the entrance channel α and in the exit channel β . It is noted that in the expansion (19) which is normally used for stripping reactions J signifies the angular momentum given to the target while J' is the angular momentum "stripped" from the projectile. The angular momentum $\vec{\lambda} = \vec{J'} - \vec{J}$ is therefore the total angular momentum given to the orbital motion ($\vec{\lambda} = \vec{l}_\rho - \vec{l}_\alpha$).

The tensor $f^{JJ'}_{\lambda m}$ in the laboratory system may now be expressed in terms of the corresponding tensor $\tilde{f}^{JJ'}_{\lambda\mu}$ in the intrinsic frame $\tilde{S}$ with z-axis along $\vec{r}$ and with the vector $\vec{K}$ in the y,z plane. We find

$$f^{JJ'}_{\lambda m}(\vec{K}.\vec{r}) = \sum_{\mu'} D^{\lambda}_{m\mu'}(\hat{r}) \tilde{f}^{JJ'}_{\lambda\mu'}(K_{\shortparallel}, K_{\perp}, r), \qquad (20)$$

where the intrinsic formfactor $\tilde{f}$ only depends on the parallel (along z) component, $K_{\shortparallel}$ and the transverse component $K_{\perp}$ of K and on the distance r between the ions. The operator $K_{\shortparallel}$ acts on the radial wavefunctions changing the coordinate r while $K_{\perp}$ which acts on the angular part of the distorted wave is related to the angular momentum of relative motion. The quantum number μ' indicates the component of $\vec{\lambda}$ along $\vec{r}$. For inelastic scattering where $\vec{K}=0$ we find $\mu'=0$.

We may now evaluate the integral over $\hat{r}$ and find the following expression for the scattering amplitude

$$f^{(c)}(\vartheta) = -\frac{m_\rho}{2\pi\hbar^2} T_{\rho\alpha}$$

$$= -\frac{\sqrt{16\pi}\, m_{\nu B}}{\hbar^2 k_\alpha k_\rho} \sum_{l_\alpha l_\rho} \langle I_A M_A JM | I_B M_B \rangle \langle I_b M_b J'M' | I_a M_a \rangle \qquad (21)$$

$$\langle \lambda m JM | J'M' \rangle \langle l_\alpha 0 \lambda m | l_\rho m \rangle$$

$$(2l_\alpha+1)^{1/2} I_{\rho\alpha}\; i^{l_\alpha - l_\rho}\; e^{i(\beta_{l_\alpha}+\beta_{l_\rho})}\; Y_{l_\rho m}(\hat{k}_\rho)$$

in a laboratory system (C) with z-axis along $\vec{k}_\alpha$.

The quantity $I_{\rho\alpha}$ is

$$I_{\rho\alpha} = \sum_{\mu'} Z \begin{pmatrix} l_\rho & \lambda & l_\alpha \\ I & \mu' & I \end{pmatrix} \int_0^\infty \chi_{l_\rho}(r)\, \tilde{f}^{JJ'}_{\lambda\mu'}(K_{\shortparallel} K_{\perp} r)\, \chi_{l_\alpha}(r)\, dr, \qquad (22)$$

where the coefficient in the front of the radial matrix element ensures the angular momentum coupling in the intrinsic frame.

In the short wavelength (classical) limit the operators $K_\parallel$ and $K_\perp$ can be substituted by the corresponding components of the average local wavenumber i.e. $K_\parallel \sim \frac{1}{2}(k_\alpha(r)+k_\beta(r))$ and $K_\perp \sim \frac{1}{2}(\ell_\alpha+\ell_\beta)$. In this limit we may furthermore substitute the radial wavefunctions in (22) by the WKB solutions

$$\chi_\ell(r) = \sqrt{\frac{k}{k(r)}} \sin\left(\int_{r_o}^r k_\ell(r')dr' + \frac{\pi}{4}\right), \qquad (23)$$

where r_o is the classical turning point defined by $k(r_o) = 0$. Utilizing furthermore that for large values of ℓ_α and ℓ_β

$$\mathcal{Z}\left(\begin{array}{ccc}\ell_\alpha & \lambda & \ell_\alpha \\ \tau & \mu' & \tau\end{array}\right) = (-1)^{\ell_\alpha - \ell_\beta + \lambda}\sqrt{2\ell+1}\; D^\lambda_{\ell_\alpha - \ell_\beta, \mu'}(0,\tfrac{\pi}{2},0) \quad (24)$$

it is seen that the summation over μ' in (22) transforms the formfactor into the frame where the x axis is along $\vec{r}$ while z points along the angular momentum $\vec{\ell}$. The component of λ along this axis is $\mu = \ell_\beta - \ell_\alpha$. If we neglect the contribution to the radial integral inside the turning points and use that

$$k_\alpha(r)-k_\beta(r) = \frac{\partial k}{\partial E}(E(\alpha)-E(\beta)) + \frac{\partial k}{\partial \ell}(\ell_a - \ell_\beta)$$
$$+ \frac{\partial k}{\partial u}(U_{aA}-U_{bB}) + \frac{\partial k}{\partial m}(m_{aA}-m_{bB}) \qquad (25)$$

one may prove that (cf. Broglia and Winther 1981, § IV.9)

$$f^{(C)}(\vartheta) = \frac{\sqrt{2\pi}}{ik}\sum_\ell (\ell+\tfrac{1}{2})^{1/2} a^{(C)}_{\beta\alpha}(\infty)e^{2i\bar{\beta}_\ell} Y_{\ell m}(\hat{k}_\beta). \qquad (26)$$

In this expression $a^{(C)}$ is the semiclassical excitation amplitude (I.25) evaluated in the coordinate system (C), m being the angular momentum transfer M'-M along the direction of the beam. The amplitude should be evaluated with an average trajectory with center of mass energy $\frac{1}{2}(E(\alpha) + E(\beta))$ and with impact parameter $\frac{1}{2}(\ell_\alpha + \ell_\beta)$ where $\ell_\alpha = \ell_\beta - \mu$, μ being the an-

gular momentum transfer perpendicular to the plane of
the orbit. The phaseshift is correspondingly
$$\bar{\beta}_l = \tfrac{1}{2}(\beta_{l_\alpha} + \beta_{l_\beta}).$$

The result (26) offers a compromise between classical and quantal description (cf. Broglia et al. 1974)
in that the radial motion is described by a classical
trajectory while the angular motion is treated quantum
mechanically. It is noted that for the case of poten-
tial scattering where $a_{\beta\alpha} = \delta(l_\alpha, l_\beta) \cdot \delta(m,o)$, (26) re-
produces the well known formula for the elastic scat-
tering amplitude. One may in the expression use the
WKB approximation for the phaseshift i.e.

$$\beta_l \underset{r \to \infty}{=} \int_{r_0}^r k_l(\dot{r}) d\dot{r}' - kr + \eta \ln(2kr) + \tfrac{1}{2}\pi l, \qquad (27)$$

The formula (26) gives a very accurate description of
the DWBA results (cf. S. Landowne et al. 1976) but also
of multiple Coulomb excitation (cf. J. de Boer et al.
1977).

By means of the formula we thus may check why the
classical results in Fig. 2 fail. To this end we study
the limit of the partial wave sum for large values of
l . For the elastic channel we find

$$f(\vartheta) = \tfrac{1}{ik} \int_0^\infty (l+\tfrac{1}{2})^{1/2} e^{2i\beta_l} a_\alpha(l) \frac{\cos\left[(l+\tfrac{1}{2})\vartheta + \tfrac{\pi}{4}\right]}{\sqrt{\sin \vartheta}} dl \qquad (28)$$

$$= \frac{1}{\sqrt{\sin \vartheta}} \sum_{\mp} \int_0^\infty dl\, (l+\tfrac{1}{2})^{1/2} a_\alpha(l) \exp\left[i\left(2\beta_l \mp (l+\tfrac{1}{2})\vartheta \mp \tfrac{\pi}{4}\right)\right],$$

where we used the asymptotic expression of $P_l(\cos \vartheta)$
for large values of l . If we assume that the amplitude for staying in the entrance channel is slowly
varying we may evaluate (28) by finding the points of
stationary phase of the exponential function i.e. the
points $l = \bar{l}$ where

$$2 \left.\frac{\partial \beta_l}{\partial l}\right|_{l = \bar{l}} = \pm \vartheta. \qquad (29)$$

For a given $\vartheta < \Theta_R$ (cf. Fig. 2(a)) there are usually two solutions corresponding to the plus sign. Expanding the exponent to second order in $\ell - \bar{\ell}$ we deform the path of integration in the complex ℓ-plane and include only the contributions around the two passes (cf. Fig. 3) to obtain

$$f(\vartheta) = \frac{\lambda}{\sqrt{\sin \vartheta}} \sum_{\bar{\ell}} (\bar{\ell} + \tfrac{1}{2})^{1/2} a_\alpha(\bar{\ell}) e^{i(2\rho_{\bar{\ell}} \mp (\ell + \frac{1}{2})\vartheta)} \left(\frac{\partial^2 \rho_\ell}{\partial \ell^2}\right)^{-1/2} \quad (30)$$

If there would be only one term we obtain the elastic cross section

$$\frac{d\sigma}{d\Omega} = |a_\alpha(\bar{\ell})|^2 \frac{(\bar{\ell} + \frac{1}{2}) \lambda^2}{\frac{\partial \Theta}{\partial \ell} \sin \vartheta} \quad (31)$$

where we have used the fact that in the WKB approximation (27) the quantity $\Theta(\ell)$ defined by

$$\frac{\partial \rho_\ell}{\partial \ell} = \frac{1}{2} \Theta(\ell) \quad (32)$$

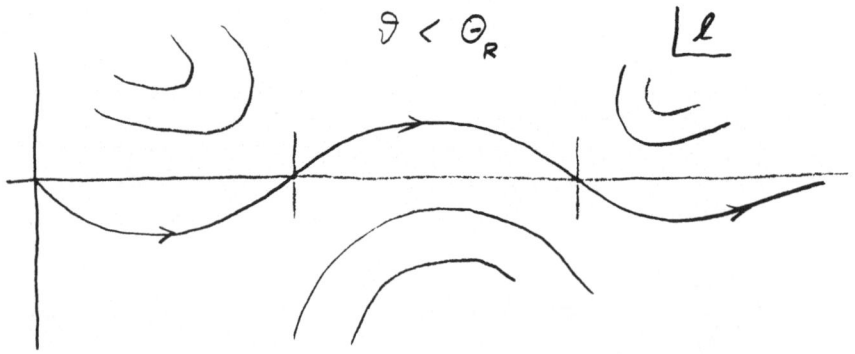

$$\vartheta < \Theta_R \qquad \ell$$

Fig. 3

is the classical deflection angle. The result (32) is then identical to the classical prescription (3). If there are several solutions $\bar{\ell}$ for one ϑ the result is

$$\frac{d\sigma}{d\Omega} = \left| \sum_{\bar{\ell}} a_\alpha(\bar{\ell}) \left(\frac{d\sigma_{geom}}{d\Omega} \right)_{\bar{\ell}} e^{i(2\beta_\ell \mp (\bar{\ell}+\frac{1}{2})\vartheta)} \right|^2 . \tag{33}$$

The different trajectories that contribute to the cross section for a given angle should thus be added coherently with a phase that is given by the action along the path. This result explains the oscillations that are seen in the experimental angular distributions for $\vartheta < \vartheta_R$. For $\vartheta > \vartheta_R$ one also finds two solutions to equation (29) with $+\vartheta$ (cf. Fig. 4). They are complex, and the path of integration from 0 to ∞ should then be deformed to go through the lower pass and one obtains (cf. Knoll and Schaeffer 1976)

$$\frac{d\sigma}{d\Omega} = \chi^2 \left| a_\alpha(\bar{\ell}_0) \frac{\bar{\ell}_0+\frac{1}{2}}{\left(\frac{\partial^2\beta_\ell}{\delta\ell^2}\right)^{1/2}_{\ell=\bar{\ell}_0}} e^{i(2\beta_{\ell_0}-(\bar{\ell}_0+\frac{1}{2})\vartheta)} + a_\alpha(\bar{\ell}) \frac{\bar{\ell}+\frac{1}{2}}{\left(\frac{\partial^2\beta_\ell}{\partial\ell^2}\right)^{1/2}_{\ell=\bar{\ell}}} e^{i(2\bar{\beta}_\ell+(\bar{\ell}+\frac{1}{2})\vartheta)} \right|^2 \tag{34}$$

The first term arises from the stationary point and gives rise to an exponentially decreasing cross section in the shadow region $\vartheta > \vartheta_R$. The second term arises from the real solution $\bar{\ell}$ to the equation (29) with minus sign,

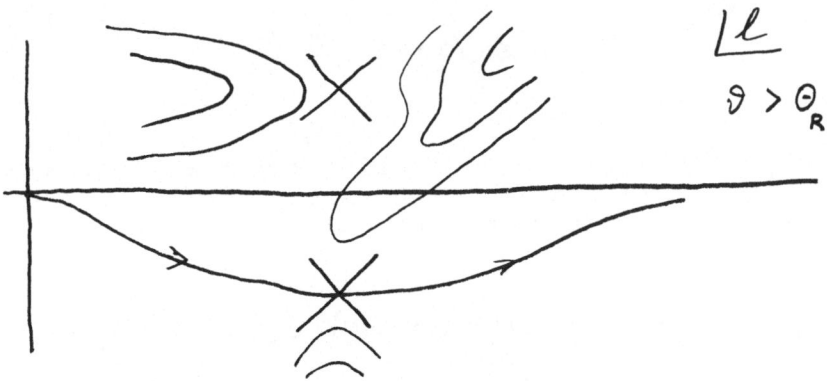

Fig. 4

which is associated with the trajectory that has been
deflected into a negative scattering angle behind the
nucleus on the far side. This term gives rise to
oscillations in the cross section in the region $\vartheta > \vartheta_R$.

We may refine the calculations by including the
fact that the function $a_\alpha (\ell)$ is usually not such a
slowly varying function of ℓ, as was assumed (cf.
Fig. 1). This fact can be included in the above considerations by redefining the phaseshift through

$$2 \bar{\beta}_\ell = 2\beta_\ell - i \ln a_\alpha$$

$$= 2\beta_\ell - i \ln |a_\alpha| + \arg a_\alpha , \tag{35}$$

where according to (6)

$$|a_\alpha| = \sqrt{P_\alpha} = e^{-\frac{1}{2}\sum_n P_n} . \tag{36}$$

We can then repeat the derivations above, but for the
fact that all stationary points $\bar{\ell}$ are now complex. We
may even reestablish the connection with a classical
deflection function (32) by defining a modified complex
scattering potential such that

$$\bar{\beta}_\ell = \int_{r_0}^{r} \left[\frac{2m}{\hbar^2} \cdot \left(E - U_{eff}(r) - \Delta U - i W(r) \right) \right]^{1/2} dr$$

$$- kr - \eta \ln(2kr) - \frac{\pi}{2}\ell . \tag{37}$$

This prescription leads to an equation for the imaginary part W of the potential which is especially simple
if W(r) is small so that we may expand

$$2 \bar{\beta}_\ell = 2\beta_\ell - \frac{1}{\hbar} \int_{-\infty}^{\infty} dt \left(\Delta U(r(t)) + i W(r(t)) \right). \tag{38}$$

In this expression we also changed to an integral over time by using

$$dt = dr/\dot{r} . \tag{39}$$

Comparing (38) with (35) we find the following equation for W(r)

$$\frac{1}{\hbar} \int_{-\infty}^{\infty} W(r(t)) \, dt = -\frac{1}{2} \sum_{n} P_n . \tag{40}$$

This equation does not determine the imaginary (absorptive) potential uniquely, since only the integral of it is known. In order to determine W we may supplement the equation by a convenient subsidiary condition. Such condition may be that we demand to have the same W(r) for all values of ℓ . The equation (40) is then to be interpreted as an identity in ℓ that can be solved (cf. Broglia et al. 1981).

The usual DWBA treatment of nuclear reactions is a first order perturbation theory based on elastic scattering in a complex (optical) potential. By the considerations above we have derived expressions for the optical potential which are in quite accurate agreement with experiments (cf. Pollarolo et al. 1983).

III GRAZING REACTIONS

The present discussion of the evaluation of reaction cross sections is based on the expression (II.26) where the amplitude $a^{(c)}$ is the solution of the coupled equations (I.15) with classical trajectories that have impact parameters corresponding to the angular momentum $\frac{1}{2}(\ell_\alpha + \ell_\beta)$. We discuss separately the three main types of grazing reactions.

a). Coulomb Excitation

For large impact parameters where the distance of closest approach r_o is larger than the range of the nuclear attraction the only reaction is Coulomb excitation. The coupled equations are

$$i\hbar \, \dot{a}_m = \sum_{n} \langle \psi_m^A | V^E | \psi_n^A \rangle e^{\frac{i}{\hbar}(E_m - E_n)t} a_n , \tag{1}$$

where for simplicity we included only excitations of
the target nucleus A. The electric interaction is

$$V^E = \sum_{\lambda\mu} \frac{4\pi z_a e}{2\lambda+1} \mathcal{M}_A(E\lambda-\mu)(-1)^\mu Y_{\lambda\mu}(\hat{r}) r^{-\lambda-1}, \quad (2)$$

where $\mathcal{M}_A(E\lambda\mu)$ are the electric multipole moments. The
equations (1) can rarely be solved by perturbation
theory. This is because the strength parameter

$$\chi_{\lambda\mu} = \frac{1}{\hbar} \int_{-\infty}^{\infty} \langle \psi_{\lambda\mu}^A | V^E | \psi_o^A \rangle dt$$

$$\approx \frac{1}{\hbar} \tau_{coll} \cdot \langle \psi_{\lambda\mu}^A | V^E | \psi_o^A \rangle_{r_o} \quad (3)$$

for the excitation of the low-lying collective 2+ state
for most heavy targets and projectiles is a number of
order unity or larger.

The first order expression

$$a^{(1)} = \frac{1}{i\hbar} \int_{-\infty}^{\infty} \langle \psi_{\lambda\mu}^A | V^E | \psi_o^A \rangle e^{\frac{i}{\hbar} \Delta E t} dt \quad (4)$$

which only leads to the correct amplitude for the excitation of non-collective states can approximately be
factorized as

$$a^{(1)} \approx -i \chi_{\lambda\mu} g_\lambda(\xi) \quad (5)$$

where the adiabaticity parameter ξ is given by

$$\xi = \tau_{coll} \frac{\Delta E}{\hbar} \quad (6)$$

in terms of the collision time $\tau_{coll} \approx r_o/v$. The

adiabatic cut off function $g(\zeta)$ is unity for $\zeta = 0$
and falls off exponentially when $\zeta > 1$. It ensures
that for low bombarding energies $v/c \approx 0.1 - 0.2$ only
low-lying states can be excited ($\Delta E < 3-5$ MeV). It is
perhaps interesting to notice that for higher bombarding energies both χ and ζ diminish i.e. firstly perturbation theory gradually becomes valid even for the
low-lying 2+ states, and secondly, high-lying states
can be reached. In the relativistic energy range the
Lorenz-contraction of the electric field makes the
collision time $\tau_{coll} \approx r_0/(c\gamma)$ go further down proportional to γ^{-1}. The field-strength entering in (3)
increases however proportional to γ and one is therefore for relativistic heavy ions able to excite very
high states with a strength that does not decrease with
γ (cf. Winther and Alder 1979).

Because Coulomb excitation is such a prolific
reaction for the low-lying collective states it is
useful that the coupled equations can be solved analytically for the two idealized main cases that are met
in practice, i.e. the excitation of vibrational states
and the excitation of rotational states in strongly
deformed nuclei. For the excitation of the n'th state
in a vibrational band built on the ground state of an
even nucleus one thus finds

$$P_n = |a_n|^2$$
$$= |a^{(1)}|^{2n} \frac{e^{-|a^{(1)}|^2}}{n!} \tag{7}$$

where $a^{(1)}$ is given by (4). This result is the
quantal equivalent of the classical result that a harmonic oscillator which is excited by a time-dependent
external field linear in the vibrational amplitude
keeps its initial distribution in phase-space. It is
just shifted and the quantal state is the corresponding
coherent state. For the excitation of rotational
states a simple result can be obtained in the limit
where the rotational states have a low energy so that
one may neglect the phasefactor exp($i \Delta Et$) in (1).
Also in this case the result can be obtained by a clas-

sical calculation. While the probabilities P_n in (7) have a smooth Poisson distribution as a function of n the excitation of the rotational states shows an interesting oscillatory behaviour. This is shown in Fig. 5 where the probabilities for exciting the rotational states of (even) spin I in an even nucleus are indicated by circles for a specific example. Indicated by a dotted line is the classical spin-distribution which has a characteristic maximum spin which is in fact

$$I_{max} \approx 4 \chi_{0 \to 2} \qquad (8)$$

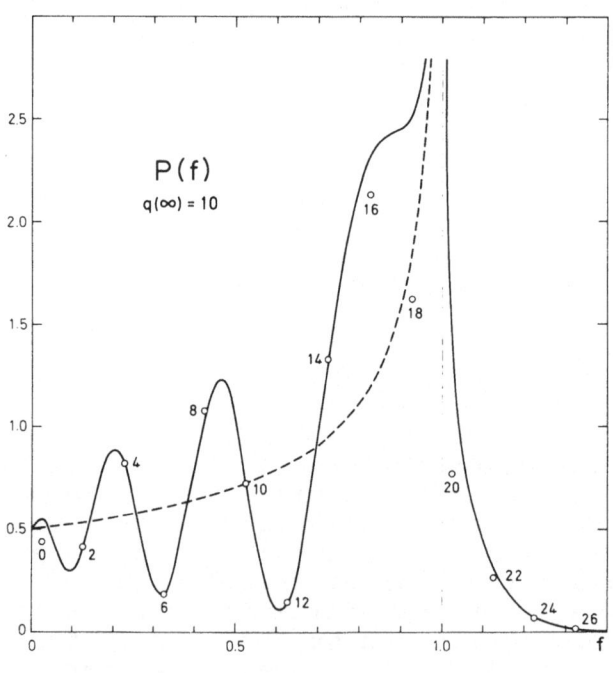

Fig. 5

The oscillatory behaviour of the quantal probabilities
around the classical probability distribution can be
understood by noting that the final angular momentum
that one achieves in a collision of a charge with a
rotor (cf. Fig. 6) is a function of the orientation of
the rotor i.e. of the angle Θ_0. If $\Theta_0 = 0$ or $\Theta_0 = \pi/2$
no torque acts on the rotor and no excitation occurs.
The maximum torque (and final angular momentum I) is
obtained for $\Theta_0 = \pi/4$. For any $I < I_{max}$ two initial
orientations will lead to the same final spin. The
situation is therefore similar to the situation we discussed in § II.c where two initial conditions occuring
in the initial quantum state with a definite phase relation interfere. In § IIc it is two different impact
parameters that occur in the plane incoming wave and
lead to the same scattering angle. Here it is two
different orientations that occur in the isotropic
wavefunction of the ground state that eventually lead
to the same final spin. The probability distribution
is

$$P_I = \left| \sqrt{P(\Theta_1)} e^{i\phi_1} - i \sqrt{P(\Theta_2)} e^{i\phi_2} \right|^2 \tag{9}$$

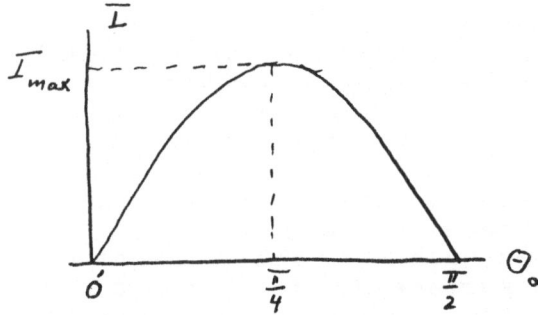

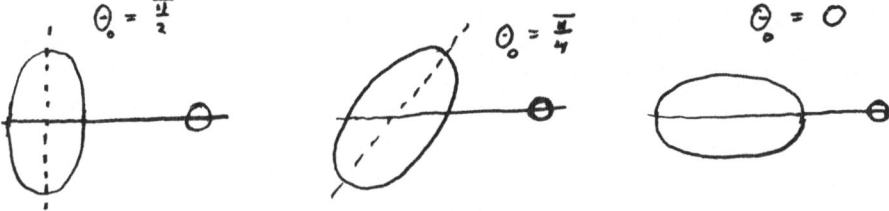

Fig. 6

where $P_I(\theta_i)$ is the classical probability corresponding to the initial orientation θ_i leading to final
spin I while ϕ_i is the action integral in units of $\hbar$.

b)   Inelastic Scattering

For smaller impact parameters inelastic scattering
may also be caused by the nuclear interaction. The
coupled equations are the same as in (1) except that
one should add the nuclear interaction i.e.

$$iha_n = \sum_m \langle n|V^E + V^N|m\rangle \, e^{\frac{i}{\hbar}(E_n - E_m)t} \, a_m(t) . \qquad (10)$$

The nuclear matrix element for the excitation of collective surface vibrational states of the target
nucleus is

$$\langle n|V^N|m\rangle = - R_A \frac{\partial U^N}{\partial r} Y_{\lambda\mu}(\hat{r}) \langle n|\alpha_{\lambda\mu}|m\rangle , \qquad (11)$$

where U^N is the ion-ion potential, R_A the radius
of the target, and $\alpha_{\lambda\mu}$ the vibrational amplitude.

In principle the strength of the nuclear excitation will be larger than the strength (3) for Coulomb
excitation for small impact parameters. However the
absorption that takes place due to transfer reactions
at such small distances, inhibits in most cases the
observation of these very strong interactions. Since
the absorption due to transfer is associated with the
position of the nuclear surface the imaginary part of
the potential also has non-diagonal matrix elements between states that describe surface modes and one should
in (11) in fact substitute the ion-ion potential by the
full optical potential i.e.

$$\frac{\partial U}{\partial r} \rightarrow \frac{\partial(U + iW)}{\partial r} \qquad (12)$$

The nuclear inelastic scattering which takes place
mainly for small impact parameters (ρ_N in Fig. 2(a))
interferes with the pure Coulomb excitation which takes
place for the large impact parameters(ρ_c in Fig. 2(a))
which lead to the same scattering angle. The cross
section is (cf. eq. (II.33))

$$\frac{d\sigma}{d\Omega} = \left| a(\ell_c) \left(\frac{d\sigma_{geom}}{d\Omega}\right)_{\ell_c}^{1/2} e^{i(2\beta_c - (\ell_c + \frac{1}{2})\vartheta)} - i a(\ell_N) \left(\frac{d\sigma_{geom}}{d\Omega}\right)_{\ell_N}^{1/2} e^{i(2\beta_{\ell_N} - (\ell_N + \frac{1}{2})\vartheta)} \right|^2 \quad (13)$$

The oscillations in the cross section from this
Coulomb-nuclear interference is similar to the oscillations in the elastic cross section except that it has
the opposite phase since the amplitudes a(ℓ_c) from
Coulomb excitation and a(ℓ_N) which arises mainly from
nuclear excitation have the opposite sign.

c)   Transfer Reactions

        In order to calculate the cross section for trans-
fer of nucleons between heavy ions we must evaluate the
matrix element (II.15). In the classical approximation
$\vec{K}$ is the average local-momentum and r is the average
relative center of mass coordinate. For a stripping
reaction,

$$a(= b + d) + A \rightarrow b + B(= A + d) \quad (14)$$

the geometry is shown in Fig. 7. For the evaluation it
is convenient to measure the position of the transferred "particle" d from the point δ signifying the
intersection of the vector $\vec{\xi} = \vec{r}_{aA} - \vec{r}_{bB}$ and $\vec{r}$. One
finds that the recoil phase may be written

$$\sigma_{\rho\alpha} = \vec{K} \cdot \vec{\xi} = \vec{k} \cdot \vec{r}_{d\delta} \quad (15)$$

with

$$\vec{k} = \frac{m_d}{2 m_0} \vec{K} \quad (16)$$

We shall consider single nucleon transfer where
m_d = M(= nucleon mass) and the matrix element (II.15)
is then

$$f(\vec{k},\vec{r})_{\rho\alpha} = \langle \psi_{a_1}^{(B)}(\vec{r}_{1A}) \psi^A \psi^b | (V_{aA} - U_{aA}) e^{i\sigma_{\rho\alpha}} | \psi^b \psi^A \psi_{a_i}^{(a)}(\vec{r}_{ib}) \rangle$$

$$= \langle \psi_{a_1}^{(B)}(\vec{r}_{1A}) | (U_{1A}(r_{1A}) - \langle U_{1A} \rangle) e^{i\sigma_{\rho\alpha}} | \psi_{a_i}^{(a)}(\vec{r}_{ib}) \rangle \quad (17)$$

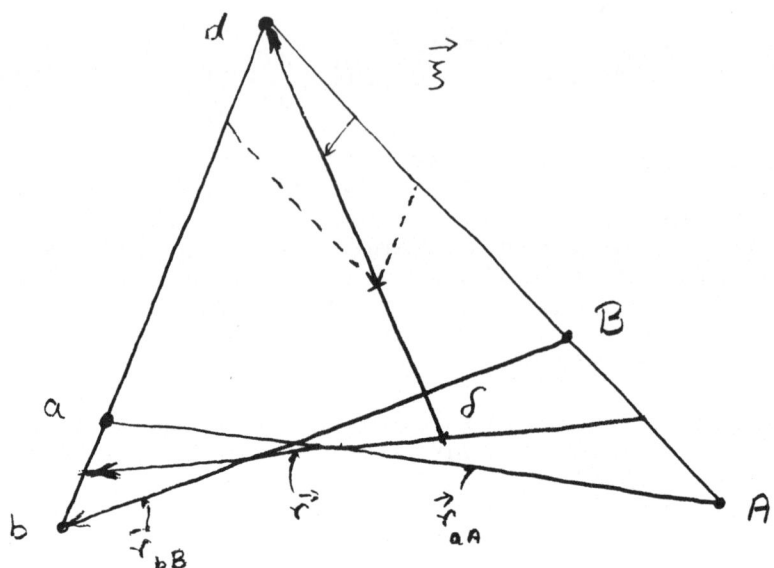

Fig. 7

where $\psi^{(B)}(r_{1A})$ and $\psi^{(a)}(r_{1b})$ are the single particle wavefunctions in target and projectile while
$U_{1A}(r_{1A})$ is the shell model potential in A. The
difference

$$\langle U_{1A} \rangle = U_{aA}(r_{aA}) - U_{bA}(r_{bA}) \qquad (18)$$

which arises from non-orthogonality and from the acceleration in the relative motion is approximately given
by

$$U_{1A} - \langle U_{1A} \rangle \approx U_{1A}^N + \left(\frac{z_d}{z_a} - \frac{m_d}{m_a} \right) \vec{r}_{1a} \cdot \vec{\nabla} U_{aA} . \qquad (19)$$

The acceleration term is thus "attractive" for neutrons
and repulsive for protons. It would be zero for a particle with charge $Z_a e/A_a$. For neutron transfer to
strongly bound states it may be neglected, while for
proton transfer the main term in (18) cancels the
Coulomb part of the shell model potential. The matrix
element (17) has in this approximation an exponential
shape

$$f(k,r) \sim e^{-\varkappa r} \qquad (20)$$

where the constant κ is determined from the slope of the wavefunction $\psi^{(a)}(r_{1a})$, i.e. $\kappa = \sqrt{2MB_a}/\hbar$ where B_a is the binding energy. The shape of the wavefunction $\psi^A(r_{1A})$ outside nucleus A is irrelevant because of the appearance of the potential U_{1A} in (17).

For the actual evaluation of the matrix element one constructs the tensor (II.19) and evaluates this formfactor in the intrinsic frame $\tilde{S}$, where the z-axis is pointing along $\vec{r}$ and $\vec{K}$ is lying in the y-z plane (cf. eq. (II.20)). In this frame the form-factor depends on the three numbers $k_{\shortparallel}$, $k_{\perp}$ and r which in the semiclassical approximation are given functions of time. One finds the result

$$\tilde{f}^{a,a'}_{\lambda\mu}(k_{\shortparallel}, k_{\perp}, r) = 2\pi (2j_a + 1)^{1/2} (2\lambda + 1)^{1/2} (-1)^{j'_a + \lambda + \frac{1}{2}} \left\{ \begin{matrix} \ell, & l, & \lambda \\ j'_a, & j_a, & \frac{1}{2} \end{matrix} \right\} J$$

$$\times \int y\,dy\,dz\, R^{(A)}_a(r_{1A}) R^{(b)}_{a'}(r_{1b}) (U_{1A}(r_{1A}) - \langle U_{1A} \rangle)\, J_{\mu'}(k_\perp y) \qquad (21)$$

$$e^{ik_{\shortparallel}(z - \frac{m_a + m_B}{2(m_a + m_A)} r)} \sum_{m, m'} \langle \ell, m, l', m', | \lambda, \mu' \rangle Y_{\ell, m'}(\vartheta_A, 0)\, Y_{\ell', m'}(\vartheta_b, 0).$$

It is noted that for small values of $k_{\perp}$ only $\mu' = 0$ contributes which means that the component of the angular momentum along the asymmetry axis $\vec{r}$ is conserved. It is the relative motion of the two ions which gives rise to Coriolis forces that violate the conservation of this (K-) quantum number. In the low-recoil approximation one may write the formfactors (21) for $\mu' = 0$ in the form

$$\tilde{f}^{a,a'}_{\lambda 0}(k_{\shortparallel}, k_{\perp}, r) = e^{i\bar{\sigma}} \sqrt{\frac{4\pi}{2\lambda + 1}}\, f^{a,a'}_{\lambda}(r) \qquad (22)$$

The phase $\bar{\sigma}$ arises from (21) by extracting the average value of the exponential function i.e.

$$\bar{\sigma} = k_{\shortparallel}\left(z_\rho - \frac{m_A + m_B}{2(m_a + m_A)} r\right) \qquad (23)$$

where z_ρ is the z-value giving the main contribution to the integral. A good approximation for z_ρ turns out to be $z_\rho = 1.0\, A_A^{1/3}$ fm. The real formfactors $f^{a,a'}_{\lambda}(r)$ can be evaluated by a simple closed expression (cf. Buttle and Goldfarb (1966) and J.M.Quesada et al. (1985).

With the formfactor (22) one may rather easily evaluate the semiclassical transfer amplitude (25) in first order perturbation theory. One finds according to (I.7-10) and (II.19-20)

$$a^{(1)} = -i \langle I_A M_A \, j, m, | I_B M_B \rangle \langle I_b M_b \, j_1' m_1' | I_a M_a \rangle$$

$$\langle \lambda \mu \, j, m, | j_1' m_1' \rangle \; I_{\lambda \mu} \tag{24}$$

with

$$I_{\lambda \mu} = Y_{\lambda \mu}(\tfrac{\pi}{2}, 0) \frac{1}{\hbar} \int_{-\infty}^{\infty} dt \int_{\lambda}^{a, a'} (r(t)) \, e^{\frac{i}{\hbar}[(E_\rho - E_\alpha)t + \gamma_{\rho \alpha} + \hbar \bar{s} + \mu \hbar \phi(t)]} \tag{25}$$

We have here evaluated $a^{(1)}$ in a laboratory coordinate system where the z-axis is along the angular momentum of relative motion while the x-axis points towards the point of closest approach of the two ions. The quantity $\phi(t)$ is the azimuthal angle of $\vec{r}$ which in this system lies in the x y plane. The phase $\gamma_{\rho \alpha}$ arises from the second term in (I.10) the first term being identical to the recoil phasefactor in (17). It may be written

$$\gamma_{\rho \alpha}(t) = \int_{0}^{t} dt' \left[U_{bB}(r(t')) - U_{aA}(r(t')) + \tfrac{1}{2}(m_{aA} - m_{bB})(\dot{\vec{r}}(t'))^2 \right]. \tag{26}$$

The orbital integral can be evaluated to a good approximation by utilizing that the formfactor $f_\lambda^m(r)$ has an exponential shape. This means that the integral receives its main contribution close to t = 0 where we may use the approximation that

$$f_\lambda^{a, a'}(r) = f_\lambda(r_0) \, e^{-\varkappa(r - r_0)}$$

$$r(t) = r_0 + \tfrac{1}{2} \ddot{r}_0 t^2$$

$$\phi(t) = \dot{\phi}_0 \, t \tag{27}$$

$$k_{\parallel}(t) = \dot{k}_{\parallel}(0) \, t \quad , \quad \gamma_{\rho \alpha}(t) = \dot{\gamma}_{\rho \alpha}(0) \cdot t$$

With this approximation we have a Gaussian integral and the result is

$$I_{\lambda\mu} = \left(\frac{2\pi}{\hbar^2 \kappa \ddot{r}_o}\right)^{1/2} Y_{\lambda\mu}^{a,a'}\left(\frac{\pi}{2},0\right) f_{\lambda}(r_o) \exp\left\{-\frac{(Q-Q_{opt}-\mu\hbar\dot{\phi}_o)^2}{2\hbar^2 \kappa \ddot{r}_o}\right\} \quad (28)$$

where $Q = E_\alpha - E_\alpha$ is the Q-value for the transfer reaction while

$$Q_{opt} = U_{bB}(r_o) - U_{aA}(r_o) - \frac{1}{2}(m_{bB}-m_{aA})v^2 + \hbar\dot{\bar{\sigma}}(o) \quad (29)$$

The quantity

$$\tau = \sqrt{\frac{1}{\kappa \ddot{r}_o}} \quad (30)$$

measures the time where the formfactor is nonvanishing, i.e. the collision time. The product of the first two factors in (28) therefore measures the strength parameter (I.19) while the exponential function gives rise to an adiabatic cut-off for Q-values which differ from the optimum Q-value, Q_{opt}. For inelastic scattering where γ and $\bar{\sigma}$ vanish the parameter in the exponential function can be compared to the adiabaticity parameter ξ in (6).

One can give a simple physical interpretation of the Q-value dependence, based on a classical picture. We thus envisage that the transfer of mass from a to A gives rise to a "frictional" force $\vec{F}$. The effect of this force is

$$\vec{F} \cdot \vec{v} = \frac{d}{dt}(E-U_{aA}) + \frac{1}{2}\frac{dm_{aA}}{dt} v^2 \quad (31)$$

where the last term is the change in kinetic energy because of the change in reduced mass caused by the transfer. The torque of the force is

$$(\vec{r} \times \vec{F})_z = \frac{dL_z}{dt} \quad (32)$$

Now we may separate (31) into radial and transverse parts and we find for the radial part

$$v_r \, F_r \;=\; \vec{v} \cdot \vec{F} - v_\varphi \, F_\varphi$$

$$\qquad = \frac{d}{dt}(E - U) + \tfrac{1}{2}\frac{dm_o}{dt} v^2 - \frac{v_\varphi}{r}\frac{dL_z}{dt} \tag{33}$$

where we used (32). The integral of this radial effect is small because $v_r = 0$ at t = 0 where the interaction is largest. Classically we thus find that

$$\int F_r \, v_r dt \;=\; E_\beta - E_\alpha - U_{bB} + U_{aA} + \tfrac{1}{2}(m_{bB} - m_{aA})(\dot{\vec{r}}(o))^2$$
$$\qquad - \dot{\phi}(0)\,\mu\,\hbar = 0 \tag{34}$$

which is essentially equivalent to the condition that the argument of the exponential function in (28) vanishes. The recoil correction due to the phase $\bar{\sigma}$ may also be reproduced by taking into account the change in the relative center of mass coordinate $\vec{r}$. (cf. also Brink (1972)).

The expression (28) offers a simple yet rather accurate result by means of which one may calculate the imaginary part of the optical potential.

References

Aküyz, Ö and Winther A, 1981, Proceedings of the International School of Physics "Enrico Fermi Varenna, 1979, p. 492, North-Holland Publ. Co. Amsterdam.

Buttle, P. A. J., and Goldfarb, L. J. B., 1966, Nucl. Phys. 78:409.

Blocki, I., Randrup, J. and Swiatecki, W., 1977, Ann. of Phys. 105:427.

Brink, D. M., 1972), Phys. Lett. 40B:37.

Broglia, R. and Winther, A., 1972, Phys. Rep. 4C, 153.

Broglia, R. and Winther, A., 1981, Heavy Ion Reactions Vol. I, Bejnamin/Cummings Publ. Co. Reading, Mass.

Broglia, R. A., Landowne, S., Malfliet, R. A., Rostokin, V. and Winther, A.,1974, Phys. Rep.11c:1.

Broglia, R. A., Pollarolo, G. and Winther, A., 1981, Nucl. Phys. A361:307.

de Boer, J., Dannhäuser, G., Massmann, H., Roesel, F. and Winther, A., 1977, Journ. of Phys. G3:889.

Knoll, J. and Schaeffer, R., 1976, Ann. of Phys.97:307

Landowne, S., Dasso, C. H., Nilsson, B. S., Broglia, R. A. and Winther, A., 1976, A259:99.

Pollarolo, G., Broglia, R. A. and Winther, A., 1983,
    Nucl. Phys. A406:369.
Winther, A. and Alder, K., 1979,
    Nucl.Phys. A319:518.
Quesada, J. M., Broglia, R. A. and Winther, A., 1985,
    (in press).

TRANSFER OF NUCLEONS BETWEEN NUCLEI

W. Von Oertzen

Hahn-Meitner-Institut für Kernforschung
Berlin, Germany
and GANIL, Caen, France

I. NUCLEAR STRUCTURE AND DYNAMICS OF NUCLEON TRANSFER CLOSE TO THE COULOMB BARRIER

a) Semi-classical Considerations

The most intuitive way to the understanding of transfer between nuclei is obtained in the semi-classical approach, which has had far reaching successes in quantitative and qualitative descriptions of heavy ion reactions (Coulomb excitation, inelastic scattering, transfer and fusion reactions, see in particular ref.1 - 4).

For large values of the Sommerfeld parameter η (Z_i-charges of the nuclei, v their relative velocity),

$$\eta = \frac{Z_1\,Z_2\,e^2}{\hbar v} = 0.16\ Z_1\,Z_2\,(E/u)^{-1/2} \tag{1}$$

and large values of the wave numbers K or $K_{eff}(r)$

$$K = \left(\frac{2\mu\ Ecm}{\hbar^2}\right)^{1/2}\ ;\ \ K_{eff}(r) = \left(\frac{2\mu\ (Ecm - V(r))}{\hbar^2}\right)^{1/2} \tag{2}$$

the motion of the centers of the colliding nuclei can be described by classical orbits. The condition for a semi-classical description implies that no interference of amplitudes occurs (this can be introduced explicitly afterwards, e.g. ref. 3). Further, for energies above the Coulomb barrier, the variation of the wave length $\lambda(r) = 1/K_{eff}(r)$ should be small due to the change of the potentials and also, of the form factors as function of r. This is expressed as (ref.4),(the gradient is taken at the "Coulomb Barrier" with radius R_B) :

$$\left| \text{Grad } \chi \ (r) \right|^2_{R_B} \ << \ 1 \tag{3a}$$

We neglect the variation of the Coulomb potential V_c at $E > V_c(R_B)$, because it is slow as compared to the nuclear part. For large r, we use the approximate form of a Woods-Saxon potential

$$V(r) = V_o \cdot e^{R_o/a} \cdot e^{-r/a} \quad \text{(with the usual notation) and obtain for}$$

condition (3a) :

$$\frac{(E_{cm} - V_a)^3 \ 8 \ \mu/\hbar^2}{(V_o/a)^2 \ e^{-(R_B - R_o)/a}} \ >> \ 1 \tag{3b}$$

Typical values of V_o/a are 100 MeV/fm and $(R_B - R_o)/a \cong 5$. Inserting these values we obtain (u is now in units of amu) for the condition (3a) to be fulfilled :

$$(E_{cm} - V_c)^3 \ \mu \ >> \ 3.4.10^2 \quad [\text{MeV}] \tag{3c}$$

This condition if often well fulfilled at energies above the Coulomb barrier (10 MeV and $\mu = 50 - 100$). The conditions are actually equivalent to demanding that the phase shifts δ_L associated with a given partial wave L change slowly. We thus obtain in the semi-classical approach the relation for the scattering angle θ :

$$2 \frac{d}{dL} \ \delta_L \ = \ \theta(L) \tag{4}$$

The angular momentum is simply given by Kb = L or for Coulomb fields

$$L = KR_m \ (1 - \frac{2 \ \eta}{KR_m})^{1/2} \tag{5}$$

or L = η ctg $\theta/2$

We have introduced here the impact parameter b and the distance of closest approach R_m (or minimum distance). The latter is the most important quantity, because in transfer the form factor is scanned by the scattering orbit as function of R_m. We use a reduced quantity d_o, where the size of the nuclei is divided out and a scaling with respect to the true extension of the wave functions of bound nucleous is obtained

$$R_{min} = R_m = \frac{\eta}{K} \ (1 + \frac{1}{\sin \theta/2}) = \frac{Z_1 Z_2 e^2}{2E} \ (1 + \frac{1}{\sin\theta/2}) \tag{6}$$

and

$$d_o = R_m/(A_1^{1/3} + A_2^{1/3}) \ ; \tag{6a}$$

d_o measures the overlapp of the two nuclei; generally a value of $d_o > 1.40$ fm has to be assured if absorption and competition with complicated channels is to be avoided. Instead of plotting $\sigma(\theta)$ we will always transform θ to d_o.

The wave length of relative motion is thus generally very small as compared to the variation of probabilities for transfer and excitation with R_m or r; this generally also true for absorption into more complicated channels. This allows the use of the following formula :

$$\sigma_{el}(\theta) = \sigma_R(\theta) (1 - P_{abs}(\theta)) \tag{7}$$

for elastic scattering and

$$\sigma_{tr}(\theta) = \sigma_{el}(\theta) P_t(\theta) F (Q,L) \tag{8}$$

for a transfer or excitation processes.

The semi- classical orbits demand that any change of the dynamical quantities (η, K and L) preserves the minimum distance R_m. If this is not the case the cross section has to be multiplied by the factor F(Q,L), which corrects for the loss of particle flux due to the mismatched orbits. Modern computer codes (ref.5) allow the calculations of the DWBA transition amplitude also for heavy systems, where up 500 or 1000 partial waves can contribute. Fig. 1 (ref. 6) shows as an example the variation of the two-neutron transfer cross section for the case of $^{112}Sn(^{120}Sn, {}^{118}Sn)^{114}Sn$ for a Q-value range from - 6 MeV to + 6 MeV. For this system the relevant values are :

$$\eta = 187.8 \; ; \quad K = 26.9 \text{ fm}^{-1} \; ; \quad \bar\lambda = 0.037 \text{ fm}.$$

The minimum distance at E_L = 544 MeV and θ = 160° is R_m (160°) = 14.08 fm (d_0 = 1.444 fm). A change of E_{cm} by 2 MeV (Q-value) changes R_m by 0.1 fm. These semi-classical restrictions to the matching of orbits lead to the "Q-value window" and matching conditions with Q $\neq$ 0 for charge transfer (see e.g. ref.2).

For energies above the Coulomb barrier absorption from the elastic channel diminishes the scattering probability ($\sigma_e/\sigma_R < 1$). Fig. 2 shows again for the $^{120}Sn + {}^{112}Sn$ the measured ratio σ_e/σ_R as function of d_0. The elastic channel contains all inelastic excitations, the absorption is thus only due to transfer and fusion processes ($\sigma_e(\theta) = \sigma_R(\theta) (1 -P_a(\theta))$. Here P_a stands for the absorption probability. The values of $(1 - P_a (d_0))$ are independent from the incident energy, as the variation of d_0 can be achieved by a change of the angle θ or E_{cm} in eq. 6.

The absorption can also be seen as the action of an imaginary potential. In the semi-classical approximation this give in first order[1] a damping coefficient from integration along the orbit using the time t as integration variable.

$$P_a(\theta) = \left| \exp\left[- \frac{1}{\hbar} \int_{-\infty}^{+\infty} W(r(t))dt\right] \right|^2 \tag{9}$$

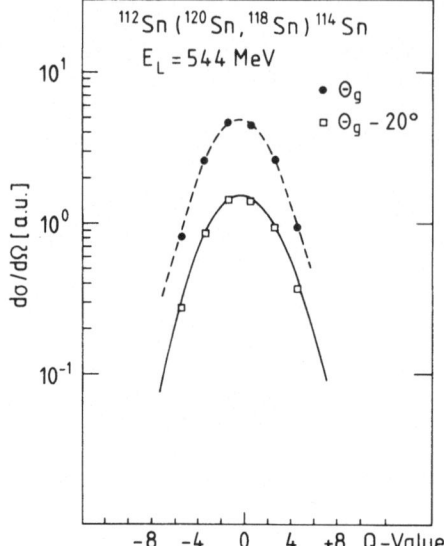

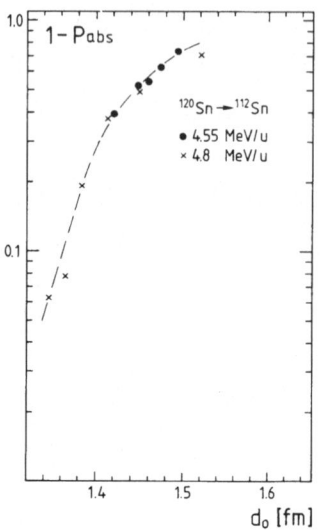

Fig. 1 Q-value dependence for
a two-neutron transfer
cross-section between Sn
isotopes.

Fig. 2 Elastic scattering cross-
section (absorption pro-
bability) $P_a(\Theta)$ plotted as
function of distance of
closest approach (parame-
ter d_0) for the system
$^{120}Sn + ^{112}Sn$ at two incident energies.

One has, however, to be aware of the fact that the imaginary
part can produce refraction, i. e. a change of the classical trajectory[7].

The probability for transfer is similarly given by an integral
of the form factor $f_t(r)$ along the scattering orbit

$$P_t(\Theta) = \left[\frac{1}{\hbar} \int_{-\infty}^{+\infty} dt\, f_t(r(t)) \right]^2 \qquad (10)$$

The form factor is obtained[2] in DWBA as the overlap of the
nuclear wave functions of the initial $\psi_A\, \psi_a$ and final $\psi_B\, \psi_b$ states.
The reaction is depicted as $A + a \rightarrow b + B$, and we have

$$f_t(r) = \langle\, \psi_B\, \psi_b\, |V_{eff}|\, \psi_A\, \psi_a\, \rangle \qquad (11)$$

Each individual overlap gives the structural factors $\theta_{\ell_i j_i}$ and a bound state wave functions of the transferred nucleon $\phi_{\ell_i j_i}(r_i)$

$$< \psi_B | \psi_A > = \phi_{\ell_2 j_2}(\vec{r}_2) \cdot \theta_{\ell_2 j_2} \qquad (12a)$$

$$< \psi_b | \psi_a > = \theta_{\ell_1 j_1}(\vec{r}_1) \cdot \theta_{\ell_1 j_1} \qquad (12b)$$

In a quantitative analysis using DWBA, the calculation of the bound state wave function $\phi_{\ell_j}(r)$ becomes one of the intricate points. Actually, at energies close or below the Coulomb barrier with transitions where the θ_{ℓ_j}'s are known, the measurement of the cross section yields rather $\phi_{\ell_j}(r)$ or the $< r^2 >$ of the bound state wave function[8].

Because of the large effective wave length at energies at the Coulomb barrier, the coordinate shifts introduced by the mass transfer generally can be neglected and the no recoil approximation is applied[2]. This approximation makes $\vec{r}_i$ and $\vec{r}_f$ (see fig.3) parallel

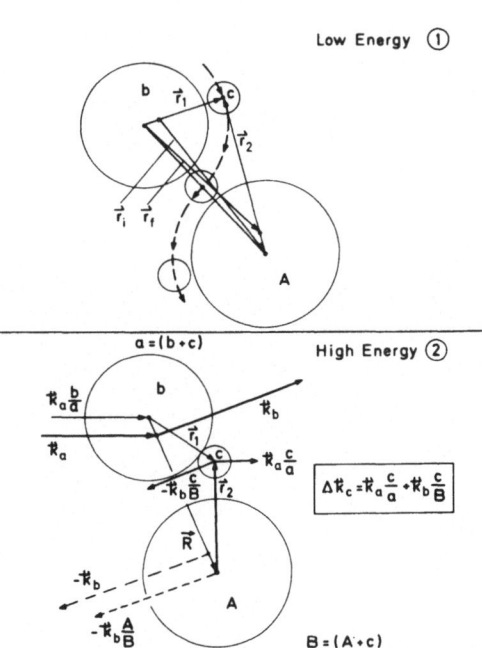

Fig. 3 a) b). Coordinates and momenta for nucleon transfer at low and high relative velocities.

as in inelastic scattering. This gives a local form of the transition form factor and the factorisation given in eq. 8 corresponds
to an independent integration of the distorted waves, which have
a much faster variation ($\lambdabar$) as compared to the decay constant of
the form factor.

It can be easily shown that the overlap, eq.(11), can be reduced for large distances to an exponential of the form (normalised
with the factor $N_{\ell j}$)

$$f_t(r) = N_{\ell j} \; e^{-\alpha r}/\alpha r \qquad\qquad (13)$$

Here the decay constant α is given by

$$\alpha = \sqrt{2\mu \; E_B /\hbar^2} \qquad\qquad (14)$$

with E_B binding energy and μ the reduced mass of the bound particle. The normalisation $N_{\ell j}$ contains all factors which are involved
in the internal matrix element (11) and the integration over
the internal variables which leaves us just with the distance r.
With this form factor the integration along the Rutherford orbit
can be done analytically (ref.9).

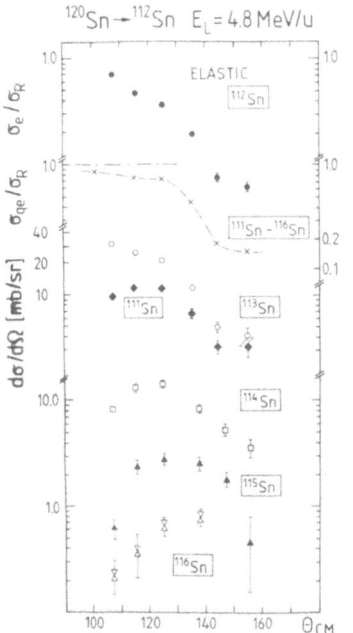 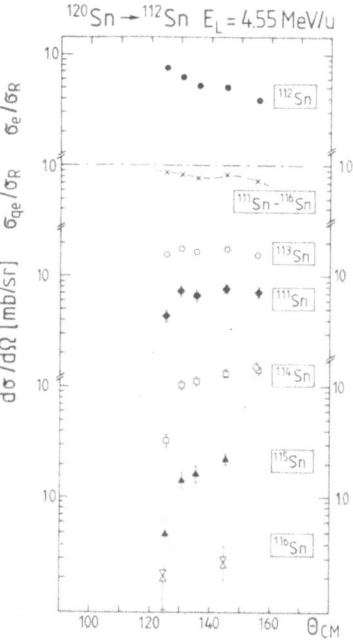

Fig. 4   Angular distributions for neutron transfer in the system
^{120}Sn + ^{112}Sn at energies close to the Coulomb barrier.

If we plot the ratio of the transfer cross section devided by the elastic cross section according to eq. 8 in a logarithmic scale we obtain a linear function which represents the transfer form factor. This is shown in fig. 4 and 5 , where angular distributions and transfer probabilities are shown for various channels in the ^{120}Sn + ^{112}Sn system. The slopes agree very well with the expected values of α from eq.14. These data correspond to a sum of low lying transitions with $|Q|<10$ MeV. The energy spread is generally much smaller. Further discussion of these probabilities will be given in sub-section b).

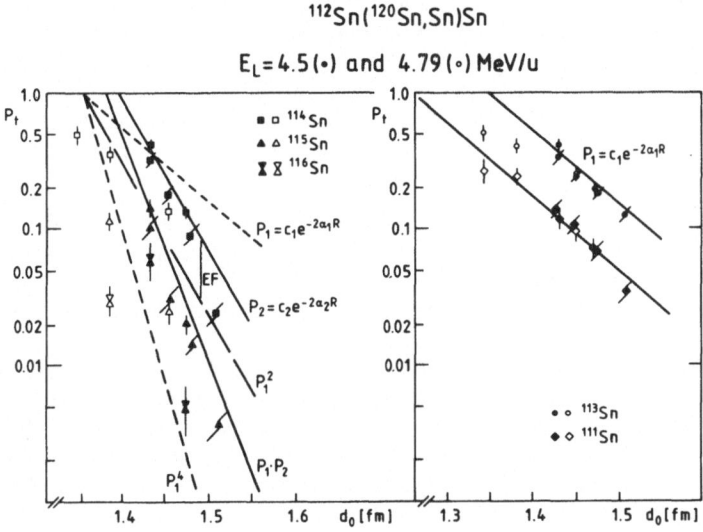

Fig. 5   Probabilities obtained from the data of fig. 4 for the transfer of one up to four neutrons.

Generally there is little or no dependence on the angular momentum transfer ℓ. The selection rules for angular momentum ℓ (multipolarity of the transition) obtained for the transfer just reflect the vector coupling of angular momenta and spins of the initial and final states,

$$\vec{\ell} = \vec{\ell}_1 + \vec{\ell}_2 \text{ and } \vec{\ell} = \vec{j}_1 + \vec{j}_2 \tag{15}$$

At low energies the internal motion of the nucleons completely dominates. By matching the internal momenta of the initial and final state (given by λ_1/R_1 and λ_2/R_2) the condition (here λ_i is the projection of ℓ_i perpendicular to the reaction plane)

$$\Delta K = \frac{\lambda_1}{R_1} + \frac{\lambda_2}{R_2} \simeq 0 \tag{16a}$$

is obtained (see fig. 3a). The dominant transitions are thus those, where $\ell = \ell_1 + \ell_2$. This produces the well known j-selectivity with transitions being favoured if j_1 and j_2 have opposite spin orbit coupling schemes

$$j_1 = \ell \pm 1/2 \quad \to \quad j_2 = \ell \mp 1/2 . \tag{16b}$$

b) <u>Transfer between very heavy nuclei</u>

For reactions between heavy nuclei apart of the fact that the semi-classical conditions are well fulfilled, the nucleon mass is negligibly small compared to the masses of the cores. The shifts of the centers of gravity due to the transfer are small, however, the wavelengths of relative motion are also small (they decrease at least $\sim 1/A$ because of the increasing energy to reach the Coulomb barrier). The use of the no-recoil approximation is thus often less justified for very heavy systems.

There are other evident changes if very heavy ions are used as projectile and target. Taking closed shell (or half-closed) nuclei, the two nuclei will be at least to 50 % in the first excited 2^+ state before contact and transfer takes place. In addition, the experimental energy resolution will not allow to separate low lying states. Thus the measured quantities defined as quasi-elastic events are summed over several MeV of excitation - a quasi-elastic channel being then defined by a change of mass and/or charge with an energy in the optimum Q-value window. The original flux contained in the ground states is recovered in the elastic and transfer channels by summing over an appropriate Q-value range (see ref. 6 and 10 and fig. 6 respectively).

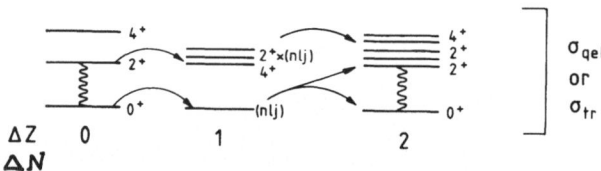

Fig. 6   Definition of elastic scattering and transfer cross sections for heavy nuclei (with vibrational spectrum)

At incident energies below the Coulomb barrier the experimental energy spectra with 2-5 MeV resolution show one "peak," which shows a width of $\sim$ 5 MeV (ref.6).

As an example for neutron transfer at energies well above the barrier we take the case of ^{86}Kr + ^{208}Pb (ref.11) at E_L = 695 MeV. In the ^{87}Kr spectrum the sum over ca10 MeV has been taken similar to the case of ^{120}Sn+^{112}Sn and the angular distribution is shown in fig.7. The angular distributions become very narrow for heavy systems. This fact can easily be deduced, if we start from the fact that localisation in r-space,ΔR (or Δb), for the transfer to occur, is independent of the masses and energies involved. We calculate the quantity $d\theta/dR$ or $\dfrac{d\theta}{dL}$ and we obtain

$$\frac{d\theta}{dL} = \frac{\sin^2\theta/2}{\eta} \quad \text{or} \quad \frac{d\theta}{db} = \frac{K}{2\eta}\,\sin^2\theta/2\,\Big|_{\theta_{gr}} \tag{17}$$

The width is inversely proportional to η but proportional to K for a given width db. For a chosen grazing angle in the classical limit (e.g. Pb + Pb) and small grazing angle ($\sin^2\theta/2 \cong 0.1$) the width can be become as narrow as a few degrees (HW).

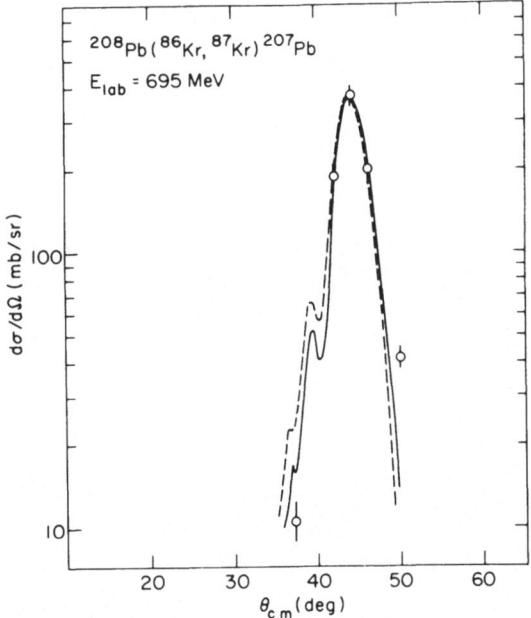

Fig.7   Angular distribution of the neutron pick-up in the ^{86}Kr + ^{208}Pb reaction. The sum over all single particle transitions is taken. The curves are DWBA calculations normalised arbitrarily (ref. 11).

A further interesting point is the question of the different components which contribute to the observed cross section. Pick-up from ^{208}Pb yields typically six different single particle states in ^{207}Pb ($3p_{1/2}$, $3p_{3/2}$, $2f_{5/2}$, $2f_{7/2}$, $h_{9/2}$, $i_{13/2}$), they combine to populate ca 5 states in ^{87}Kr ($3s_{1/2}$, $2d_{3/2}$, $d_{5/2}$, $g_{7/2}$, $i_{11/2}$); in total we have 30 ! single nucleon transfer transitions (see fig.10). Actually, we should add those transitions, where either projectile or target are in vibrational states and the final nuclei in a single-hole (or particle) core-excited state. If we consider that each transition carries at least two angular momenta ℓ, we obtain a total of 100 to 200 individual transitions (incoherent), which give the cross section observed (which is 380 mb at θg for the case of Kr + Pb).These transitions are certainly not completely independent. In table 1 we show from ref. 11 the calculated cross sections for some transitions. Due to the strong favouring of the maximum $\ell=(\ell_1 + \ell_2)$ only ca10 cross sections give 90 % of the cross section.

Table 1. Configurations and cross sections (in mb/sr) contributing to the total transfer probability observed over a width of ca5.0 MeV in the ^{208}Pb(^{86}Kr, ^{87}Kr)^{207}Pb reaction(DWBA) calculations, ref. 11 .(See also fig. 8,10a).The numbers in paranthesis are the ℓ-values.

^{87}Kr / ^{207}Pb	s $1/2^+$	d $3/2^+$	d $5/2^+$	g $7/2^+$	i $11/2^+$
p $1/2^-$	8.5(1) 13.5(1)	3.5(2) 2.6(2)	29. (3)	1.45(4)	12.7(6)
p $3/2^-$	11.4(1)	28.1(3) 2.2(2) 1.8(1)	20.6(3) 6.9(2) 22.9(1)	25.2(5) 1.5(4) 1.8(3)	7.91(6) 1.5 5.1
f $5/2^-$	5.5(3) . .	2.0(4) . . .	23.6(5) 5.1(4) . .	1.44(6) < 1.0 (1-5)	19.9(8) . .
f $7/2^-$	2.7(3)	13.5(5) 1.5 <1.0	6.95(5) 2.8 . .	20.8(7)	5.11(8) . .
h $9/2^-$	0.2(5)	<1.0	3.26(7) . . .	0.13(7) . .	7.8(10) . .
i $13/2^+$	0.3(6)	2.02	1.25(8) .	5.56(10) .	1.8(11) .

Similar data have been obtained in the interaction of ^{58}Ni + ^{208}Pb (ref.12). In figure 8 we show the relevant transfer probabilities deduced from ref.12 and those for the case of ^{86}Kr + ^{208}Pb. We notice that the correct slope of the exponential form is observed and that the same absolute magnitude of $P_t \approx 0.1$ is observed at $d_0 = 1.55$fm for both cases. Surprisingly the same value is obtained for the

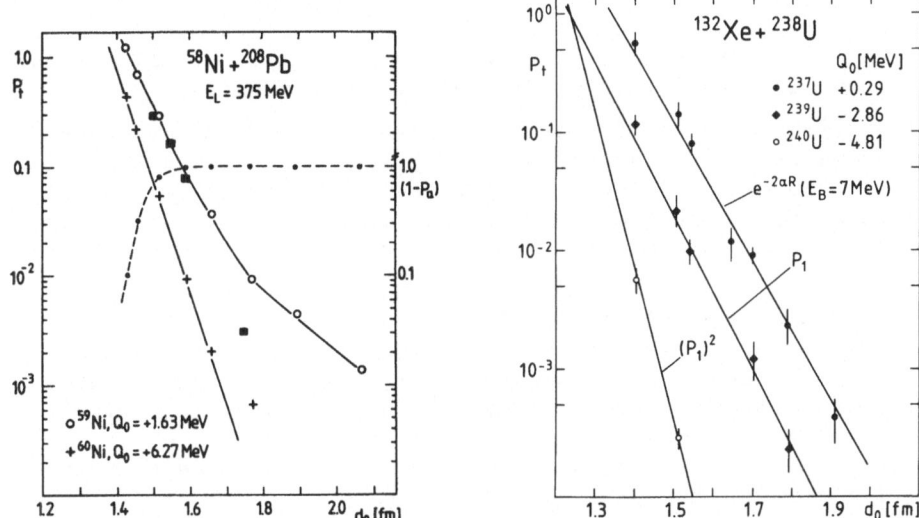

Fig. 8 and 9 Transfer probabilities in the interaction of ^{58}Ni + ^{208}Pb (ref. 12),^{132}Xe + ^{238}U (ref.13) et ^{86}Kr + ^{208}Pb (ref.11, black squares).

case of Sn + Sn (fig. 5) and also for the case of Sub-Coulomb transfers observed in the System ^{132}Xe + ^{238}U (ref.13). The latter is shown in fig.9.

We thus find that in the limit of the rather macroscopic definition of the one-neutron transfer probability, we observe, indepent of the nuclear system, approximately (within the errors of the definition of the cross section) the same transfer probability (see ref.14 for a further discussion of this point). This probability most likely represents the tunneling of a nucleon between two fermi surfaces, where the splitting into configurations is smeared out by the definition of the measured quantities.

c) Two-nucleon transfer

In the data shown in fig. 5, 8 and 9, the two neutron transfer is also shown in terms of a transfer probability as function of minimum distance. As expected from the semi-classical picture, the slope of these probabilities is twice that of the one neutron case.

This is obtained either by writing for α for a dineutron

$$\alpha_{2n} = \sqrt{\frac{2\mu \cdot 2E_B \cdot 2'}{\hbar^2}} \tag{18}$$

or if we consider sequential transfer, the probability to transfer x neutrons is simply

$$P_{xn} = (P_{1n})^x \simeq e^{-x(2\alpha R)} \tag{19}$$

For an increasing number of transferred nucleons a fan of straight lines crossing at a fixed value of d_o with $P_t = 1$ for purely sequential transfer is predicted (see also fig.11).

Previous studies of two-nucleon transfer induced by heavy ions have shown that the transfer is mostly sequential and not a pair transfer (ref. 15,17). However, the pairing interaction will still influence the two-nucleon transfer by introducing a constructive interference of intermediate steps and thus lead to an enhancement.

It has to be emphasized that a behaviour of the form factor or of $P_t(d_o)$ according to eq.(19) is only observed if transitions along low lying bound states are observed. In multi-nucleon transfer for energies well above the Coulomb barrier, the cross section often is localised at much higher excitation. However, even if a narrow Q-value window is chosen in the finally observed channel, the intermediate steps may correspond to unbound states and the form of equ.(19) main not be observed.

II. TWO-STEP INTERACTIONS IN TRANSFER REACTIONS

a) Two-neutron transfer and pairing correlations

The transfer of two nucleons between two heavy nuclei has been the subject of numerous experimental and theoretical studies[2,17]. There are two major questions which arise :

i) what is the relative importance of one-step pair and the sequential (two-step) transfer,

ii) the occurence of enhancement due to the pairing correlations and the possibility of a nuclear Josephson effect[18].

The first question can not be answered by inspection of the data. The general experience with the most recent calculations[17] is that the sequential transfer dominates by factors of 10 to 100 in the absolute cross-section. In the limit of no pairing interaction the one-step transfer of two nucleons does not exist, in the calculations this term is then completely cancelled by the non-orthogonality term[15].

The sequential transfer can be written in the following form (it is a second order term)

$$T_{ba}^{(2)} = \Sigma_m < \chi_b^- \psi_B \psi_b \mid V_{eff} \mid \bar{\Phi}_m > G_m^{(+)} < \bar{\Phi}_m \mid V_{eff} \mid \psi_a \psi_A \chi_a^+ > \qquad (20)$$

$$\text{with } G_m = (E_m - T_m - < V_m > \pm i\varepsilon)^{-1}$$

SEQUENTIAL TRANSFER OF
TWO NEUTRONS

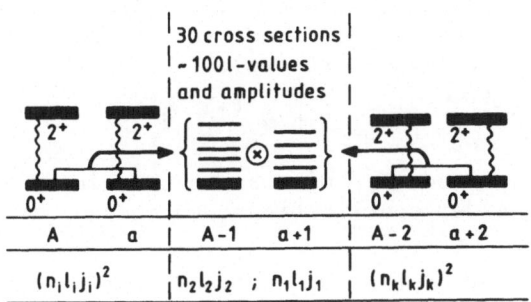

Fig. 10 a)   Schematic representation of intermediate states contributing in a two-step transfer process.

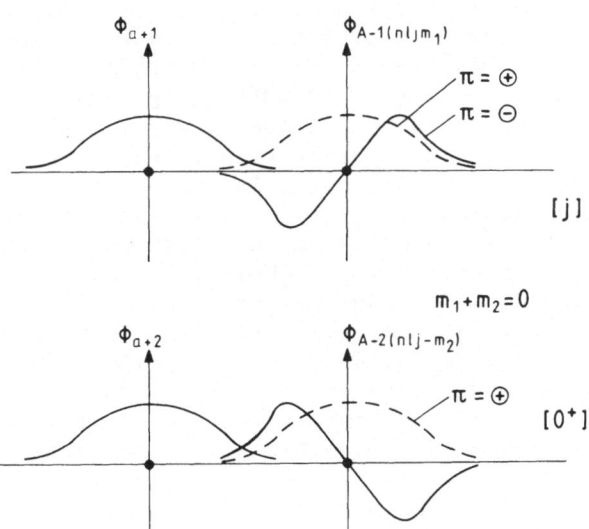

Fig. 10 b)   Overlapps of single nucleon wave functions in the first (top) and second (bottom) steps in a two nucleon transfer populating a $(n\ell j)^2 0+$ state. The sign of the overlapp changes with the parity π.

The sum is taken here over all possible intermediate configurations (states) which contribute to the final state. These contributions are coherent and contain the configurations, which are
present in the wave functions due to the pairing interaction. This
applies to the projectile and the target. For the case of
^{86}Kr + ^{208}Pb we have seen that as many as 30 intermadiate states
are possible, the enhancement is thus obtained via the sum of these
intermadiate states due to the configurational mixing in the initial
and the final state and the combinations of these states (see tab.1,
fig. 10a). The first order term has the usual form of the DWBA
transition matrix element (Φ_j stands for the product of wave functions of initial and final states).

$$T_{ba}^{(1)} = < \chi_b^{(-)} \Phi_f \mid V_{eff}^i \mid \chi_a^{(+)} \Phi_i > \tag{21}$$

The total amplitude T_{ba} will again contain an interference of
first and second order contributions, which can give a constructive
or also destructive effect[17].

The enhancement of the two-neutron transfer can be well defined
in the case of semi-classical conditions, because we have a direct
measure of the form factor and the structural factor through the
transfer probability defined in eq.10, and shown for several cases
in fig. 5, 8 and 9. The measured transfer probabilities must exhibit
the right slope as function d_o (or R_m) to insure that transitions to
lowest lying states are measured (see in particular ref.10 and 14).
The multiplication of the probabilities for the multistep transfer
produces a fan of curves crossing at $P_t = 1.0$ and increasing slope.
An enhancement is than seen as a parallel shift relative to the
prediction of pure sequential transfer (fig.11). The enhancement
factor EF actually should be defined by reference to the form factors for a chosen single particle transition (one) and that of
a $0^+ \rightarrow 0^+$ ground state transition. In view of the experimental
(and principal) difficulties, it is only possible to define an average enhancement, $\overline{EF}$, by deviding the measured quantities which
correspond to a sum of levels, by their relevant phase-space factors
N_{1n} and N_{2n}. The data for the case of ^{120}Sn + ^{112}Sn show for the
two-neutron transfer a parallel shift (fig. 5) which amounts to ca
factor 3. For the ^{114}Sn channel, we can assume that, apart from the 0^+
transition, the 2^+ and 3^- states contribute to the measured cross
section. In the single nucleon transfer, we have at least one
third of the (6 x 6) = 36 transitions with appreciable cross sections (see eg.table 1). In addition, the average multipolarity
(spin multiplicity) of the single nucleon transfer is much higher
(factor 3). We may thus derive $\overline{EF}$ from the measured value EF

$$\overline{EF} = \frac{P_{2n}}{(P_{1n})^2} \frac{(N_1)^2}{N_2} = EF \frac{(N_1)^2}{N_2} \tag{22}$$

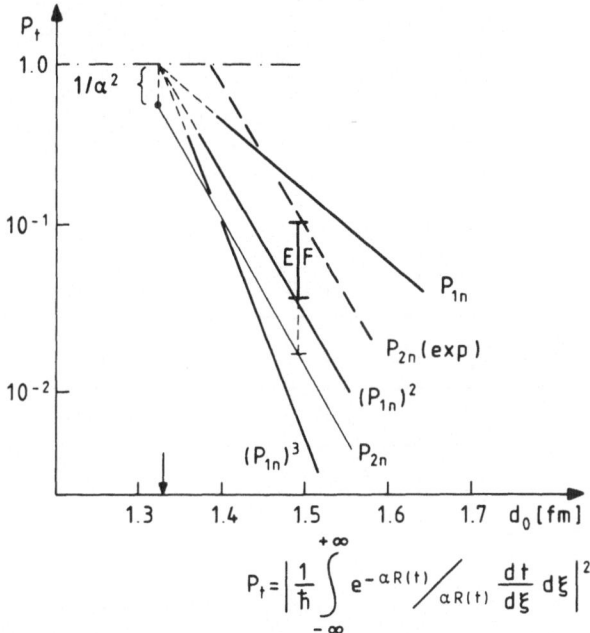

$$P_t = \left| \frac{1}{\hbar} \int\limits_{-\infty}^{+\infty} e^{-\alpha R(t)} \Big/ \alpha R(t) \frac{dt}{d\xi} d\xi \right|^2$$

Fig. 11   Semi-classical transfer probabilities for the sequential
          transfer of nucleons and the definition of enhancement.

        A value for the average enhancement $\overline{EF}$ of $30(\frac{(3 \times 13)^2}{3}) = 1600$
is obtained.

        We have used here a slightly different approach as compared to
that given in ref.6, where a value for $\overline{EF} = 40$ is given.

        We go back to the case of lighter ions and discuss the problem
of the calculation of the two-nucleon transfer in the context of
second order DWBA. There has been considerable progress in the calculation of the absolute cross sections[17,19]. In the most recent
calculations, e.g. $(^{16}O, {}^{18}O)$ on ^{208}Pb, agreement is generally
achieved within a factor of 1.5 to 2. Configuration mixing is for
the one-step pair transfer, as well as for sequential transfer, the
most important source of uncertainty. In the two cases (ref.15, 16),
where configurations extending into the next major shell of different parity are used, it was found, that there occurs a destructive
interference between states of different parity. In table 2, we
give the result of Kammuri, where the inclusion of the next shell
$(g_{9/2})$ reduces the two-step cross section and increases the one-step
contribution. The origin of this cancellation is found in the change
of the sign of the overlapp of single particle wave functions   in
the second step if a $0^+ \to 0^+$ transition is considered. The two

angular momenta have to be coupled with (ℓ_1, m_1) and $(\ell_2, -m_2)$ in order to form a pairing state $(\ell_1 = \ell_2)$! Fig. 10 b gives a schematic illustration of this circumstance.

Table 2.   Cross sections(in μb/sr) calculated for two neutron transfer ^{60}Ni $(^{18}$O, ^{16}O$)^{62}$Ni with and without the $(g_{9/2})^2$ configuration in ^{62}Ni (ref.16).

$$^{60}\text{Ni} \; (^{18}\text{O}, \; ^{16}\text{O}) \; ^{62}\text{Ni} \; (f_{7/2}, \; p_{3/2}, \; g_{9/2})$$

	all conf	without $g_{9/2}$
$\sigma(1)$ one-tep	194	143
$\sigma(2)$ two-step	556	596
$\dfrac{\sigma(2)}{\sigma(1)}$	2.87	4.17 !

In the limit of complete configuration mixing due to a very strong pairing interaction, a complete cancellation of the sequential transfer is expected and we would be left with a one-step pair transfer alone. In this context, it is intriguing to find that the spatial correlation of a neutron pair is in fact increased by the mixing of configurations of different parity. This has recently been shown by Catara et al[43]. We show in fig.12 the result of their calculation of the probability distribution $P(r,R)$ of two neutrons

$$P(r,R) = | \psi \, (\vec{r}_1 \, \chi_1, \; \vec{r}_2 \, \chi_2) | \; r^2 \; R^2 \; d\hat{r}d\hat{R}d\chi_1 \; d\chi_2 . \qquad (23)$$

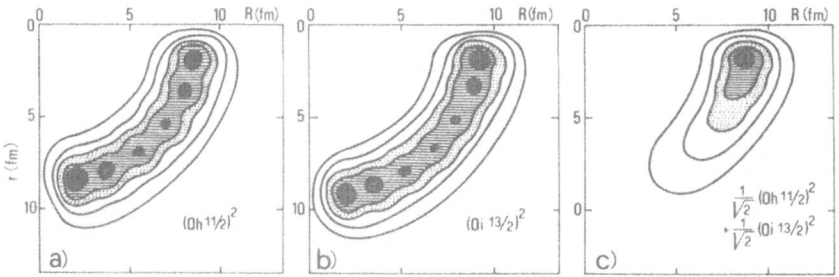

Fig.12   Contour plots of the probability distributions for two neutrons (relative distance r, distance of their center of mass to core, R) for two configurations $(h_{11/2})^2$ and $(i_{13/2})^2$ and for the mixed configuration with $h_{11/2}$, $i_{13/2}$, amplitudes equal to $1/\sqrt{2}$.

Here, $\psi(1,2)$ is the two-neutron wave function and the coordinate R describes the motion of the center of the pair with respect to the core, and r the relative motion of two neutrons. The density profile of each configuration shows clearly the high number of nodes expected for a state with intrinsic quantum numbers $\ell = 0$, $S = 0$ for the pair moving relative to the core with quantum numbers N,L (the well known relation for the transformation $2 \times (2 n + \ell) = 2 N + L$ gives N=5 for $(h_{11/2})^2$ and $N = 6$ for $(i_{13/2})^2$. The mixed state gives a very strong concentration of the density for small values of r and large R (fig. 12), i.e. a correlated neutron pair.

These two facts actually suggest that a dominant one-step pair transfer between very heavy nuclei is conceivale if nucleides can be made to interact,which are placed with their Fermi-levels close to major shells. The observation of a nuclear Josephson effect is conceivable in such cases. The presently available data on one and two-neutron transfer, $^{120}Sn + ^{112}Sn$ (ref. 6, 14 and fig. 5), and $^{136}X + ^{238}U$ (ref. 13 and fig. 9) are mostly dominated by sequential transfer, however large enhancements due to the pairing correlations are seen as expected from (p, t) and (t,p) results.

b) Two-step interactions involving inelastic transitions and transfer

Another important source of two-step interactions are inelastic transitions preceeding or following a transfer process. This has been studied extensively (see ref.20). We want to address ourselves here to the more recent studies connected with the spin-orbit interaction of heavy ions. It has been shown that the origin of the large values of the V_{LS}-potential is due to the second order (and higher order) coupling of inelastic channels and/or transfer channels[21]. In this frame, it was possible to explain the different signs of the analyzing power of ^6Li and ^7Li. The main difference in the coupling scheme to the inelastic projectile excitation for the two ions is the fact, that for ^7Li the spin sequence is: larger spin (3/2, Q = 0, and lower spin (1/2), Q < 0, and opposite to that in ^6Li: (1), Q = 0 and (3+), Q < 0.

For a further illustration of second order coupling effects, I will discuss the $^{12}C + ^{13}C$ system.

The states (single particle states in ^{13}C) which are included in the coupling scheme are shown in fig. 13. This system is particularly suited to illustrate certain effects introduced by higher order interactions[22]. We have marked in fig. 13 the transitions which lead to the strong coupling scheme (iterative methods do not converge for the coupled reaction channel calculations). The most important ones are the transfer transitions between the mixed orbits of $p(1/2^+)$ and Sd, which is called hybridization in atomic

physics[23],[24]. Actually the favourable conditions for the mixing of
single particle orbits of different parity occur for the same spe-
cies (carbon) in atomic and nuclear physics in view of the shell
structure of the quantum systems. Similar to the case already

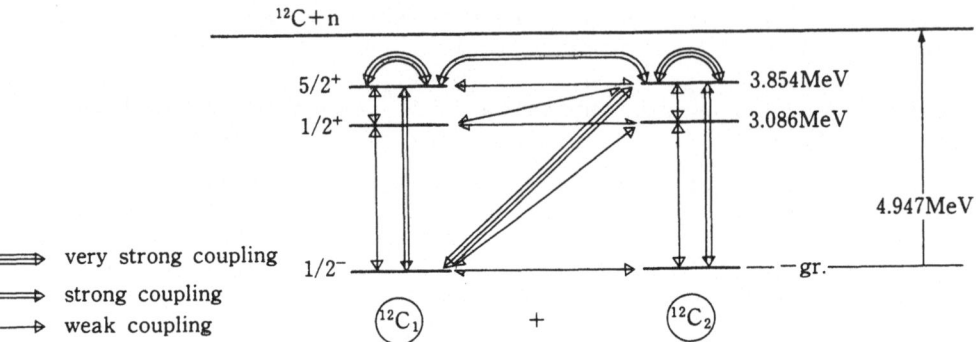

Fig. 13   Coupling scheme for transfer and inelastic transitions in
          the interaction of ^{12}C + ^{13}C.

discussed in the case of spatial correlations of the neutron-pair
in fig. 12 in the previous section, the mixing of parities leads
to an extremely polarized orbit. The density distribution of such
a mixed configuration is shown in schematic way in fig. 14. In a

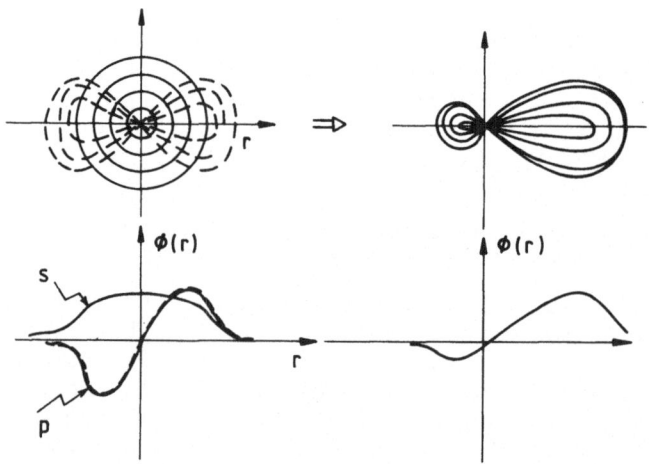

Fig. 14   Polarised single particle orbits due to mixing (hybridi-
          zation) of p-sd orbits in ^{13}C.

general way we can state that inelastic transitions could be incorporated as polarised orbits in the description of transfer processes. These polarised orbits are states which contain several configurations (with the same m-quantum number) and will generally
lead to an enhanced cross section. This reminds us of the mixing
due to pairing in the two-nucleon case, where mixing of different
parties introduces an enhancement of pair transfer.

Concerning the spin-orbit interaction, the ^{12}C + ^{13}C system
offers a perfect example. Fig. 15 shows the polarization of ^{13}C in

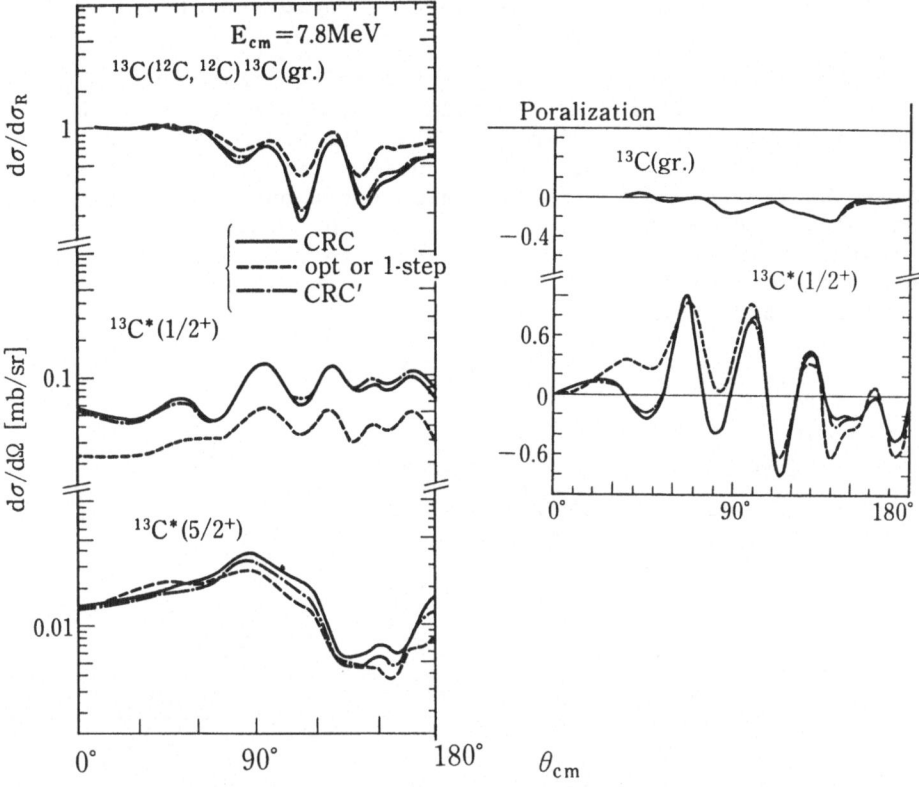

Fig. 15   Vector polarization of ^{13}C in the interaction of
^{12}C + ^{13}C. The polarization is zero in the elastic channel
in a first order calculation and is produced by a second
or higher order coupling effect.

the elastic and inelastic channels[25]. Whereas no polarization is
observed in the elastic channel in a first order calculation (absence of any explicit V_{LS}-potential) the inelastic channel shows
polarization just due to angular momentum matching. The second or
higher order coupling calculation introduces a polarization in the

elastic channel which can be described in first order calculations
by an V_{LS}-potential.

c) Charge exchange as a two-step transfer

The two-step transfer of nucleons can be conceived not only as
a sequential transfer of two nucleons in the same direction but
also as pick-up + stripping of the same nucleon (second order transfer coupling in the elastic and inelastic scattering[26],[27]) or that
of a proton and a neutron. The latter case turns out to be the
dominant mechanism for charge exchange at energies below 30 MeV/u
(see also chapter III).

As an example, we shall discuss the (^{13}C, ^{13}N) reaction on
^{58}Ni. Similar to the case of two-neutron transfer, we have the intermediate configurations, namely $^{12}C + ^{59}Ni$ and $^{14}N + ^{57}Co$, as
shown in fig. 16 and 17, of the one-nucleon transfers (they are
observed as real cross sections), which contribute coherently to
the population of a specific final state. The mixing of several
configurations is again important in order to reproduce the absolute cross sections in a CRC calculation. Fig. 16 lists the
transitions used in this calculation.[28]

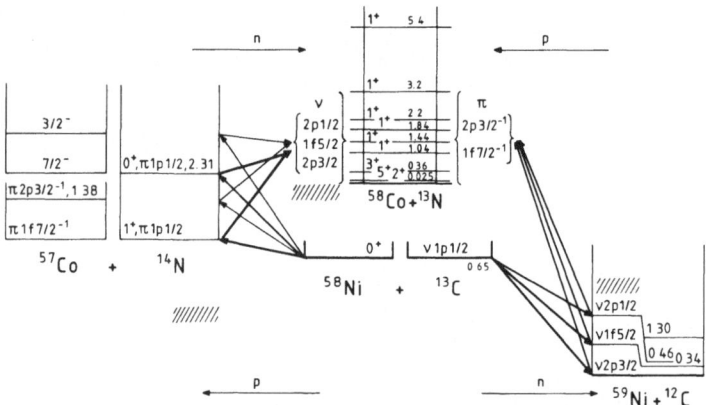

Fig. 16   Reaction scheme for the charge exchange reaction
          (^{13}C, ^{13}N) on ^{58}Ni. The proton pick-up and neutron strip-
          ping reactions are seen also in fig. 17, 18.

The energy spectra observed for the different reaction
routes   are shown in fig. 17. The final calculations show very
good agreement with the data concerning the general shape as well
as the absolute magnitude (fig. 18).

Further discussion of charge exchange will be given in
Chapter III for higher relative velocities.

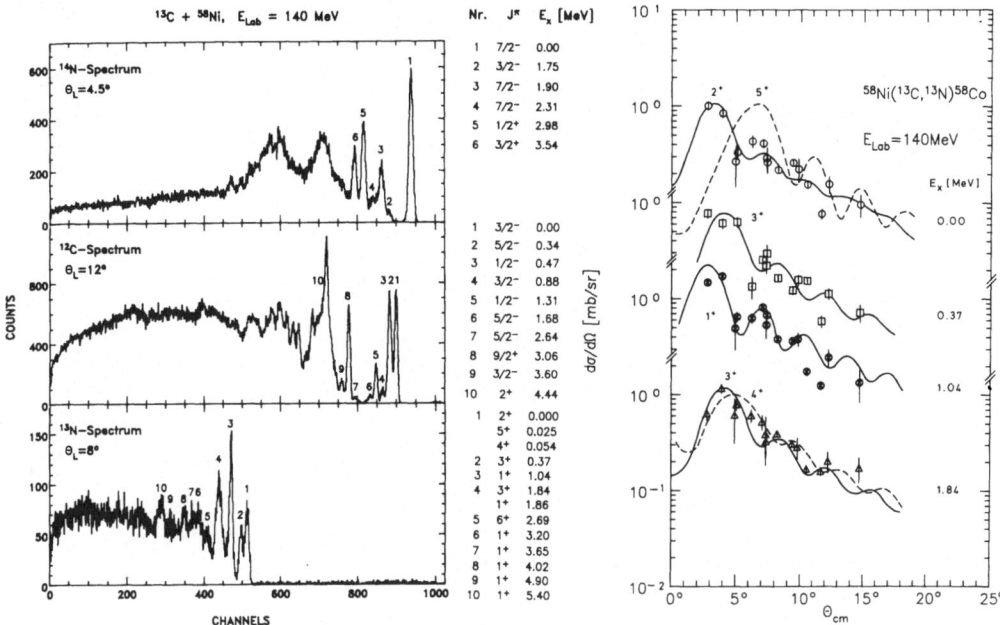

Fig.17 Spectra of dominant reaction channels observed with the charge exchange (^{13}C, ^{13}N) on ^{58}Ni.

Fig. 18 Angular distributions of the (^{13}C, ^{13}N) reaction. The curves are CRC calculations for a n and p exchange.

III. HIGHER RELATIVE VELOCITIES.

a) Diffractive and refractive scattering

At low relative velocities, the nuclear attraction generally is so strong (provided that $Z_1 Z_2$ is not too large (<2000)), that the projectile is always captured. The deflection function gives a branch of negative angles, which extends well beyond $\theta = -180°$.

The total potential V_{eff} (r)

$$V_{eff}(r) = V_c(r) + V_N(r) + \frac{\ell(\ell+1)\hbar^2}{2\mu \, r}$$ (24)

has still a pocket. At a given energy E_{crit}, the pocket will disappear and the centrifugal forces are strong enough to produce a finitie maximum negative deflection angle (see fig. 19). With this shape of the deflection function an appreciable cross section, which corresponds to the negative angle branch will be due to a refractive interaction, because the particles have to traverse a region of the real potential which must be sufficiently strong. This can be expressed by the refractive index, n

$$n = \sqrt{(E - V)/E}$$ (25)

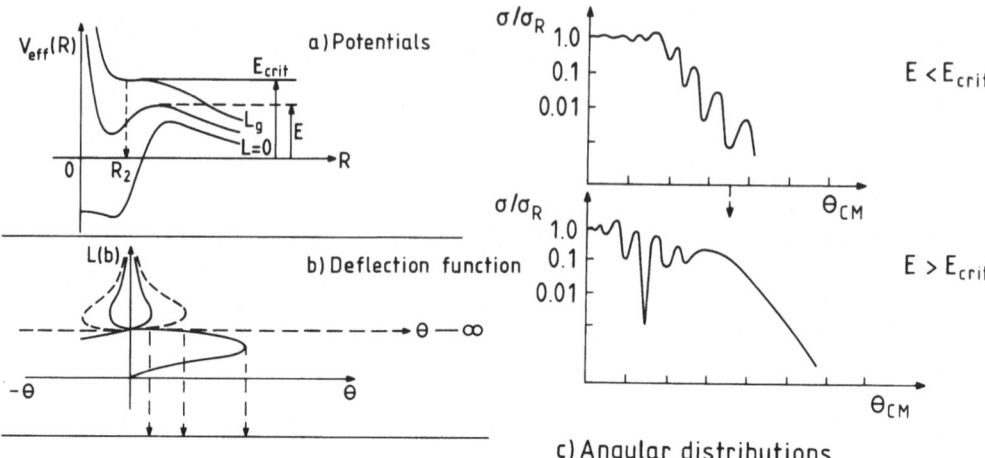

Fig. 19   Conditions for diffractive and refractive scattering :
a) effective potentials, b) deflection function,
c) angular distributions.

Figure 19 shows also schematically the evolution of the elastic scattering, when going beyond E_{crit}. The angular distribution exhibits a damping of the diffractive structure followed by a smooth fall-off at large angles ; it is this part at large angles, which deserves particular attention.

As an illustration, we show the case of $^{12}C + ^{12}C$ scattering (fig. 20) in the energy range between 20 MeV/u to 80 MeV/u. As discussed by Satchler[29] and by Bohlen et al[30], the large angle part of the differential cross section is extremely sensitive to the real potential. The formation of a true rainbow effect in this angular range is hindered by the absorption at small impact parameters (or L-values). Figure 21 shows the deflection function and the transmission coefficients for the case of 25 MeV/u $^{12}C + ^{12}C$. At the maximum negative angle (nuclear rainbow angle) the amplitude has decreased to 0.005. If we choose different absorptions for the interior, but the same at the surface, the diffractive part of the angular distribution is not changed, however, the region of the nuclear rainbow angle is strongly enhanced for weaker absorption (fig.22). For the transfer process, which is governed by a form factor, which has large values at small distances, a similar sensitivity as in the elastic scattering on the real potential is expected. Figure 23 illustrates this for the case of neutron transfer (^{13}C, ^{12}C) on ^{12}C. Two different transitions, the unique ($\ell=1$) transition to the $1/2^+$ state, at E_L = 55 MeV/u, and the ground state transition ($1/2^-$ state) at 23 MeV/u are shown. The two choices for the real potentials, taken from ref.30, give distinct differences at large angles, where the refractive part of the cross section is localised. The angular distributions can again be

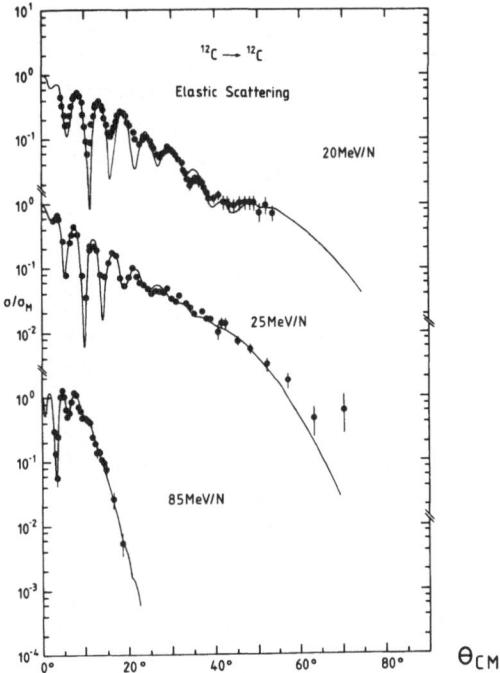

Fig. 20 Elastic scattering of ^{12}C on ^{12}C at three different ener-
        gies.

interpreted as an interference of near and far side contributions
with different slopes in the corresponding angular range and an
interference region where both are equal (see also fig.20).For
inelastic scattering the refractive effects at large angles are
barely visible because the form factor is generally peaked at the
surface (typically the collective form factor is given by the first
derivative of the real potential).

        The most drastic change, as compared to inelastic scattering,
is observed in transfer reactions, because at high relative velocities the internal motion becomes comparable or even negligible
as compared to the motion of the centers. The transfer process can
be seen as a jump from one train (TGV, ref.31) to another travel-
ling in the opposite sense. In order to succeed, the advice is to
run in the respectively opposite directions before and after the
jump.

b) Recoil effect and TGV

        In the first discussion (chapter I), we have already indicated
that the finite mass of the nucleon induces a shift in the center
of mass coordinates between projectile and ejectile, which has to
be small relative to the value of λ, if the "no-recoil" approximation is to be applied.

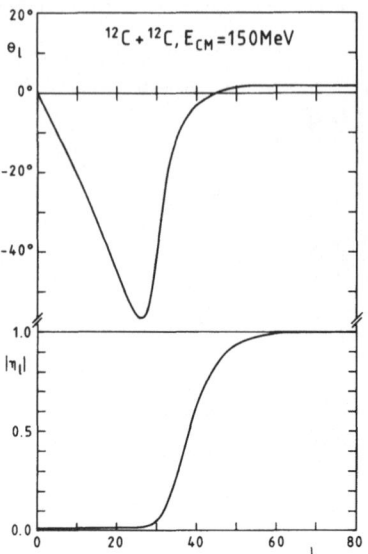

Fig. 21   Deflection function $\theta(\ell)$ and transmission coefficients $|\eta_\ell|$ for the scattering of ^{12}C on ^{12}C at E_L = 25 MeV/u.

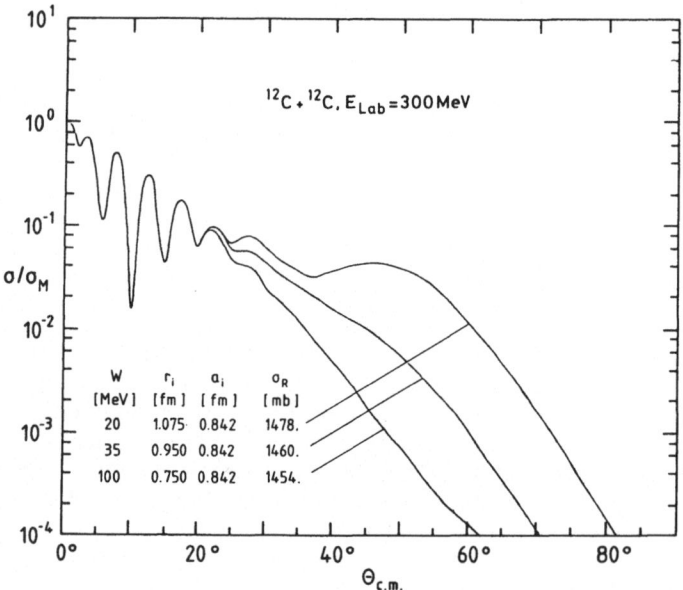

Fig. 22 Angular distributions of the elastic scattering of ^{12}C on ^{12}C, E_L = 25 MeV/u calculated with different absorption in the interior. The parameters of $W(r)$ give the same surface absorption and thus the same values of σ/σ_R up to θ = 25°.

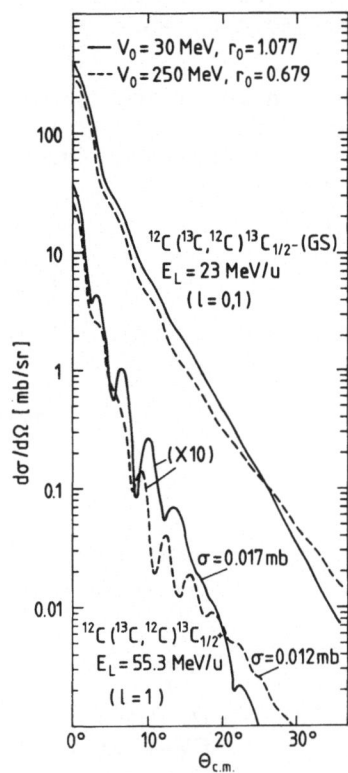

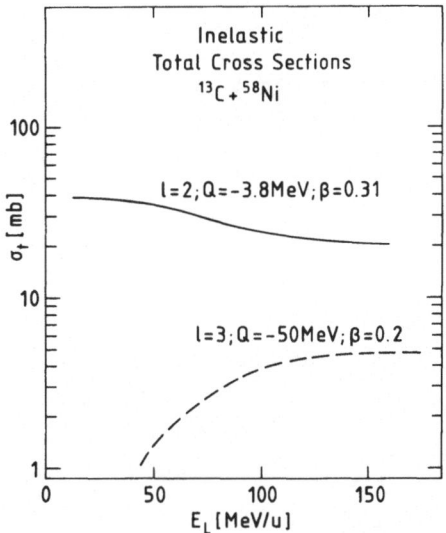

Fig. 23 Refractive (dashed
curves) and diffractive
(full curve) potentials
in a neutron transfer re-
action $^{12}C(^{13}C,\ ^{12})\ ^{13}C$
at two incident energies.

Fig. 24 Dependence of the cross section of inelastic transitions
(two Q-values) on incident
energy.

Actually, inelastic scattering reflects (apart from a different form factor) the features of a no-recoil reaction (see fig.24).

For inelastic excitations, the total cross section saturates
for a given Q-value and multi polarity. This is shown in fig. 24
for inelastic excitation (one-step) for ^{13}C on ^{58}Ni for the energy
range between 20 MeV/u and 150 MeV/u. The differential cross section thus rises strongly if the relevant grazing angle is taken.

The situation changes drastically if we consider transfer of
a nucleon. For the same system $^{13}C + ^{12}C$ calculations using the
finite range code Ptolemy[5] show a steep decrease of the total cross
section for a transfer between well defined bound states.

In figure 25, the evolution of the differential cross section
as function of energy is shown for the $p_{1/2}$ (groundstate $\ell = 0,1$)

and $S_{1/2}$ $(\ell = 1)$. The optical model potentials used in this DWBA calculation are those of ref.30 (V_o = 250 MeV, as also shown in figure 23).

The decrease of the total cross section can be understood by considering the DWBA-matrix element (discussed in chapter I) in a slightly different form (see in particular the original treatment of Greider[32], who discussed the problems of recoil in 1969).

We approximate the distorted waves by :

$$\chi_j(\vec{r}) = B_j(r) \exp\left\{i\vec{K}_j \vec{r}_j\right\} \quad ; \quad j = i,f; \tag{26}$$

and for the integration variables we choose $\vec{r}_1$ and $\vec{r}_2$ (fig. 3) by using the relations

$$\vec{r}_i = \vec{r}_1 \frac{b}{a} - \vec{r}_2 \quad \text{and} \quad \vec{r}_f = \vec{r}_1 - \frac{A}{B} \vec{r}_2 \tag{27}$$

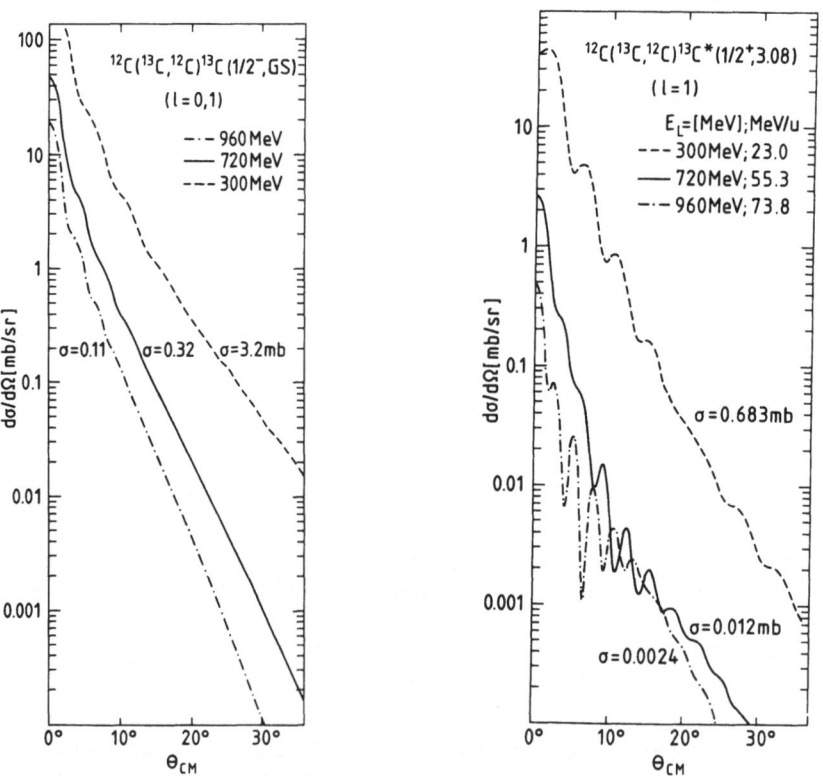

Fig. 25 Differential cross sections calculated using DWBA and a refractive potential (see fig.23) for the neutron transfer $^{12}C(^{13}C, {}^{12}C)$ ^{13}C at three incident energies.

Here, the mass ratios are represented by their symbols. The dependence of the amplitudes on the distance is assumed to be small. The transition amplitude then takes the form :

$$T_{ba} = B_i B_f \int \exp\left\{ -i\left[\vec{K}_i \frac{b}{a} \vec{r}_1 - \vec{K}_f \vec{r}_1\right]\right\} \phi_2^*(\vec{r}_2) \, V_{eff}(\vec{r}_1) \cdot$$

$$\cdot \phi_1(\vec{r}_1) \exp\left\{ i\left[\vec{K}_f \frac{A}{B} \vec{r}_2 - \vec{K}_i \vec{r}_2\right]\right\} d\vec{r}_1 \, d\vec{r}_2 \tag{28}$$

We know define the momenta imparted onto the cores b and A due to the transfer of the particle c, $q_b = \vec{q}_1$, $\vec{q}_A = -\vec{q}_2$

$$\vec{q}_1 = \vec{K}_f - \vec{K}_i \frac{b}{a} \tag{29}$$

$$\vec{q}_2 = \vec{K}_i - \vec{K}_f \frac{A}{B} \tag{30}$$

For a further discussion, it is useful to transform the integral (see also ref. 33)

$$< \phi_1(\vec{r}_1) \mid V(\vec{r}_1) \mid e^{i\vec{q}_1 \vec{r}_1} > =$$

$$= (2\pi)^{-3} \int d\vec{q} \, \phi_1^*(\vec{r}_1) \, e^{i\vec{q}\vec{r}_1} \int e^{-i\vec{q}\vec{r}_1} \, V(r_1) \, e^{i\vec{q}_1 \vec{r}_1} \, d\vec{r}_1 \tag{30}$$

which takes the form of a Fourier-tranform of the bound state $\phi(\vec{r}_1)$ multiplied by a transition amplitude $Tq \, q_1$

$$T_{ba} \cong \int dr_2 \, \phi_2^*(\vec{r}_2) e^{-i\vec{q}_2 \vec{r}_2} \int d\vec{r}_1 \, \phi_1(\vec{r}_1) \, e^{i\vec{q}\vec{r}_1}$$

$$\int d\vec{q} \, e^{-i\vec{q}\vec{r}_1} \, V_{eff}(\vec{r}_1) \, e^{i\vec{q}_1 \vec{r}_1} \tag{31}$$

The first term is the Fourier-transform of the bound state $\phi_2(\vec{r}_2)$ than the Fourier-tranform of state $\phi_1(r_1)$ is taken with a momentum q which is obtained by a transition from q_1 to q mediated by the interaction $V_{eff}(r)$. The total change of the momentum is given by $\vec{Q} = \vec{q}_b + \vec{q}_A$

$$\vec{Q} = \vec{K}_f \frac{c}{B} + \vec{K}_i \frac{c}{a} \tag{32}$$

The probability for the transition therefore depends on the overlap of the two momentum distributions separated in momentum space by Q. This is illustrated in a schematic way in figure 26.

In consideration of the internal momenta, it is necessary to take into account the fact that the momenta are smaller at the nuclear surface. This is best seen in the form of the Wigner function as calculated for example by Durand et al[34]. The Wigner function gives the momentum distribution as function of a spatial coordinate. For peripheral collisions as considered in the transfer process, the local momenta are in the average much smaller,

therefore limitations due to the finite momenta are felt at much
smaller relative velocities (1/5 of $v_F \simeq 10$ MeV/u).

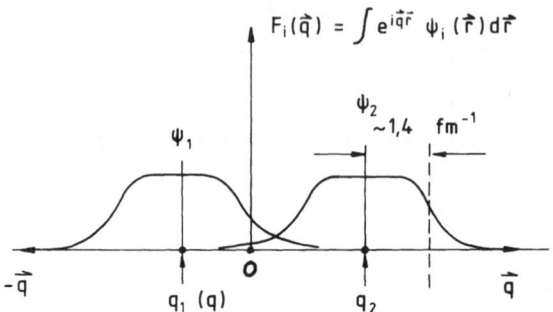

Fig. 26 Schematic illustration of the overlap of two momentum
distributions separated by $\vec{Q} = \vec{q}_1 + \vec{q}_2$

Any kind of DWBA calculations, which calculates the complete
finite range intergral will contain the features discussed in this
subsection. Fig. 27 shows such calculations for the total transfer
cross sections for the case of $^{13}C + ^{12}C$ as already shown in the

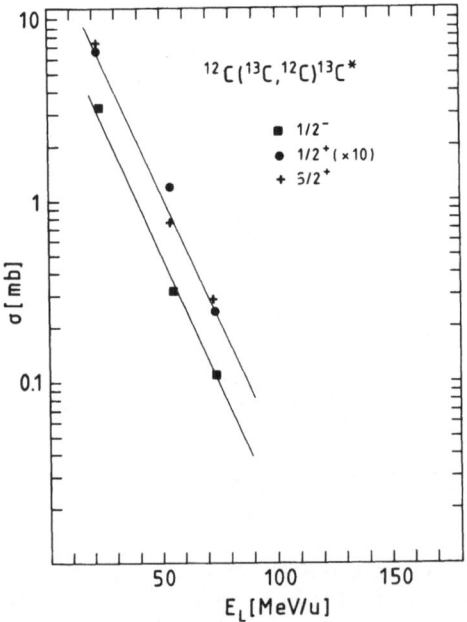

Fig. 27   Energy dependence of the neutron transfer as calculated
with the DWBA (see also fig. 24 and 25).

angular distributions. The exponential decrease of the cross section reflects the diminishing overlap of the momentum distributions (compare also with fig.24).

In a more direct way, the semi-classical model of Brink[35] gives a matching condition which is analogous to the situation described here. Assuming that the reaction takes place in the reaction plane between states with angular momenta ℓ_1 and ℓ_2, the projections normal to the reaction plane will be λ_1 and λ_2. They will give a measure for the internal momenta by the quantities :
λ_1/R_1 and λ_2/R_2 ; the matching condition thus reads :

$$\Delta K = \frac{\lambda_1}{R_1} + \frac{\lambda_2}{R_2} - \left| \vec{K}_i \frac{c}{a} + \vec{K}_f \frac{c}{B} \right| \approx 0$$

We will have in addition maximum polarization of the spins due to the conditions on λ_1 and λ_2.

The total change of "external" momentum has to be furnished by the internal motion. Thus, in contrast to the rule at low relative

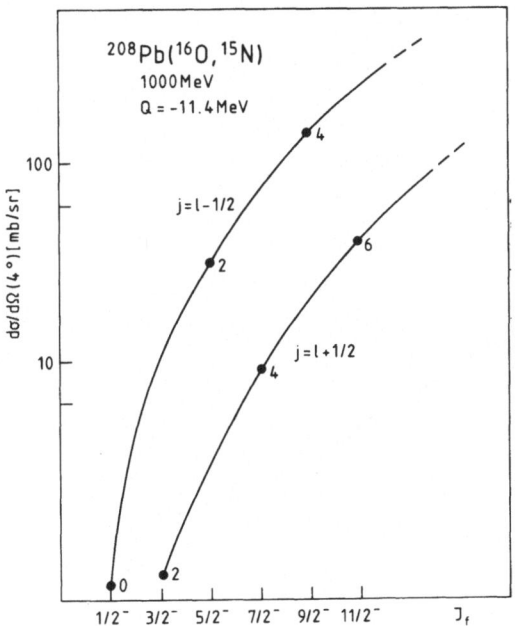

Fig. 28 Spin selectivity of a proton stripping reaction at high relative velocities.

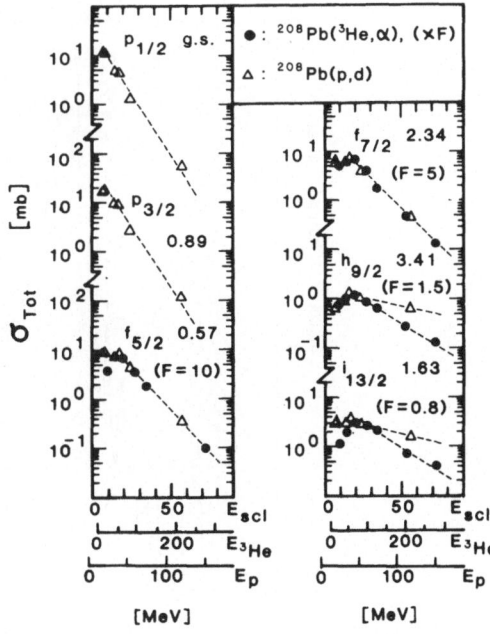

Fig. 29 Energy dependence of one nucleon transfer reactions (^{3}He, α) on ^{208}Pb according to Kurihara and Sakai (ref.36).

velocities (chapter I) the angular momenta ℓ_1 and ℓ_2 will be parallel and the smallest value of $\ell_1 + \ell_2 = \ell$ will be favoured $\ell = |\ell_1 - \ell_2|$. Thus the j-selectivity observed at low velocities (namely the dominance of $j_1 = \ell \pm 1/2 \rightarrow j_2 = \ell \mp 1/2$) is reversed and we favour now transitions with <u>equal</u> spin-orbit coupling scheme (fig. 28, ref.39).

Experimental information on the energy dependence of transfer reactions has recently been discussed by Kurihara and Sakai[36]. They found that total cross sections for (α, ^{6}He) and (α, ^{3}He) reactions extending up to ca 55 MeV/u exhibit an exponential decrease. Figure 29 shows an exemple of the systematics of this work. The slope of the decrease does show some dependence on the total angular momentum transferred.

From this discussion, it is evident that transfer of nucleons at high relative velocities, will have small probabilities. The measurement of these cross sections can yield rather new information on the high momentum components of wave functions. Alternatively, the transfer process will populate dominantly high spin states (maximum ℓ_2 and λ_2). One example is given in figure 30 (ref.37).

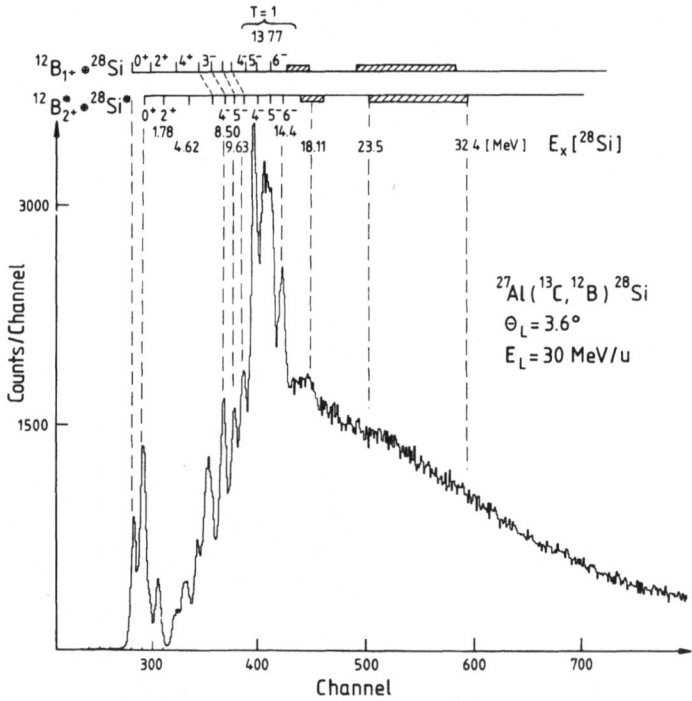

Fig. 30 Population of particle-hole states $(d5_{/2}^{-1} \otimes f7_{/2})_{j^\pi}$ with high spin with T = 0 and T = 1 in the ^{27}Al(^{13}C, ^{12}B) ^{28}Si reaction.

proton stripping at 30 MeV/u (^{13}C, ^{12}B) on ^{27}Al. In this case, the
d5/$_2$ - hole is preserved and (d5/$_2^{-1}$ $\otimes$ f7/$_2$) configurations (4 ,
5 , 6) in ^{28}Si with T = 0 as well as with T = 1 are strongly populated. The isobaric analog states are well seen ; the structures at
higher energies are probably due to higher subshells as partially
observed in high energy (α,t) reactions by Gales[38].

c) Charge exchange

   From the discussion in the previous section we realise, that
charge exchange with heavy ions must loose the two-step character
(neutron and proton exchange) at energies larger ca 50 MeV/u. For
heavy ions beams, as they become available now, the variety of
choices of projectiles offers unique possibilities, which are complementary to (p,n) studies and which allow in addition the study
of (n,p) reactions and reactions with ΔT = 2. We show in fig. 31

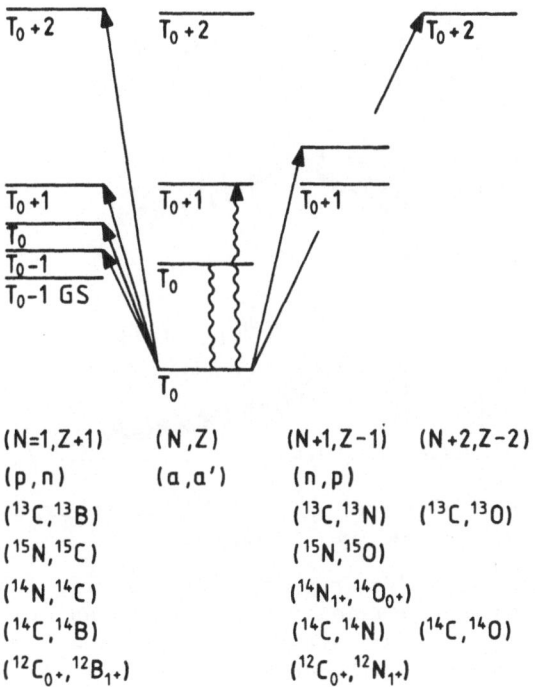

Fig. 31   Illustration of the different charge exchange reactions,
          which can be induced by heavy ions.

a scheme of the possible reactions. By choosing projectiles with neutron excess it is possible to populate the $T_o + 2$ states either via the "(p,n)" reaction (here it will be suppressed strongly due to Clebsch-Gordon coefficients, similar to the $T_o + 1$ state) or by double charge exchange in the other direction. In addition the reactions will differ with respect to the properties concerning spin flip.

As a final example, we will discuss the $(^{13}C, \ ^{13}B)$ and $(^{13}C, \ ^{13}N)$ reaction on ^{12}C. The residual nuclei produced by this reaction are ^{12}N and ^{12}B, which are mirror nuclei. The comparison of the two spectra thus immediatly focusses our attention on the different selectivity of the two reactions (ref.40) (fig.32a, b).

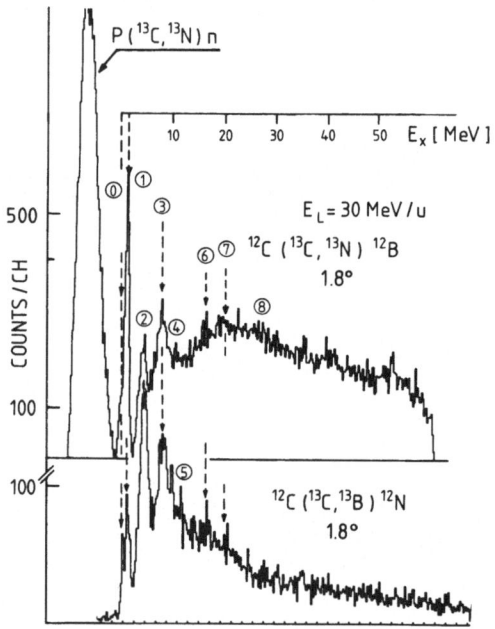

Fig. 32 a) Spectrum of the "(n,p)" and "(p,n)" charge exchange reactions induced by ^{13}C on ^{12}C .

There are states which carry unnatural parity like 1^+, 2^- and 4^-. These states have magnetic properties and correspond to a spin-flip transition. The two states (2^-, 4^-) $\cong$ 4.5 MeV (not separated) are much stronger populated in the $(^{13}C, \ ^{13}B)$ reaction than in the $(^{13}C, \ ^{13}N)$. The $^{13}B_{3/2}$ - groundstate is a spin flip transition relative to $^{13}C_{1/2}$. The heavy ion charge exchange contains two vertices, one for the projectile-ejectile, the other for the target-residual nucleus.

The transitions at these vertices are not independent, the spin flip
transition in the two vertices can thus be complementary and the
spin-flip in the target can be enforced by the choice of the
projectile-ejectile spin configurations. Some choices are given in
figure 31.

Further inspection of the spectra hows a strong peak at the
position of the Giant Dipole resonance at 7.6 - 8.0 MeV. This group
is generally also seen in (p,n) and (^{3}He,t) reactions[41], on
^{12}C, however, with varying shape if spectra at different angles
are compared. The concentration of particle hole strength with and
without spin-flip is thus observed in a narrow region of excitation.
At excitation energies beyond 15 MeV, we see in both spectra
strength which is not seen in the (p,n) and (^{3}He, t) reactions.
These groups correspond to 30 and 35 MeV excitation in ^{12}C, where
high lying Octopule strength has been suggested based on inelastic
scattering[42]. The position of states in ^{12}B, ^{12}N and ^{12}C is summarised in figure 32 b).

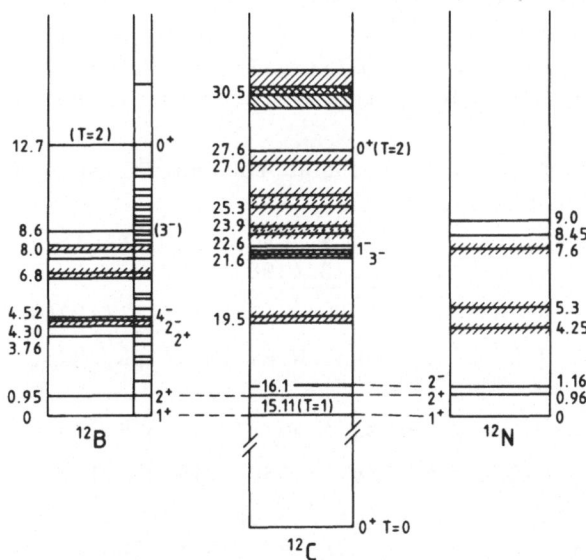

Fig. 32 b) States in the nuclei ^{12}B, ^{12}C, ^{12}N relevant to the charge exchange reaction.

An important difference for the two reactions finally has to
be mentioned which is connected to the background. In the case of
^{13}B ejectile, pick-up plus decay contributions are negligible
because of the thresholds in mass (^{14}C → ^{13}B) is placed far off the
stability valley and is therefore "protected" from sequential

decay of heavier reaction products. For ^{13}N the surplus in the cross section at excitation energies beyond $E_x > 20$ MeV is most likely due to the particle emission of ^{14}N and ^{14}C unbound states populated by pick-up reactions. The same argument holds for the ^{12}B spectra in figure 30.

Spectra on the $(^{13}C, ^{13}N)$ on heavier nuclei show continuous yields, with steps similar to the spectra observed in (π^-, π^0) studies. The heavy ion reaction, however, will populate rather states with larger ℓ and a comparison between (π^-, π^0) and "(n,p)" spectra with heavy ions could help to study the isovector strength a high excitations in nuclei.

REFERENCES

[1] R.A. Broglia, S. Landowne, R.A. Malfliet, V. Rostokin and A. Winther., Phys. Rep. CII (1974).

[2] L.J.B. Goldfarb and W. von Oertzen in Heavy Ion Collisions, ed. R. Bock, North Holland, 1979, vol. 1.

[3] Classical and Quantum Mechanical Aspects of Heavy Ion Collisions ed. H.L. Harney et al., Lecture Notes in Physics, vol. 33, 1975, Springer.

[4] W. Nörenberg, H.A. Weidenmüller, Introduction to the theory of Heavy Ion Collisions, Lecture Notes in Physics, vol. 51 (1980) Springer.

[5] Code Ptolemy by S.C. Pieper and M.H. Macfarlane, Argonne National Laboratory.

[6] W. von Oertzen, B. Gebaure, A. Gamp, H.G. Bohlen, F. Busch and D. Schüll, Z. f. Physik, A313 (1983) 189.

[7] J. Knoll and R. Schaeffer, Phys. Lett. 52B (1974) 131

[8] M.A. Franey, J.S. Lilley and W.R. Phillips, Nucl. Phys. A324 (1979) 193.

[9] W. von Oertzen, Nucl. Phys. A148 (1970) 529.

[10] H. Siekmann et al., Z.f. Physik A307 (1982) 113.

[11] A.J. Baltz, P.D. Bond, O. Hansen et al., Phys. Rev. C29 (1984) 2392. See also Phys. Rev. Lett 47 (1981) 1039.

[12] K.E. Rehm et al., Phys. Rev. Lett 51 (1983) 1426.

[13] G. Franz, J.V. Kratz, W. Brüchle, H. Folger and B. Haefner Z. f. Physik A291 (1979) 167.

[14] W. von Oertzen, XXII Winter Meeting on Nuclear Physics, Bormio 1984, p. 289

[15] U. Götz et al, Phys. Rep. 16C (1975) 115

16   T. Kammuri, Nucl. Phys. A259 (1976) 343

17   B.F. Bayman and J. Chen, Phys. Rev. C26 (1982) 1509 and
     ref. cited here.

18   K. Dietrich, Ann. Phys. (N.Y.) 66 (1971) 480,
     and H. Weiss, Phys. Rev. C19 (1979) 834 and references.

19   D.H. Feng, T. Udagawa and T. Tamura,   Nucl. Phys. A274
     (1976) 262

20   K.S. Low, Proceed. Europ. Conf. on Nuclear Physics
     with Heavy Ions, Caen (1976)
     J. de Physique C5 (1976) 15 and
     J. Phys. Soc. of Japan 44 (1978).

21   D. Fick, Ann. Rev. Nucl. Part. Sci. 31 (1981) 53 and ref.

22   W. von Oertzen and B. Imanishi, Nucl. Phys. A424 (1984) 262,
     see also H. Fröhlich et al, Nucl. Phys. A420 (1984) 124

23   B. Imanishi and W. von Oertzen, Phys. Lett. 87 B (1979) 188.

24   L. Pauling, The Nature of the Chemical Bond.

25   B. Imanishi and W. von Oertzen, Phys. Lett. 118 B (1982) 273

26   For a discussion of second order coupling see
     S. Landowne, Nucl. Phys. A409 (1983) 53 C, also W. Bohne et al.,
     Nucl. Phys. A332 (1979) 501

27   A. Etchegyoen et al., Nucl. Phys. A397 (1983) 343.

28   H.G. Bohlen et al. Annual Report HMI, Berlin,1983, and to be
     published.

29   G.R. Satchler, Nucl. Phys. A409 (1983) 3C.

30   H.G. Bohlen et al., Z. f. Physik A308 (1982) 121.

31   Guide de voyageur TGV (3 juin 1984 au 29 Septembre 1984) SNCF
     1984, Paris. This advise is not given here. Also, W. von
     Oertzen, to be published.

32   K.R. Greider, Proceedings Int. Conf. on Nuclear reactions
     induced by heavy ions, 1969, Heidelberg, eds. R. Bock and
     W.R. Hering (North Holland),
     also L.R. Dodd and K.R. Greider, Phys. Rev. 180 (1969) 1187

33   D.F. Jackson, "Nuclear Reactions", Methuen, London 1970 p.186.

34   M. Durand, U.S. Ramamurthy and P. Schuck, Phys. Lett. 113B
     (1982) 116

35   D.M. Brink, Phys. Letters 40B (1972) 37 also N. Anyas-Weiss
     et all. Phys. Rep. 12 C (1974) 203

36   T. Kurihara and M. Sakai, Phys. Letters 133B (1983) 157

[37] W. von Oertzen, M. Buenerd et al, ISN Grenoble,to be published.

[38] S. Gales, International Symp. on Highly excited States and
Nucl. Structure, Orsay, 1983.
J. de Physique 45 (1984) C4-39

[39] W. Mittig, GANIL Caen, private communication.

[40] W. von Oertzen, M. Buenerd et al, ISN Grenoble, to be published.

[41] W.A. Sterrenburg et al, Nucl. Phys. A405 (1983) 109.

[42] M. Buenerd et al, Nucl. Phys. A286 (1977) 377

[43] F. Catara, A. Insolia, E. Maglione and A. Vitturri, Phys. Rev.
C 29 (1984), 1091.

NONRELATIVISTIC THEORY OF HEAVY ION COLLISIONS

G. Bertsch

Department of Physics
University of Tennessee
Knoxville, TN 37996

and

Oak Ridge National Laboratory[*]
Oak Ridge, TN 37831

INTRODUCTION

A wide range of phenomena is observed in heavy ion collisions,
calling for a comprehensive theory based on fundamental principles of
many-particle quantum mechanics. At low energies, the nuclear dynamics is controlled by the mean field, as we know from spectroscopic
nuclear physics. We therefore expect the comprehensive theory of
collisions to contain mean-field theory at low energies. The mean-
field theory will be the subject of the first lectures in this chapter. This theory can be studied quantum mechanically, in which form
it is called TDHF (time-dependent Hartree-Fock), or classically,
where the equation is called the Vlasov equation.

At high energies the mean field becomes insignificant in comparison with the effects of nucleon-nucleon collisions. A proper
theory needs to include collision effects, as is done in the classical theory of gases with the Boltzmann equation. The lectures will
go on from mean-field theory to the derivation of theories incorpor-
ating collisions. Again, there are both quantum and classical extensions of mean-field theory. The classical equation turns out to

---

[*]Operated by Martin Marietta Energy Systems, Inc. under contract
DE-AC05-84OR21400 with the U.S. Department of Energy.

be just the Boltzmann equation with a Pauli blocking factor in the collision integral, and is called the Uehling-Uhlenbeck equation. However, the collisional theories must be regarded as tentative, because the assumptions made in their derivation cannot be given firm justification, as will be seen. Also, the evaluation of such theories and comparison with experiment is still in an early stage. In the last year, considerable progress has been made in the numerical solution of the Uehling-Uhlenbeck equation, and the techniques will be described in the last lectures. The physics of the heavy-ion collision looks quite different with and without nucleon-nucleon collisions; even at moderate energies, the collisions tend to bring a rapid approach to local equilibrium in the nuclear medium.

DERIVATION OF MEAN-FIELD THEORY

This lecture will present a derivation of mean-field theory that is readily extended to collisional theories. The starting point is the Hamiltonian which we will assume, for the present, contains only kinetic energy and two-particle interactions. We write the Hamiltonian in second quantized notation as

$$\hat{H} = \sum_{i,j} \langle i \mid T \mid j \rangle \, a_i^\dagger a_j + \frac{1}{2} \sum_{\substack{i,j \\ k,\ell}} \langle ij \mid V \mid k\ell \rangle \, a_i^\dagger a_j^\dagger a_\ell a_k. \tag{1}$$

The indices i,j label a convenient complete set of single-particle states. The second quantized formalism is used because it is the easiest way to deal with the antisymmetry of the wavefunction, which will be nontrivial in the later development of the theory. The wavefunction satisfying the Hamiltonian (1) will be denoted by $\psi(t)$; if we knew ψ, the problem would be solved. We shall set ourselves the more modest goal of determining the one-body density matrix, ρ. This is defined by its matrix elements

$$\rho_{ij} = \langle \psi \mid a_i^\dagger a_j \mid \psi \rangle \tag{2}$$

The equation of motion for the density matrix is the relation satisfied by its time derivative; this is evaluated using the equation of motion for ψ and ψ^*,

$$\hat{H} \mid \psi \rangle = i \frac{\partial}{\partial t} \mid \psi \rangle; \quad \langle \psi \mid \hat{H} = -i \partial/\partial t \, \langle \psi \mid. \tag{3}$$

The result is the usual commutator formula,

$$\dot{\rho}_{ij} = \frac{1}{i} \langle \psi \mid [a_i^\dagger a_j, \hat{H}] \mid \psi \rangle$$

$$= \frac{1}{i} \langle \psi \mid [a_i^\dagger, \hat{H}] a_j + a_i^\dagger [a_j, \hat{H}] \mid \psi \rangle \tag{4}$$

The commutator of the kinetic energy operator with a creation or annihilation operator is a linear combination of such operators,

$$[a_i, \hat{T}] = \sum_j \langle i | T | j \rangle \, a_j$$

$$[a_i^\dagger, \hat{T}] = -\sum_j \langle j | T | i \rangle \, a_j^\dagger$$

(5)

The kinetic energy contribution to the equation of motion is then

$$\dot{\rho}_{ij}\bigg|_{\text{kinetic}} = \frac{1}{i} \langle \psi | [a_i^\dagger a_j, \hat{T}] | \psi \rangle$$

$$= \frac{1}{i} \langle \psi | -\sum_k \langle k | T | i \rangle \, a_k^\dagger a_j + \sum_k \langle j | T | k \rangle \, a_i^\dagger a_k | \psi \rangle$$

$$= \frac{1}{i} \sum_k \{ \langle j | T | k \rangle \, \rho_{ik} - \langle k | T | i \rangle \, \rho_{kj} \}.$$

(6)

Note that the only information required about the wavefunction in Eq. (6) is the single-particle density matrix.

Equation (6) can be simplified further by making a suitable choice of representation. If we use momentum eigenstates for our basis, the kinetic energy will be diagonal, and the density matrix obeys the simple equation,

$$\dot{\rho}_{ij}\bigg|_{\text{kinetic}} = \frac{1}{i} \left(\frac{p_j^2}{2m} - \frac{p_i^2}{2m} \right) \rho_{ij}$$

(7)

The commutator of the potential with the density operator is more complicated, requiring the expectation value of operator products with quadruple Fock space operators,

$$\langle \psi | [a_i^\dagger a_j, \hat{V}] | \psi \rangle = \frac{1}{2} \sum_{k, \ell, m} (\langle jk | V | \ell m \rangle - \langle kj | V | \ell m \rangle) \langle \psi | a_i^\dagger a_k^\dagger a_m a_\ell | \psi \rangle$$

$$+ \frac{1}{2} \sum_{k, \ell, m} (\langle \ell m | V | ik \rangle - \langle \ell m | V | ki \rangle) \langle \psi | a_\ell^\dagger a_m^\dagger a_k a_j | \psi \rangle$$

At this point, we are ready to derive mean-field theory. The fundamental approximation is to evaluate the expectation of the four-particle operator by a product of single-particle densities,

$$\langle \psi | a_i^\dagger a_j^\dagger a_k a_\ell | \psi \rangle \approx \langle \psi | a_i^\dagger a_\ell | \psi \rangle \langle \psi | a_j^\dagger a_k | \psi \rangle$$

$$- \langle \psi | a_i^\dagger a_k | \psi \rangle \langle \psi | a_j^\dagger a_\ell | \psi \rangle$$

(9)

$$= \rho_{i\ell} \rho_{jk} - \rho_{ik} \rho_{j\ell}.$$

Equation (9) is the Hartree-Fock approximation to the two-particle density matrix. It is exact if the wavefunction is a Slater determinant of single-particle wavefunctions. The antisymmetry is explicit with the two terms on the right; if k and ℓ are interchanged, the expectation changes sign. The first term alone is the Hartree approximation; it requires that a physical distinction be made between i, ℓ states and j,k states. The approximation is good for treating the two-particle density for particles in widely separated parts of the system if there are no long-range correlations in the wavefunction.

The Hartree-Fock two-particle density matrix cannot be correct for particles close to each other when the interactions are very strong, but it might still serve as a useful approximation if the Hamiltonian is appropriately modified. In that case, the theory will be called the mean-field theory.

Inserting (9) and (6) in Eq. (4) gives a closed equation of motion for the single-particle density matrix. It is convenient to write this in terms of a single-particle potential, U, defined by the matrix elements

$$\langle j|U|\ell\rangle = \sum_{km} (\langle jk|V|\ell m\rangle - \langle jk|V|m\ell\rangle)\rho_{km} \tag{10}$$

Then the potential interaction commutator is

$$\langle\psi|[a_i^\dagger a_j, \hat{V}]|\psi\rangle = \sum_k (\langle j|U|k\rangle\rho_{ik} - \rho_{kj}\langle k|U|i\rangle) \tag{11}$$

and the equation of motion can be expressed in terms of a single-particle Hamiltonian H_{MF}

$$\dot{\rho}_{ij} = \frac{1}{i}[\rho_{ik}\langle j|H_{MF}|k\rangle - \langle k|H_{MF}|j\rangle\rho_{kj}] \tag{12}$$

where $\hat{H}_{MF} = \hat{T} + \hat{U}$. Equation (12), together with (10), defines the time-dependent Hartree-Fock approximation. There are two particular representations for Eq. (12) that are useful for numerical calculations or for further approximation. The first representation is obtained by expanding the density matrix as a sum over dyadic matrices,

$$\rho_{ij} = \sum_\alpha (b_i^\alpha)^* b_j^\alpha \tag{13}$$

Then Eq. (12) is solved if the b satisfy

$$\frac{db_j^\alpha}{dt} = \frac{1}{i}\sum_k \langle j|H_{MF}|k\rangle b_k^\alpha. \tag{14}$$

In coordinate space, the vector $\sum_{j} b_j^{\alpha} | j\rangle$ is just the single-particle wavefunction $\phi^{\alpha}(x)$. The coordinate space representation is especially useful if the potential U is approximated by a local function of position.

Another useful representation uses the Wigner function. This is defined as a certain Fourier transform of the density matrix, starting from either the coordinate space or momentum space representation,

$$f(p,r) = \int d\vec{s} \; e^{ip\cdot s} \; \rho_{r + \frac{s}{2}, r - \frac{s}{2}} \tag{15a}$$

$$f(p,r) = \int \frac{d\vec{q}}{(2\pi)^3} \; e^{-iq\cdot r} \; \rho_{p + \frac{q}{2}, p - \frac{q}{2}} \tag{15b}$$

The Wigner function has all the properties of a classical phase space distribution function while remaining a quantum-mechanical density. For example, the ordinary spatial density is obtained from f by integrating over p,

$$\int \frac{d\vec{p}}{(2\pi)^3} \; f(p,r) = \int \frac{d\vec{p}}{(2\pi)^3} \; d\vec{s} \; e^{ip\cdot s} \; \rho_{r + \frac{s}{2}, r - \frac{s}{2}}$$

$$= \int d\vec{s} \; \delta^{(3)}(s) \; \rho_{r + \frac{s}{2}, r - \frac{s}{2}} \tag{16}$$

$$= \rho_{r,r} = \rho(r).$$

It is also easy to show that the expectation of the momentum and kinetic energy are given by the integrals over f:

$$\langle \vec{p} \rangle = \int d\vec{r} \; \frac{d\vec{p}}{(2\pi)^3} \; \vec{p} \; f(\vec{p},\vec{r})$$

$$\tag{17}$$

$$\langle T \rangle = \int d\vec{r} \; \frac{d\vec{p}}{(2\pi)^3} \; \frac{p^2}{2m} \; f(p,r)$$

We now express the mean-field equation of motion in the Wigner representation. This may be done by taking the time derivative of the definition of the Wigner function, Eq. (15), and substituting the equation of motion for $\dot{\rho}$,

$$\frac{df(p,r)}{\partial t} = \frac{1}{i} \int \frac{d\vec{q}}{(2\pi)^3} e^{-iq \cdot r} \left[\rho, H_{MF}\right]_{p + \frac{q}{2}, p - \frac{q}{2}}$$

$$= \frac{1}{i} \int \frac{d\vec{q}}{(2\pi)^3} e^{-iq \cdot r} \left\{ [\rho, \hat{T}] + [\rho, U] \right\}_{p + \frac{q}{2}, p - \frac{q}{2}} \qquad (18)$$

$$= \frac{\partial f}{\partial t}\bigg|_{kinetic} + \frac{\partial f}{\partial t}\bigg|_{potential}$$

The momentum representation has been used for ρ, which allows the first term in Eq. (18) to be evaluated with Eq. (7),

$$\int \frac{d\vec{q}}{(2\pi)^3} e^{-iq \cdot r} [\rho, \hat{T}]_{p + \frac{q}{2}, p - \frac{q}{2}} = - \int \frac{d\vec{q}}{(2\pi)^3} e^{-iq \cdot r} \frac{p \cdot q}{m} \rho_{p + \frac{q}{2}, p - \frac{q}{2}}$$

$$(19)$$

The factor $\vec{q}\, e^{-iq \cdot r}$ in the above integral is replaced by $i\vec{\nabla}_r\, e^{-iq \cdot r}$. Next, the gradient operator is moved outside of the integral, leaving an integral which is proportional to the Wigner function,

$$\frac{\partial f}{\partial t}\bigg|_{kinetic} = - \int \frac{d\vec{q}}{(2\pi)^3} e^{-iq \cdot r} \frac{p \cdot \nabla_r}{m} \rho_{p + \frac{q}{2}, p - \frac{q}{2}}$$

$$= -\nabla_r \cdot \int \frac{d\vec{q}}{(2\pi)^3} e^{-iq \cdot r} \frac{\vec{p}}{m} \rho_{p + \frac{q}{2}, p - \frac{q}{2}} \qquad (20)$$

$$= -\nabla_r \cdot \frac{p}{m} f(p,r)$$

The second term in Eq. (18) can only be simplified under restricted assumptions. Let us first assume that the mean-field potential is local in coordinate space

$$\langle \vec{r} | U | \vec{r}' \rangle = \delta(\vec{r} - \vec{r}') U(\vec{r}) \qquad (21)$$

In general, this is not true for the exchange potential. It will be true if the interaction is zero range, and it is a reasonable approximation in any case. The effects of the nonlocality can be incorporated into the theory in other ways, such as with the effective mass, which will not be discussed in these lectures. The potential term in Eq. (18) is evaluated in the coordinate space representation, using Eq. (21),

$$\frac{\partial f}{\partial t}\bigg|_{\text{potential}} = -\frac{1}{i} \int d\vec{s}\ e^{i p \cdot s}\ \left(U\left(r + \frac{s}{2}\right) - U\left(r - \frac{s}{2}\right) \right)\ \rho_{\ r+\frac{s}{2},\ r-\frac{s}{2}}$$

(22)

The lefthand side can be expressed in terms of the Wigner function by using the inverse Fourier transform of Eq. (15a), and the result is an integral over the Wigner function for all values of p. With one additional approximation, the theory simplifies much further. We assume that U is a sufficiently smoothly varying function of r to permit truncation of the Taylor series for the U factor by the lowest nonvanishing term,

$$U(r+s) - U\left(r - \frac{s}{2}\right) \approx s \cdot \nabla_r U$$

(23)

We insert this in Eq. (22) and use the same trick to convert the s factor in the integrand to a gradient which can be taken out of the integral. The result is identical to the potential term in the classical Liouville equation for the single-particle distribution function in the field U. The final equation is that Liouville equation,

$$\frac{\partial}{\partial t} f + \frac{p}{m} \cdot \nabla_r f - \nabla_r U \cdot \nabla_p f = 0$$

(24)

With U the self-consistent field associated with f, Eq. (24) is known as the Vlasov equation. It is remarkable that quantum physics only plays a role in the initial conditions on f. The initial f must respect the Pauli principle, e.g. be based on a Slater determinantal many-particle wavefunction. In ground state wavefunctions, f will be close to 1 in occupied regions of phase space and close to zero outside. But the identity of the particles plays no role beyond the initial conditions.

Conservation Laws

Conservation laws are very important in areas of physics for which the dynamic equations are difficult to solve. When one develops a simplified theory in such areas, the requirement that the theory respect the conservation laws is often sufficiently stringent to be helpful in formulating the theory. For the physics of heavy-ion collisions, the most important conservation laws are for particle number, momentum, and energy. Mean-field theory passes the test of satisfying these conservation laws. This is easy to show for the case of particle number, which is defined in the three representations as

$$N = \sum_i \rho_{ii} = \sum_\alpha \int d^3r\ \phi_\alpha^*(r)\phi_\alpha(r) = \int \frac{d^3p}{(2\pi)^3}\ d^3r\ f(p,r)$$

(25)

The proof that N is conserved proceeds by taking its time derivative and evaluating the righthand side using the equation of motion. In the density matrix representation, this yields

$$\frac{dN}{dt} = \sum_i \dot{\rho}_{ii} = \frac{1}{i} \left[H_{MF}, \rho \right]_{ii} = 0 \tag{26}$$

The last step follows by the symmetry of the commutator. In the particle wavefunction representation, the conservation law proof relies on the Hermitian character of H_{MF},

$$\frac{dN}{dt} = \frac{1}{i} \sum_\alpha \int d^3r \left[\phi_\alpha^* H_{MF} \phi_\alpha - \phi_\alpha H_{MF} \phi_\alpha^* \right] = 0 \tag{27}$$

Finally, in the Wigner representation, the conservation law is

$$\frac{dN}{dt} = \int \frac{d^3p}{(2\pi)^3} d^3r \left[-\frac{p}{m} \cdot \nabla_r f + \nabla_r U \cdot \nabla_p f \right] \tag{28}$$

$$= -\int \frac{d^3p}{(2\pi)^3} \int f \frac{p}{m} \cdot d^2r + \int d^3r \int \frac{f}{(2\pi)^3} \nabla_r U \cdot d^2p = 0$$

with the last step obtained by evaluating the surface integrals at a large distance or momentum, where f=0.

The conservation law for momentum depends on the translational invariance of the potential interaction in the Hamiltonian. We made certain approximations in the treatment of the interaction when we derived mean-field theory. The theory will conserve momentum as long as those approximations preserve translational invariance. To illustrate this, let us assume a generalized interaction,

$$\hat{V} = \frac{1}{2} \int d^3r \, d^3r' \, \hat{\rho}(r)\hat{\rho}(r') \, v_2(r,r') \tag{29}$$

$$+ \frac{1}{3} \int d^3r \, d^3r' \, d^3r'' \, \hat{\rho}(r)\hat{\rho}(r')\hat{\rho}(r'')v_3(r,r',r'') + \dots$$

Neglecting exchange terms, the mean field associated with the above interaction is

$$U(r) = \frac{1}{2} \int d^3r' \, \rho(r')\left(v_2(r,r')+v_2(r',r)\right) \tag{30}$$

$$+ \frac{1}{3} \int d^3r' \, d^3r'' \, \rho(r')\rho(r'')\left(v_3(r,r',r'')+v_3(r',r,r'')\right.$$

$$\left. + v_3(r',r'',r)\right) + \dots$$

We now examine the equation of motion for the total momentum. In the particle wavefunction representation this is

$$\frac{d\vec{p}}{dt} = \frac{d}{dt} \sum_\alpha \int d^3r \ \phi_\alpha^* \left[\frac{\vec{\nabla} - \overleftarrow{\nabla}}{2i} \right] \phi_\alpha$$

$$= \frac{1}{i} \sum_\alpha \int d^3r \ \phi_\alpha^* \left\{ \left[\frac{\vec{\nabla} - \overleftarrow{\nabla}}{2i}, \ T \right] + \left(\frac{\vec{\nabla} - \overleftarrow{\nabla}}{2i}, \ U \right) \right\} \phi_\alpha \tag{31}$$

The commutator of the momentum operator with the kinetic energy vanishes, and the commutator with U is proportional to the gradient of U,

$$\frac{d\vec{p}}{dt} = - \int d^3r \ \rho(r) \ \vec{\nabla}_r U \tag{32}$$

Translational invariance of the interaction implies that

$$\left(\nabla_r + \nabla_{r'} + \dots \right) v_m(r, r', \dots) = 0 \tag{33}$$

The integral in Eq. (32) can be shown to vanish using Eqs. (30) and (33).

The mean-field theory with the generalized interaction Eq. (29) also conserves energy. To prove this, we start by expressing the energy as the expectation of the Hamiltonian, using the mean-field density matrix and neglecting exchange terms,

$$E = \sum_\alpha \int d^3r \ \phi_\alpha^* \ T\phi_\alpha + \frac{1}{2} \int d^3r \ d^3r' \ \rho(r)\rho(r')v_2(r,r') + \frac{1}{3} \int \dots \tag{34}$$

The equation of motion for E becomes

$$\frac{dE}{dt} = \frac{1}{i} \sum_\alpha \int d^3r \ \phi_\alpha^* [T, H_{MF}] \phi_\alpha + \int d^3r \ d^3r' \ \overset{.}{\rho}(r)\rho(r')v_2(r,r')$$

$$+ \int \underbrace{d^3r \ d^3r' \ d^3r'' \dots}_{\int d^3r \ \overset{.}{\rho}(r) \ U(r)} \tag{35}$$

The second and later terms in this equation can be combined, and the coefficient $\overset{.}{\rho}$ is equal to the mean field. Finally, $\overset{.}{\rho}$ is evaluated by the equation of motion, yielding

$$\frac{dE}{dt} = \frac{1}{i} \sum_\alpha \int d^3r \ \phi_\alpha^* \ [T, H_{MF}] \phi_\alpha + \frac{1}{i} \sum_\alpha \int d^3r \ \phi_\alpha^* \ [U, H_{MF}] \phi_\alpha$$

$$= \frac{1}{i} \sum_\alpha \int d^3r \ \phi_\alpha^* \ [H_{MF}, H_{MF}] \phi_\alpha = 0 \tag{36}$$

Having disposed of the global conservation laws, we now inquire about the existence of local conservation laws, in which the density of a conserved quantity is related to its flux. In that form the conservation laws are of quite direct value, because one of the objects of theory is to calculate the transport of the various conserved quantities from one nucleus to another. To establish a local conservation law, we first define the density whose integral is the globally conserved quantity

$$Q = \int d^3r \; q(r) \tag{37}$$

Then a local conservation law will exist if a current $\vec{j}(r)$ can be found so that the equation of continuity is satisfied,

$$\frac{\partial q}{\partial t} + \vec{\nabla} \cdot \vec{j} = 0 \tag{38}$$

In the law for the conservation of the number of particles, the density and current have their usual quantum-mechanical definitions,

$$\rho(r) = \sum_{\alpha} \phi_{\alpha}^{*} \phi_{\alpha} \tag{39}$$

$$\vec{j}(r) = \sum_{\alpha} \phi_{\alpha}^{*} \left(\frac{\vec{\nabla} - \overleftarrow{\nabla}}{2im} \right) \phi_{\alpha}(r) = \int \frac{d^3p}{(2\pi)^3} \; \frac{\vec{p}}{m} \; f(p,r) \tag{40}$$

The equation of continuity is easily derived from the equation of motion for ρ. The local conservation law for momentum is less obvious. The momentum density is defined by

$$\vec{p}(r) = m\vec{j}(r), \tag{41}$$

and the momentum flux will be determined from the equation of motion. Evaluating the time derivative of $\vec{p}(r)$ in the usual way, we find

$$\frac{\partial \vec{p}(r)}{\partial t} = \frac{1}{i} \sum_{\alpha} \left[\phi_{\alpha}^{*} \left(\frac{\vec{\nabla} - \overleftarrow{\nabla}}{2i} \right) H_{MF} \phi_{\alpha} - \left(H_{MF} \phi_{\alpha}^{*} \right) \left(\frac{\vec{\nabla} - \overleftarrow{\nabla}}{2i} \right) \phi_{\alpha} \right] \tag{42}$$

$$= -\nabla_{\mu} \cdot \sum_{\alpha} \phi_{\alpha}^{*} \left(\frac{\vec{\nabla} - \overleftarrow{\nabla}}{2} \right) \left(\frac{\vec{\nabla} - \overleftarrow{\nabla}}{2m} \right)_{\mu} \phi_{\alpha}^{*} - \rho(r) \; \vec{\nabla} U(r) \tag{43}$$

Only the first term in Eq. (43) has the required form as a divergence of a vector. The second term can be manipulated into the proper form under certain assumptions. If the interaction is short range compared to the scale of variation of the density, then $U(r)$ may be taken to depend only on the density at r, i.e. $U(r) = U(\rho(r))$. The potential energy is then expressible as an integral over the potential energy density given by

$$V(r) = \int_0^{\rho(r)} d\rho \, U(\rho) \tag{44}$$

We now define the tensor

$$\tilde{\tilde{W}} = \delta_{\mu\nu} \, (\rho U - V) \tag{45}$$

and evaluate its divergence

$$\vec{\nabla} \cdot \tilde{\tilde{W}} = \vec{\nabla}(\rho U - V) = U\vec{\nabla}\rho + \rho\vec{\nabla}U - \vec{\nabla}\int_0^{\rho(r)} d\rho \, U(r) = \rho\vec{\nabla}U.$$

This is just the second term in Eq. (43) showing that the required momentum flux from the potential interaction is given by (45). The total momentum current is then

$$\Pi_{\mu\nu} = \Pi_{\mu\nu}^{particle} + \delta_{\mu\nu}(\rho U - V) \tag{46}$$

where $\Pi^{particle}$ is the particle contribution to the momentum flux,

$$\Pi_{\mu\nu}^{particle} = -\sum_\alpha \phi_\alpha^* \left(\frac{\vec{\nabla} - \overleftarrow{\nabla}}{2}\right)_\mu \left(\frac{\vec{\nabla} - \overleftarrow{\nabla}}{2m}\right)_\nu \phi_\alpha$$

$$= \int \frac{d^3p}{(2\pi)^3} \frac{p_\mu p_\nu}{m} f(p,r). \tag{47}$$

In the next lecture, Eq. (46) will be applied to the calculation of momentum transfer between colliding nuclei.

Effective Hamiltonians

Because of the strength and spin dependence of the interaction, mean-field theory cannot be applied directly to the nuclear Hamiltonian. Correlations are induced between the particles at short distances which may be important for a fundamental understanding of nuclear properties but which play a minor role in the physics at low excitation energies. In principle, there are theories such as the Brueckner theory that allow one to derive an effective mean-field description starting from a fundamental interaction. However, this program has not been entirely successful; the predicted binding energy and nuclear matter density are not in agreement with the empirical saturation properties. The present-day philosophy is to bypass the steps to get from a fundamental interaction to an effective mean field and simply postulate an effective Hamiltonian which satisfies the known empirical requirements. Some guidance as to the form of the effective Hamiltonian comes from many-particle perturbation theory which expresses the binding energy (per particle) as a power series in the Fermi momentum or the cube root of the density,[1]

$$E = ak_F^2 + bk_F^3 + ck_F^4 + \dots$$

$$= a'\rho^{2/3} + b'\rho + c'\rho^{4/3} + \dots \tag{49}$$

The first term is the kinetic energy of a free Fermi gas having the coefficient $a = 3/5 \; h^2/2m$. The second and higher terms are due to potential interactions and correlations. The main empirical constraints on the function (49) are that it have a minimum at $\rho = 0.17$ fm or $k_F = 1.34$ fm, and that the binding energy at the minimum be E = -15 MeV. Another constraint that may be invoked is that the compressibility coefficient of nuclear matter be consistent with the empirical vibrational frequency of the giant monopole vibration. Having three conditions to satisfy, the next two terms in the power series expansion are overdetermined. We can make a three-parameter potential model by using the power of ρ as a free parameter in the third term. Thus, we consider a mean field of the form

$$U(\rho) = A\rho + B\rho^\sigma \tag{50}$$

The associated potential energy density is

$$V(\rho) = \frac{1}{2} A\rho^2 + \frac{1}{\sigma+1} B\rho^{\sigma+1} \tag{51}$$

The choice $\sigma=2$ gives a density-dependent mean field that can be directly associated with a potential of the form Eq. (29). The proper compressibility is obtained for a choice $\sigma=7/6$. The two mean-field functions are

$$\text{Stiff:} \quad U(\rho) = -124 \left(\rho/\rho_o\right) + 70.5 \left(\rho/\rho_o\right)^2 \text{ MeV}$$

$$\text{Soft :} \quad U(\rho) = -356 \left(\rho/\rho_o\right) + 303 \left(\rho/\rho_o\right)^{7/6} \text{ MeV} \tag{52}$$

The designations stiff and soft refer to the compressibility coefficients which are K=380 MeV and K=200 MeV, respectively. In Figs. 1 and 2 are plotted the binding energy and the mean field associated with these two functions. Figure 1 shows that the functions reproduce the proper binding energy and saturation density. The two functions are quite similar at subnuclear densities and only differ substantially at strong compressions. The mean field is close to -50 MeV for both functions at normal density, in agreement with other kinds of empirical information about the potential. Notice that the soft function is rather flat at higher-than-normal densities. This observation will simplify the discussion of the compressed matter dynamics.

Heavy-Ion Collisions in Mean-Field Theory

The dominant physics of close collisions between nuclei is associated with the mutual exchange of particles between the nuclei. Nucleons pass freely from one nucleus to the other when they come together. Particularly with the soft potential field, the field does not change very much in the overlap region and the motion of the

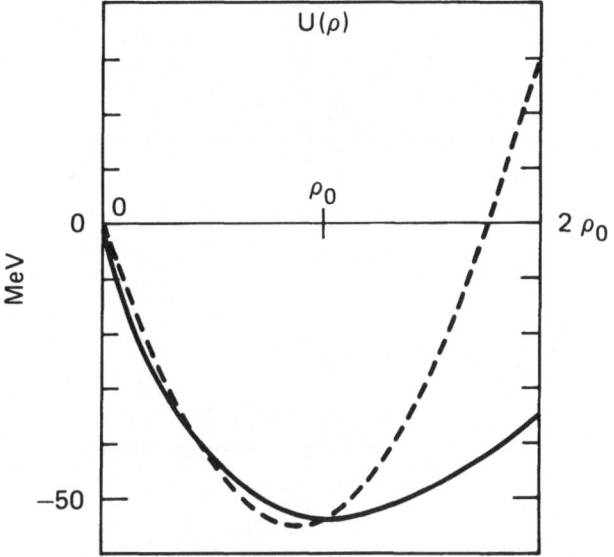

Fig. 1.   Mean-field potential $U(\rho)$ for the two functions of Eq. (52). The stiff and soft potential functions are shown by dashed and solid lines, respectively.

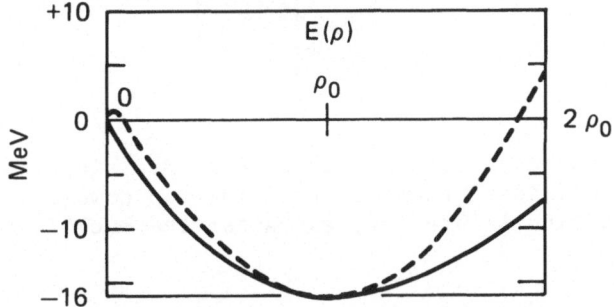

Fig. 2.   Energy per nucleon as a function of density. The soft and stiff functions are shown by solid and dashed lines, respectively.

nucleons is only affected by the external surfaces of the two nuclei. The evolution of the density matrix is shown schematically in terms of the Wigner function in Fig. 3. The dependence of f on the longitudinal coordinate and momentum is shown with the shaded areas representing regions with f $\cong$ 1. Of course, the actual Wigner function will have smooth variation instead of the sharp boundaries sketched. The boundaries in coordinate space represent the physical surfaces of the nuclei; the momentum boundaries are at the Fermi momentum. Since

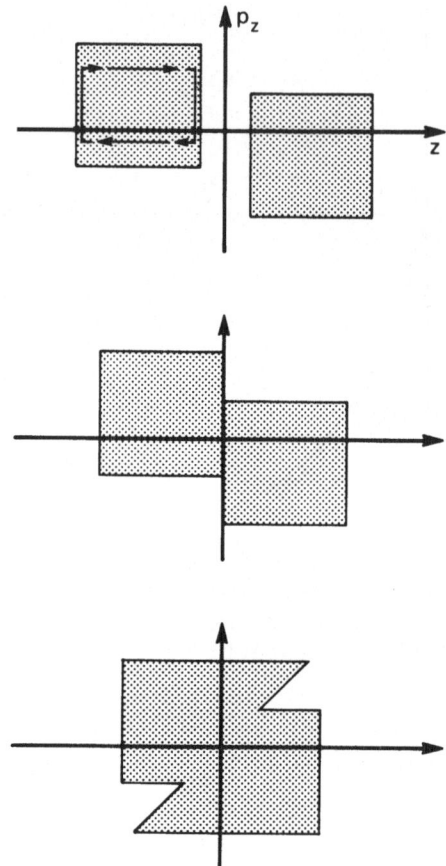

Fig. 3.   Schematic picture of phase space distribution of nucleons in
          two colliding nuclei. From the top down, picture shows dis-
          tribution before, at, and after the nuclei come into con-
          tact.

the nuclei are moving toward each other, the distribution functions
for each nucleus have a displacement in momentum. In the classical
picture of the motion, particles are moving within the phase space
boundaries and are reflected at the nuclear surface to stream in
closed loops, as shown by the arrows. Once the nuclei touch, there
is no inner surface to reflect from, and the nucleons pass over to
the other side. This is indicated in the lower two drawings.

The qualitative physics described here is born out quite well by
numerical studies of the quantum and the classical mean-field
theories.[2,3] In Fig. 4 is shown the comparison of one-dimensional
TDHF with the corresponding Vlasov theory from Ref. 3. The two

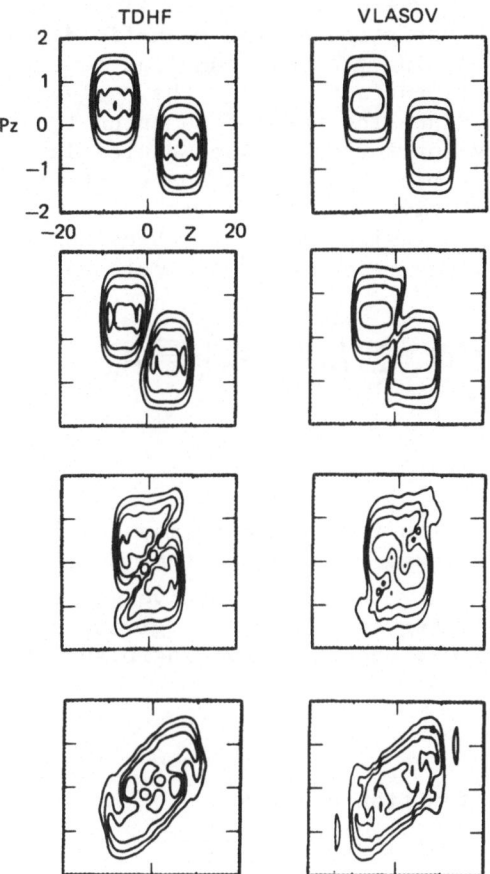

Fig. 4.   Comparison of TDHF and Vlasov equation for the evolution of
          the phase space distribution function in one dimension, from
          Ref. 3.

theories are very similar, with the only qualitative difference from
the schematic picture of Fig. 3 being the presence of a hole at p=0
and r=0.

    For three-dimensional geometry, the quantum mean-field theory is
quite difficult numerically, but enough calculations have been done
to survey the range of collision conditions possible. Some reviews
of the theory and comparison with experiment are given in Refs. 4 and
5. In general, the theory is successful in describing the energy and
momentum transfer at low energies.

Reduction of Mean-Field Theory to Force Dynamics

    Because of the difficulty in treating three-dimensional aspects realistically, it is useful to consider simplified treatments. A semiquantitative description of the mean-field dynamics can be developed by approximating the momentum flux tensor. This can be evaluated at the contact surface between the nuclei to calculate the momentum transfer. Since the rate of momentum transfer is a force, we will have a theory for forces between nuclei in close contact. We begin by choosing a plane to divide space into two regions and define the conserved quantities associated with each nucleus by integrating over the appropriate region of space. Labeling the regions A and B, the momentum, position, and number of nucleons in nucleus A are given by

$$
\vec{p}_A = \int_A d^3r \ \vec{p}(r) = \sum_\alpha \int_A d^3r \ \phi_\alpha^* \left(\frac{\vec{\nabla} - \overleftarrow{\nabla}}{2i}\right)\phi_\alpha
$$
$$
\vec{r}_A = \int_A d^3r \ \vec{r}\rho(r)
$$
$$
N_A = \int_A d^3r \ \rho(r).
$$
(53)

The equations of motion for r_A and p_A may now be determined using the local conservation laws. Taking the time derivative of the equation for r_A, we replace $\dot{\rho}$ by the negative divergence of the current. The divergence can be integrated by parts, giving a surface integral which vanishes if we choose the dividing plane so that

$$
\frac{dN_A}{dt} = 0
$$
(54)

The result is the familiar equation

$$
\frac{d\vec{r}_A}{dt} = \frac{\vec{p}_A}{m}
$$
(55)

The force equation is obtained from a similar set of manipulations,

$$
\frac{d\vec{p}_A}{dt} = \int dS \cdot \Pi
$$
(56)

Since the momentum flux could only be defined for short-range interactions, Eq. (56) does not contain the Coulomb interaction, which must be added separately. Note that Eq. (56) is in the form of a surface integral on the area of contact between the two nuclei. This form is convenient and physically sensible for discussing the macroscopic force dynamics. We next evaluate the pressure tensor Π using macroscopic limits. In infinite nuclear matter, the particle flux contribution to the pressure can be calculated from the geometry of the Fermi surface. If the potential field remains constant in the colliding nuclei, the Fermi surface will be in the shape of two intersecting Fermi spheres centered at the momenta of the two nuclei. The overlap region of the two spheres has the same density as the other parts of the distribution--refer to Fig. 3 to see how the nucleon distributions merge without violating the Pauli principle. The particle momentum flux for the intersecting sphere geometry was calculated by Randrup[11] who keeps the lowest terms in a power series in the relative momentum of the spheres. The result is

$$\Pi_{\mu\nu}^{particle} \cong \delta_{\mu\nu} \cdot constant + \frac{3}{16} \rho_0 V_F m \left(\vec{v}_A - \vec{v}_B\right)_\mu \hat{n}_\nu + \hat{n}_\mu \left(\vec{v}_A - \vec{v}_B\right)_\nu \tag{57}$$

where $\hat{n}$ is the normal to the dividing surface, and v_F is the Fermi velocity. The case $v_A - v_B = 0$ should represent nuclear matter in equilibrium, which has zero pressure. Thus the constant term in Eq. (57) must be cancelled by the potential contribution to the pressure at normal nuclear matter density. The cancellation will not be perfect for nonzero relative velocities. However, for the soft potential function, the potential contribution is small and Eq. (57) is a reasonable approximation to the entire momentum flux.

The infinite matter Fermi gas approximation will break down near the surface of a nucleus. In fact, the pressure tensor is nonzero in the surface even for nuclear matter in equilibrium, giving rise to the surface tension. Rather than treat the full quantum mechanics of the density matrix in the surface region, we shall adopt a macroscopic description, and augment the bulk momentum flux with a surface contribution. The empirical surface tension, $\sigma=1$ MeV/fm^2, fixes the integral of the pressure tensor across the surface. We shall describe the contact area of the two nuclei as a circle of radius R_n, which reduces the dynamic equation to

$$\frac{d\vec{P}_A}{dt} = \pi R_n^2 \frac{3}{16} \rho_0 V_F m \left(\left(\vec{v}_A - \vec{v}_B\right) + \hat{n}\left(\vec{v}_A - \vec{v}_B\right) \cdot \hat{n}\right) + 2\pi R_n \sigma$$
$$+ \text{Coulomb interaction} \tag{58}$$

The solution of this equation requires knowledge of the evolution of R_n and of the relative velocities of the Fermi spheres. Randrup makes the assumption that the Fermi sphere velocity is equal to the instantaneous velocity of the nuclear centers of mass, $v_{A,B} = \dot{r}_{A,B}$. This results in a linear friction force. The situation is more complicated in mean-field physics. The velocity of the Fermi sphere is determined by the physical surface velocity at the points where the particles in question changed direction. There is a time delay of the order of the nuclear transit time before these particles reach the contact zone. The motion of the farther nuclear surfaces is additionally delayed with respect to the center-of-mass motion. Thus, in mean-field physics a better description of the velocity of the Fermi surfaces is to relate it to the center-of-mass velocity at an earlier time,

$$\left(v_A - v_B\right)_f \approx \left(\dot{r}_A - \dot{r}_B\right)_{t-\tau_D} \tag{59}$$

with the delay time τ_D having the order of magnitude

$$\tau_D \sim (1-2) \cdot \frac{2R}{V_F} \qquad (60)$$

where R is the radius of one of the nuclei. Let us estimate this numerically for medium heavy nuclei. Typical parameters are R~5 fm, v_F~1/4 c, giving

$$\tau_D \sim (1-2) \cdot \frac{(2)(5 \text{ fm})}{1/4 \text{ c}} \sim 40\text{-}80 \text{ fm/c} \qquad (61)$$

We next consider the evolution of the contact zone or neck region between the two nuclei. In typical TDHF collisions, the neck forms quickly once the nuclei touch, and grows in size to a substantial fraction of a nuclear radius. If the nuclei come apart again, due to the Coulomb or the centrifugal forces, the neck becomes elongated and shrinks rather slowly in size. This is illustrated by a TDHF calculation, shown in Fig. 5 from Ref. 12, of a collision between two Pb nuclei. The nuclei touch when they approach within 14 fm of each other. The neck grows to about 5.5 fm radius at the distance of closest approach, and then shrinks at about 1/3 of the rate at which the nuclei draw apart.

These features were put in a completely geometric description of the neck radius by A. Bonasera.[6] It is then possible to make a self-contained model for the mean-field collisions. A comparison of the mean-field dynamics and the macroscopic model is shown in Fig. 6,

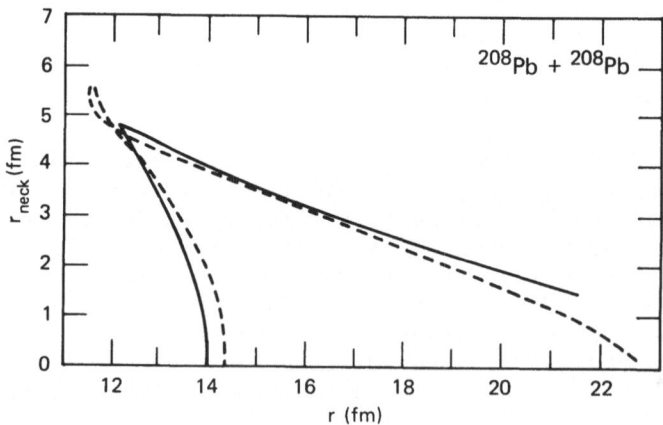

Fig. 5.   Neck evolution in TDHF[12] compared with geometric model of Ref. 6.

with the relative velocity of the two nuclei plotted as a function of
separation. The nuclei approach with a negative velocity and are
slowed down by the Coulomb force. Between 7 and 9 fm separation, the
nuclear potential attraction becomes significant, impeding the decel-
eration. At 7 fm separation, the nuclei touch, and particle transfer
becomes important. The associated repulsive force rapidly reduces
the velocity to zero. In fact, there is an overshoot, and the nuclei
slightly rebound. For the particular collision conditions in Fig. 6,
the two nuclei remain fused. If the memory effect in the particle
momentum flux is eliminated in favor of a linear friction, the collision does not show a rebound.

Observables

As mentioned earlier, mean-field theory is very successful in
describing the main features of low-energy collisions associated with
energy loss and momentum transfer. Let us first consider the case of
complete momentum transfer, i.e. fusion of the two nuclei. As a
function of impact parameter and initial energy, there will be a domain of fusion, which may be defined operationally by requiring that
the system hold together for a certain (finite) length of time. Such
a fusion region is shown in Fig. 7. There are three distinct physical processes determining the boundaries of fusion. For moderately
charged nuclei and low energies and angular momentum, the physics is
very simple. If the nuclei surmount the potential barrier of the
combined Coulomb and external nuclear potentials, they touch and the
attractive force from the surface tension in the neck holds them together. The low energy edge of the fusion domain is determined by
this potential barrier physics. At larger impact parameters and
therefore higher angular momenta, the centrifugal force is larger
than the attraction from the surface tension, and the nuclei scission by a process of elongation and thinning out the neck. For a
fixed geometry, the balance of forces occurs at a critical angular
momentum,

$$\hbar^2 \ell^2 / \mu r^3 \approx 2\pi R_N \sigma \tag{62}$$

For the example in the figure, appropriate parameters are $r\sim 6$ fm,
$R_n \sim .5$ fm, giving $\ell \sim 35\hbar$. The scission boundary from TDHF actually occurs at this value of the angular momentum. That angular momentum
is also close to the maximum angular momentum sustainable in a liquid
drop of nuclear matter of that size. Indeed, we see that the only
physics that is important for the scission in the mean-field treatment, namely the surface tension and the inertia, is also present in
the liquid-drop model. The third boundary in Fig. 7, at small impact
parameter and high energy, is known as the fusion window. Physically, the nuclei flow through each other under these conditions, and
the attractive bulk forces are insufficient to hold them together in
the end. In the macroscopic description, this comes about because
of the memory effect in the particle momentum flux. This

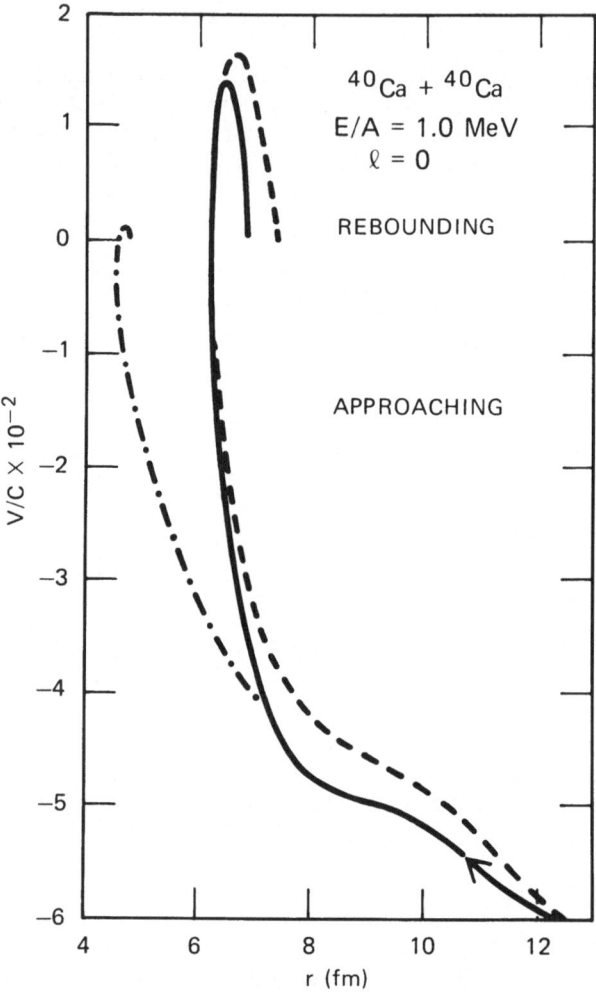

Fig. 6. Relative velocity of collision partners as a function of
separation distance. Dashed line shows a TDHF calculation,
and solid line shows the simulation by macroscopic force
dynamics.[8] The dot-dashed line shows the evolution when
the delay time τ_D is set to zero.

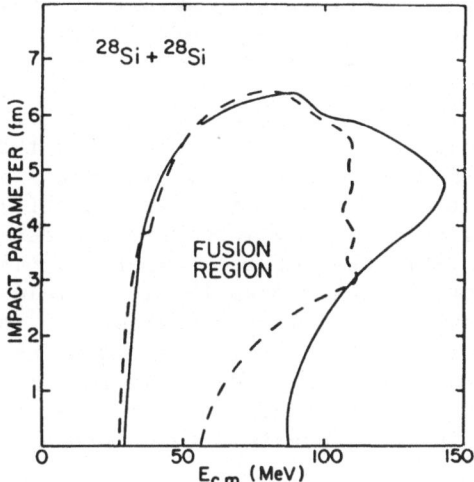

Fig. 7. Fusion region in collisions of ^{28}Si + ^{28}Si. The dashed boundary is the TDHF result,[7] and the solid line is the simulation by macroscopic force dynamics.[6]

contribution to the force is strongly repulsive, because of the high velocity of the nuclei. The force persists for a significant time after the centers of mass have been brought to rest, so the nuclei begin to separate with a substantial velocity. The Fermi surface of the combined spheres is then smaller than an individual sphere and the density falls below nuclear matter density in the neck's region. The pressure tensor becomes negative, i.e. there is an attractive force between the two nuclei. However, there is a maximum sustainable negative pressure in nuclear matter, the tensile strength. When this is exceeded, the neck will break up into regions of nuclear matter density separated by void regions. We call this mechanism neck snap. However, the macroscopic treatment does not describe this boundary of TDHF as well as the other boundaries. Since the nuclear matter is under a stress extreme here, the details of the Hamiltonian should play some role.

Recently, Davies and collaborators made a study of the fusion window threshold with a number of Hamiltonians, and found a

substantial dependence.[8] The biggest variation was with the
Hamiltonians Skyrme II and IV, which gave thresholds of 62.5 MeV and
42.5 MeV, respectively for the window in collisions between ^{16}O
nuclei. It should be possible to understand these differences in
terms of the properties of the pressure tensor with the various interactions, but so far this has not been done.

On the experimental side, the measured fusion cross sections in
medium heavy systems agree with the mean-field theory at low energies. An example is shown in Fig. 8. The cross section increases
with energy until the scission begins at the critical angular momentum. The window should cause a more dramatic decrease at higher energies, but so far this has not been observed. The fusion window may
well be nonexistent, in which case the mean-field theory would be
limited in validity to the lowest energies. An inadequacy of mean-
field theory also is found for very light systems, such as $^{12}C + ^{12}C$,
where there are pronounced wiggles in the fusion excitation function.
These are manifestations of resonances or other quantum effects that
are not contained in the mean-field description.

For heavy systems, the Coulomb force becomes significant in determining the fusion boundary. The scission boundary moves down as
the charge increases, and at some point it extends to zero impact

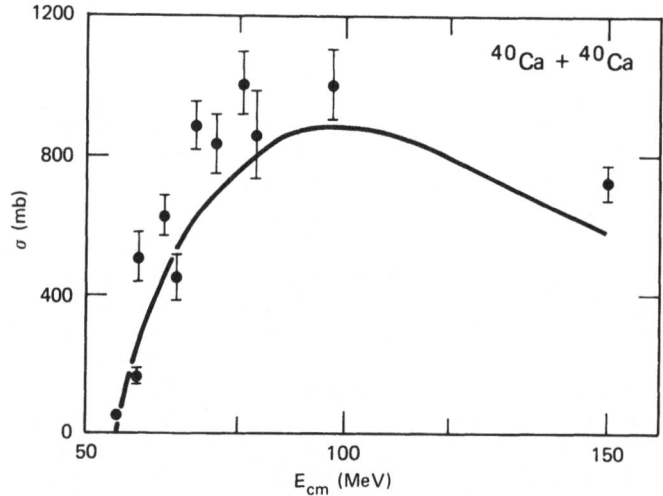

Fig. 8. Experimental fusion cross section in collisions of
$^{40}Ca + ^{40}Ca$, compared with macroscopic force model.[6] Experimental data are from Refs. 9 and 8.

parameter. We illustrate this with the macroscopic calculation for ^{120}Sn + ^{120}Sn shown in Fig. 9. The fusion boundary has a higher threshold than the potential barrier, which may be understood quite simply. After the nuclei surmount the barrier, they are accelerated toward each other by the external nuclear field. This force can be quite strong; in the proximity model it has a maximum magnitude

$$F_{proximity} \sim 2\pi R\sigma \tag{63}$$

As soon as the nuclei touch the repulsive particle exchange comes into play. The net nuclear force at that time is the surface tension force,

$$F_{surface} \sim 2\pi R_N \sigma \tag{64}$$

which we see is always less than the proximity force. Nuclei are eager to touch each other, but less enthusiastic to remain bonded.

For the more highly charged systems, the net force will be attractive only if the neck exceeds a certain radius, requiring additional bombarding energy above the barrier. This dynamic fusion threshold was first described by Nix and Sierk;[13] Swiatecki gives it the name "extra push". An example is the system ^{209}Bi + ^{54}Cr, shown in Fig. 10. The experimental data show that the nuclei react with each

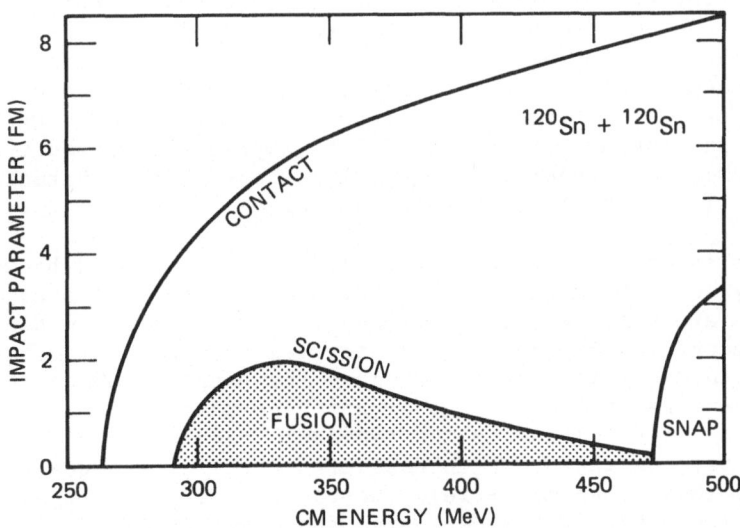

Fig. 9. Fusion region in ^{120}Sn+^{120}Sn collisions showing the dynamic fusion threshold.

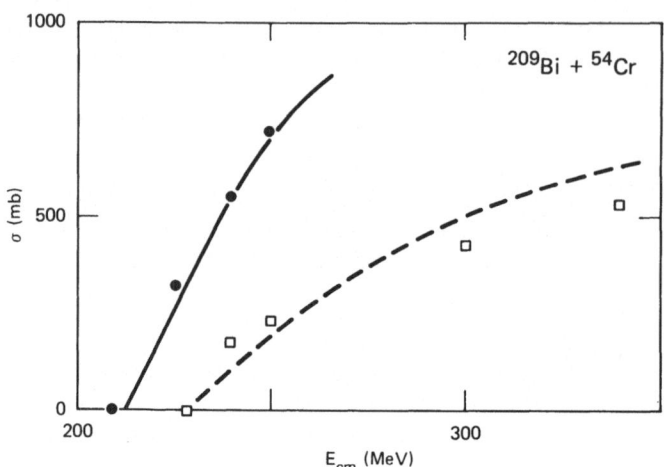

Fig. 10.   Reaction cross section and fusion cross section for
           ^{209}Bi + ^{54}Cr. The prediction of the macroscopic mean-field
           model is shown as solid and dashed lines respectively.

other starting at an energy of about 210 MeV, but do not begin to
fuse until 230 MeV or so. As with the fusion window, the precise
value of the dynamic fusion barrier seems to depend quite sensitively
on the nuclear Hamiltonian. Davies et al. found that in one system,
^{84}Kr + ^{139}La, the barrier varied from 410 MeV to 660 MeV with
Hamiltonians ranging from Skyrme III to Skyrme II, respectively.[8]
Again, there is no understanding of how the specific characteristics
of the Hamiltonians affect the threshold.

    Other aspects of low energy collisions that should be mentioned
are energy loss and angular distributions of deep inelastic scattering, which are well described by mean-field theory. The deep inelastic scattering is contained in the region between the barrier
curve and the scission curve. Because the neck shrinkage rate is
low, much energy is lost from the center-of-mass motion. The angular
distribution is sensitive to the temporal description of the elongation process; the fact that angular distributions are qualitatively
reproduced shows that mean-field theory has about the right time dependences. However, these data also show features that are beyond
the possibility to describe in mean-field theory. Namely, there are
broad dispersions in angle and in energy, while mean-field theory
predicts a unique angle and energy for each impact parameter. There

are a number of ways to go beyond mean-field theory to describe fluctuations. A recent theory of Balian and Veneroni,[14] for example, treats fluctuations associated with neighboring mean fields. More commonly, a theoretical framework is adopted in which the Hamiltonian is separated into a center-of-mass term, an internal term, and a coupling that is treated in some approximate way.

BEYOND MEAN FIELD: NUCLEON-NUCLEON COLLISIONS

At higher energy, mean-field theory gives a poor description of the single-particle density matrix because of the collisions between particles. The central approximation of mean-field theory, Eq. (9), must be replaced by something better. A natural way to approach this task is to use mean-field theory as a starting point and improve the wavefunction with perturbation theory. From a notational point of view, the perturbation development with the full two-particle interaction is cumbersome, so we will start with a simpler problem, developing the perturbations due to a one-particle external field. The generalization to two-particle interactions is merely a question of putting the extra particle operators into the equations.

Perturbation theory begins with a separation of the Hamiltonian into a part which can be solved exactly, H_0, and the remainder V, the perturbation. As mentioned above, we take H_0 to be the mean-field Hamiltonian. The unperturbed states are the eigenstates of that Hamiltonian (which change in time) and the multiparticle states built out of those eigenstates. We shall use these states for a basis and assume that the unperturbed density matrix is diagonal,

$$\langle \psi \left| a_i^\dagger a_j \right| \psi \rangle = \rho_{ij} = \delta_{ij} n_i \tag{65}$$

Equation (65) is a key assumption in the later development, but unfortunately it is not readily justified. Physically, we assume that the time scale for the evolution of the single-particle wavefunctions is long compared to the time for the perturbations to develop. Whether the assumption is justified in practice remains to be seen.

The actual perturbation is the two-body interaction minus the mean field, so that the operator V only has matrix elements off-diagonal in both particle states. For our outline of the derivation, we take V to be an external single-particle field. The first-order perturbed wavefunction at time t is

$$\left| \psi(t) \right\rangle = \left| \psi(o) \right\rangle + \frac{1}{i} \int_o^t dt' \sum_{k,k'} e^{-i\Delta\varepsilon t'} V_{kk'} a_k' a_k \left| \psi(o) \right\rangle$$

$$= \left| \psi(o) \right\rangle + \sum_{k,k'} V_{kk'} \frac{(e^{-i\Delta\varepsilon t} - 1)}{\Delta\varepsilon} a_k' a_k \left| \psi(o) \right\rangle \tag{66}$$

$$\Delta\varepsilon = e_{k'} - \varepsilon_k$$

where we have started from the mean-field wavefunction at t=0. We next substitute the perturbed wavefunction in Eq. (4) to get a new dynamic equation. The equation of motion for the diagonal density matrix in the instantaneous basis of H_{MF} depends only on the perturbation,

$$\frac{dn_i}{dt} = \frac{d}{dt} \langle \psi | a_i^\dagger a_i | \psi \rangle = \frac{1}{i} \langle \psi | [a_i^\dagger a_i, \hat{v}] | \psi \rangle$$

$$= \frac{1}{i} \langle \psi | a_i^\dagger [a_{ij} \hat{v}] + [a_i^\dagger, \hat{v}] a_i | \psi \rangle \tag{67}$$

$$= \frac{1}{i} \sum_j V_{ij} [\langle \psi | a_i^\dagger a_j | \psi \rangle - \langle \psi | a_j^\dagger a_i | \psi \rangle]$$

When the perturbed wavefunction (66) is inserted into (67), the righthand side consists of terms of first, second, and third order in V. The first-order terms may be included in the mean field and are not of interest. The third-order terms are neglected in comparison to the second-order terms. These terms required the evaluation of operator expectation values involving four operators. A typical expectation value is

$$\langle \psi(o) | a_i^\dagger a_j \, a_k^\dagger a_\ell | \psi(o) \rangle = \delta_{k\ell} \, n_i \, \delta_{jk} \, (1-n_j) \tag{68}$$

After some simplification, the result for the dynamic equation is

$$\frac{dn_i}{dt} = \sum_j \frac{2\sin\Delta\epsilon t}{\Delta\epsilon} V_{ij}^2 \, (n_j(1-n_i) - n_i(1-n_j)) \tag{69}$$

The sine function of $\Delta\epsilon$ is a representation of the delta function, so in the limit where there are enough levels so that $\Delta\epsilon$ takes on values small compared to the inverse time scales of interest, we can express the final result as

$$\frac{dn_i}{dt} \cong 2\pi \sum_j \delta(\epsilon_i - \epsilon_j) \, V_{ij}^2 \, (n_j(1-n_i) - n_i(1-n_j)) \tag{70}$$

This is recognizable as Fermi's golden rule, applied particles rather than to the system as a whole. The only consequence of the many-particle context, besides the mean field, is the presence of occupation factors in the formula.

The correponding formula for a perturbation due to a two-particle interaction is

$$\frac{dn_i}{dt} \cong 2\pi \sum_{\substack{ii' \\ jj'}} \langle ii'|V|jj'\rangle^2 \, \delta(\varepsilon_i+\varepsilon_{i'}-\varepsilon_j-\varepsilon_{j'}) \, [n_j n_{j'}(1-n_i)(1-n_{i'})$$

$$- n_i n_{i'}(1-n_j)(1-n_{j'})] \tag{71}$$

It might also be interesting to consider perturbations of the form of couplings between particles and vibrations. Labelling the vibration by q, the formula describing the effect of the vibrations on the single-particle density is

$$\frac{dn_i}{dt} = 2\pi \sum_{j,q} \langle i|V|jq\rangle^2 \left[\delta(\varepsilon_j+\varepsilon_q-\varepsilon_i)(n_j-n_i)(1+n_q) \right.$$

$$\left. + \delta(\varepsilon_i+\varepsilon_q-\varepsilon_j)(n_j-n_i) \, n_q \right] \tag{72}$$

To summarize the theory, the mean-field approximation is used as a basis of the representation of the single-particle density matrix. The off-diagonal matrix elements evolve by the mean-field Hamiltonian, while the diagonal matrix elements evolve by Eq. (71). It is easy to see that the theory conserves energy. The only difference from the mean-field theory, the changing of the diagonal density matrix elements, gives an energy change

$$\frac{dE}{dt} = \sum_i \varepsilon_i \frac{dn_i}{dt} \tag{73}$$

We substitute Eq. (71) into (73) and see that the sum vanishes, due to the symmetry and energy conservation built into (71). The theory will also conserve momentum if the interaction is translationally invariant.

The theory embodied in Eq. (71) has not yet been calculated to a point that allows a judgement about its adequacy. One attempt at a calculation was made by Wong and Davies,[15] who apply mean-field theory and Eq. (71) to collisions of ^{16}O on ^{16}O and ^{40}Ca on ^{40}Ca. Unfortunately, these authors used as a basis the time-evolved eigenstates of the initial mean-field Hamiltonian, rather than the instantaneous eigenstates of H_{MF}.

In that basis, there are very few levels crossings, and the energy-conserving delta function prevents practically all collisions from occurring. The reason why that happens may be seen in a momentum space representation. The occupied and unoccupied states are concentrated at momenta below and above the Fermi momentum, respectively, with equal probability for positive and negative momenta due to the parity symmetry of the initial state. We have seen that in the mean-field collision dynamics the momenta of the particles do not

change very much. So in the collisional geometry of the Fermi surface with two intersecting spheres, the unoccupied states will remain
at a higher energy than the occupied states.

Another approach to introduce collisional dynamics is the theory
of Nörenberg, et al.[16] Here the basis of states is the set of instantaneous eigenstates of H_{MF}, as required by (71). However, the
initial occupation numbers of these states are not taken from the
time-dependent mean-field wavefunction, but are assumed to be 0 or 1,
depending on the occupation probability in the initial state. It is
also assumed that the level crossings are distinct, occurring one at
a time. Then the transitions can be calculated by the Landau-Zener
formula [HW],

$$P = \exp\left(-2\pi \frac{v^2}{\dot{\varepsilon}_i + \dot{\varepsilon}_j - \dot{\varepsilon}_k - \dot{\varepsilon}_\ell}\right) \tag{74}$$

In constrast, the perturbation theory limit becomes valid when many
level crossings take place over the range of time required for an individual transition. Sketching the energy of a single-particle level
as a function of time or of some collective coordinate, the two
limits are shown in Fig. 11.

The number of level crossings in a heavy ion collision can be
estimated with a Fermi gas model. The state of maximum compression
is approximated by the intersecting sphere momentum distribution. In
the diabatic picture, the energy of single-particle states change
smoothly during the collision process, so that an unoccupied state
must cross all unoccupied states having lower energy. In this way,
we can examine a particular momentum state in the intersecting Fermi
sphere, and compute how many unoccupied states were crossed. If each
nucleus starts with a momentum p per particle in the center-of-mass
frame, then a particle at the tip of the Fermi surface crosses a number of single-particle states given by

$$N = \left(2\pi p/P_F\right) \frac{A}{4} + \mathcal{O}(p^2) \tag{75}$$

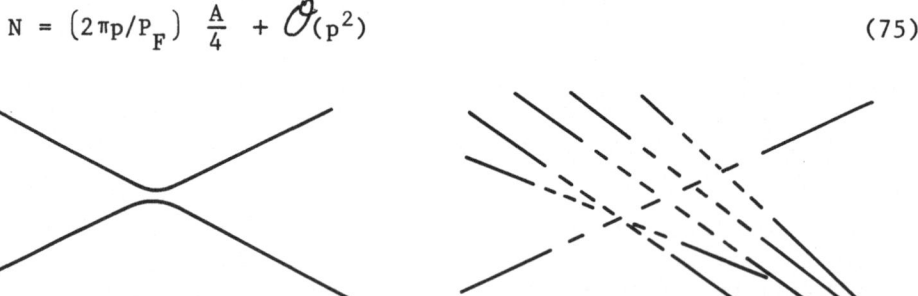

(a)   (b)

Fig. 11. Sketch of single-particle level crossing for situations in
          which Landau-Zener formula would be valid, and for which
          perturbation theory would be valid.

In this equation we count only spatial states of a given kind of nucleon; if both spin orientations are counted, the number should be doubled. The energy change of the particle during the diabatic transformation is

$$\Delta E \sim \frac{\hbar^2}{m} p P_F + \mathcal{O}(p^2) \tag{76}$$

so the energy spacing of the crossings is

$$\Delta E/N \approx \frac{P_F^2}{m} \frac{2}{\pi A} \approx 50/A \text{ MeV} \tag{77}$$

This should be compared with a typical matrix element of the two-particle interaction. The most important matrix elements are of the pairing type, involving particle degenerate energy states. These may be estimated using a zero-range interaction having a strength consistent with the mean-field potential. Calling the strength of the interaction v_o, the volume of the system $\mathcal{V}$, the estimate is

$$\langle i\bar{i} \mid v_o \delta^3(r'r') \mid j\bar{j} \rangle \approx \frac{1}{2} \frac{v_o}{\mathcal{V}} = \frac{1}{2} \frac{v_o \rho_o}{A} \approx \frac{U(\rho_o)}{2A} \approx \frac{25}{A} \text{ MeV} \tag{78}$$

This energy is smaller than the energy spacing in Eq. (76), so the approximation of separated crossings is reasonable at low energies. At higher energies the phase space increases more rapidly, and the perturbation limit is more appropriate.

The Uehling-Uhlenbeck Equation

Equation (71) can be reduced to classical physics in the treatment of large systems with slowly varying potential fields.[17] In that case the eigenstates of H_{MF} are well approximated by wavepackets localized in both momentum and position. The occupation number of the states is then just the phase space occupation probability given by the Wigner function

$$n_i \sim f(p,r) \tag{79}$$

To make the formula completely classical, we also use Born approximation for the particle-particle scattering cross section,

$$\frac{d\sigma}{d\Omega} = \frac{2\pi}{v} \int \frac{P_3^2 dP_3}{(2\pi)^3} \langle p_1 p_2 \mid V \mid p_3 p_4 \rangle^2 \delta(\varepsilon_1 + \varepsilon_2 - \varepsilon_3 - \varepsilon_4) \tag{80}$$

with v the relative velocity of the particles. Then the dynamic equation is

$$\frac{Df(p,r)}{Dt} = \int \frac{dp^3_1 \; d^3p_2 \; d^3p_3 \; d\Omega_3}{(2\pi)^9}$$

$$\times \; v \frac{d\sigma}{d\Omega} \left[f(p_1,r)f(p_2,r)\bigl(1-f(p_3,r)\bigr)\bigl(1-f(p,r)\bigr) \right.$$

$$\left. - f(p,r)f(p_3,r)\bigl(1-f(p_1,r)\bigr)\bigl(1-f(p_2,r)\bigr) \right]$$

$$(2\pi)^3 \; \delta^{(3)} \bigl(p_1+p_2-p_3-p\bigr) \qquad\qquad (81)$$

where the total derivative D/Dt represents the entire lefthand side of Eq. (24). This form was first proposed by Uehling and Uhlenbeck.[17] It is the Boltzmann equation with the collision integral modified by the Pauli blocking factors (1-f). This equation is attractive for consideration in the theory of intermediate energy collisions, because it encompasses both the mean-field physics valid at low energies, and the independent collision dynamics applicable at high energies.

The numerical solution of the Uehling-Uhlenbeck equation is difficult, but is undoubtedly within reach of present-day computational techniques. I shall describe one method for solving the equation that has been pursued by the Michigan State University group.[18] The method is based on the Particle-in-Cell technique of numerical hydrodynamics.[19] The system is described by a set of test particles, each having specified momentum and position. The density of these particles in phase space represents the distribution function f(p,r). If the particles obeyed Newtonian mechanics within the mean field, the distribution function would evolve according to the Vlasov equation. The Newtonian equations are integrated with the particle positions calculated at discrete time steps, t+Δt, t+2Δt, etc. The particle positions are updated by the equation

$$\vec{r}_i(\Delta t) = \vec{r}_i(o) + p_i(\Delta t/2)\Delta t/m \qquad\qquad (82)$$

where p(Δt/2) is the momentum of the particle at the midpoint of the time interval.* To find the momentum vectors at the half time steps, we use the equation

$$\vec{p}_i(\Delta t/2) = \vec{p}_i(-\Delta t/2) + \vec{F}_i(o)\Delta t \qquad\qquad (83)$$

where $F_i(o)$ is the force at t = 0. The force is calculated by dividing space up into cells. The density in each cell is determined by counting the number of test particles n in that cell,

$$\rho(r) = \frac{n}{N(\Delta x)^3} \qquad\qquad (84)$$

---

*The algorithm is an order of Δt more accurate using the momentum at the midpoint rather than at the initial time.

where N is the number of test particles used to represent one physical particle. The potential field of the cell is specified by some function of the density, and the force is determined from the difference in potentials in adjacent cells,

$$F_x(r) = \frac{U\big(\rho(i,j,k0)\big) - U\big(\rho(i+1,j,k)\big)}{\Delta x}, \text{ etc.,} \qquad (85)$$

where the point r lies between the midpoints of cells (i,j,k) and (i+1,j,k).

The collision integral in Eq. (81) is calculated by simulating nucleon-nucleon collisions between the test particles. Each test particle is permitted to interact with 1/N of the other test particles in a way that produces the required scattering cross section for isolated nucleons. The Pauli blocking factors are included by accepting or rejecting each collision in a probabalistic way. We examine the final state of each collision, constructing a sphere about the phase space coordinate of each particle. The phase space density f is determined by counting the number of test particles in that sphere. The collision is accepted with a probability $(1-f_p)(1-f_p')$, using a random number generator to make the probabalistic decision.

The numerical parameters that enter the calculation are the cell size, the number of test particles to represent one physical particle, and the time step. We have not yet made a systematic study of the parameter requirements to achieve a definite accuracy, but have only done some exploratory calculations. In Ref. 18 we used cells of 2 fm on a side, which seems adequate for short time intervals, but becomes faulty after times of the order of 100 fm/c: the nuclei do not remain spherical but acquire the square shape of the cells. This situation is rectified by using cells 1 fm on a side, which allows propagation of nuclei to at least 120 fm/c. The number of test particles to represent a physical particle should be large to reduce numerical fluctuations. We found that a nucleus could be described reasonably well with about 10 test particles per occupied cell. Then the numerical fluctuations in the density are of the order of 30%, which seems tolerable in the calculation of the forces. Thus with cells 1 fm on a side, the number of test particles is 50 times the number of nucleons. The requirements on the time step are that the particle travel only a small fraction of cell length and that the collision probability be small during one step. In practice, t=0.5 fm/c seems adequate, and we have used that value. In our present studies, we assume an isotropic cross section of 4 fm^2.

An important test of the numerical scheme is to study the behavior of a nucleus by itself. Besides maintaining its overall shape and density, it should not leak particles, and the momentum distribution should be a sharp-edged Fermi distribution. With the above numerical parameters, we find that less than one nucleon leaks out of an ^{16}O nucleus in a time 120 fm/c, which should be satsifactory to

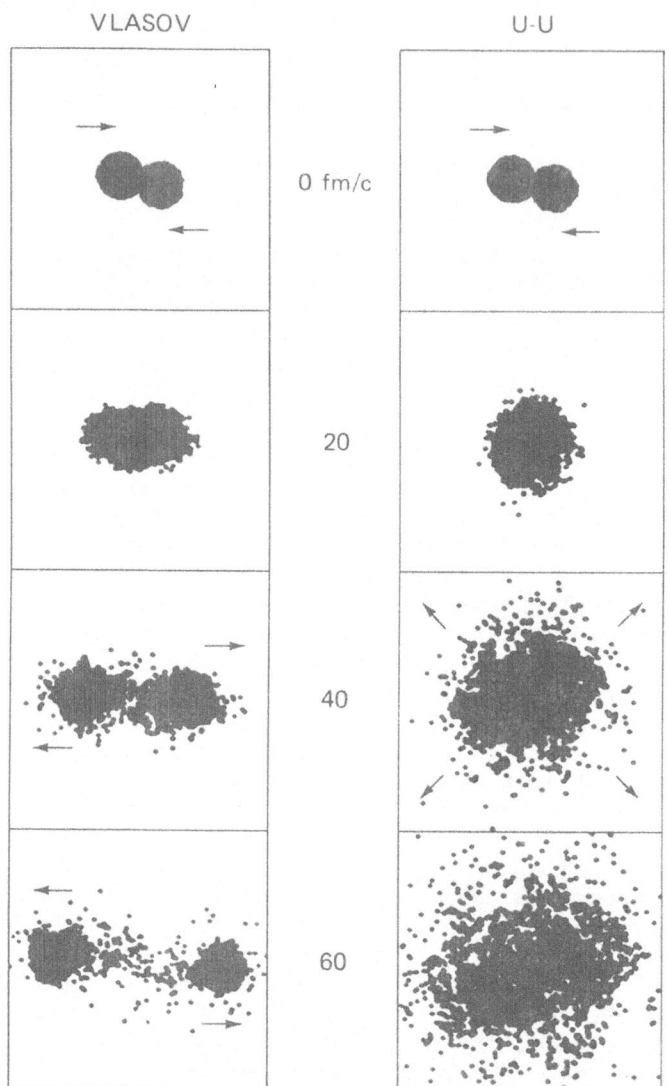

Fig. 12. Collisions of ^{12}C + ^{12}C at 84 MeV/n, calculated by J.
          Aichelin.[21] The lefthand column shows the positions of
          the test particles, evolved according to the Vlasov
          equation with the soft potential function, Eq. (52). In
          the righthand column, the collision integral is included
          in the evolution equation. The numerical parameters in
          these calculations are Δt=0.5 fm/c, cell size Δx=1 fm,
          and N=100 test particles/nucleon.

treat large cross section phenomena for energies down to 25 MeV/n. One check on the effectiveness of the Pauli blocking algorithm is to record the fraction of collisions that are blocked in a nuclear ground state. We find that for ^{16}O, 90% of the collisions between particles are blocked.

We next examine collisions between nuclei. Figure 12 shows the projection of test particles on the x,z plane for collisions of ^{12}C on ^{12}C at 84 MeV/n. The first set of figures shows the theory without collisions, i.e. the Vlasov equation. In that theory, the nuclei go through each other, as they do in TDHF. The prediction of the U-U equation is shown in the second column. The nucleon-nucleon collisions are extremely important, converting the directed motion into a spherically expanding distribution. There is some suggestion[20] that thermal equilibrium may come about very rapidly in collisions at intermediate energy; the study here gives very rough qualitative support to that view.

One observable quantity that can be studied rather easily is the single-particle distribution in the final state. In the laboratory frame, the energy distribution of the emitted particles is quite flat for forward angles, changing to a distribution that falls off steeply with energy at backward angles. This behavior is qualitatively reproduced by the numerical calculations with the Uehling-Uhlenbeck equation.[21] However, it is not completely straightforward to make a quantitative comparison, because the theory provides only a single-particle distribution function, and experiments measure free particles separately from clusters such as alphas which are emitted copiously at the lower energies. Kruse et al. have also studied the single-particle distribution,[22] using a somewhat different technique for particle acceleration in the Uehling-Uhlenbeck equation. They make a clustering correction to the single-particle distribution to obtain the free proton spectrum. They find qualitative agreement between theory and experiment, shown in Fig. 13, with some differences between the theory with and without the mean-field term.

The inclusive single-particle momentum distribution averages over impact parameter and orientation of the collision, losing possible dependences on these parameters. It is possible to regain some of this information with a global analysis of momentum flow. In this technique, the momentum of a large number of particles are measured simultaneously, and the quadrupole tensor of the distribution is determined. One popular definition of this quantity is

$$F_{\mu\nu} = \sum_i \frac{P_\mu(i)P_\nu(i)}{\partial m_i{}^1}$$

where i labels the different particles emerging from the collision. The tensor tends to be close to isotropic for collisions between

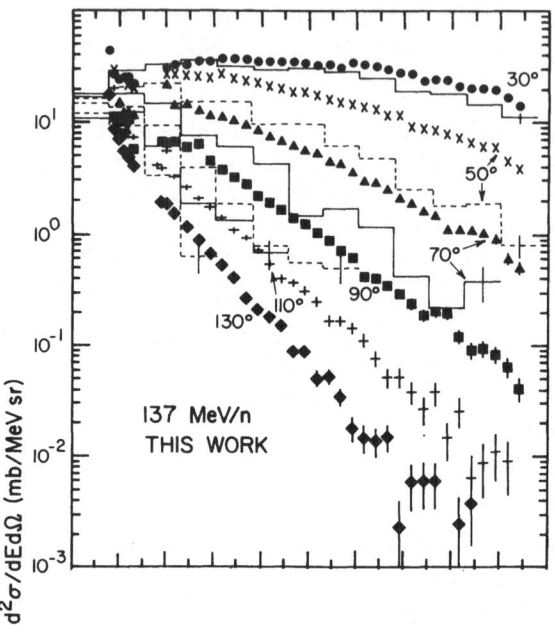

Fig. 13. Proton spectrum from collisions of Ar + Ca at 137 MeV/n.
The data[23] are indicated by points, and the theory[22] by
histograms.

equal mass target and projectile, but off-diagonal anisotropies are
predicted to occur at finite impact parameter in models for which the
pressure has other sources besides nucleon-nucleon collisions. Although the anisotropy is rather small, it has a large effect on the
orientation of the principal axes of the tensor $F_{\mu\nu}$. The orientation
of the longest principal axis, called the flow angle, is thus quite
sensitive to the underlying dynamics. Flow angles were recently
measured[24] for collisions of $^{40}Ca + ^{40}Ca$ and $^{93}Nb + ^{93}Nb$ at energies
of 400 MeV/n. In the lighter system, the distribution of flow angles
peaked at zero degrees, showing that the anisotropy was small compared to the difference $F_{zz}-F_{xx}$. However, for the heavier system,
the most central collisions, defined by the highest multiplicity of
secondaries, showed a peaking at a flow angle of 30°. This result is
reproduced in the theory based on the Uehling-Uhlenbeck equation. In
Fig. 14 is shown the predicted distribution of flow angles for the
soft and stiff potential functions.[25] We see that there is some

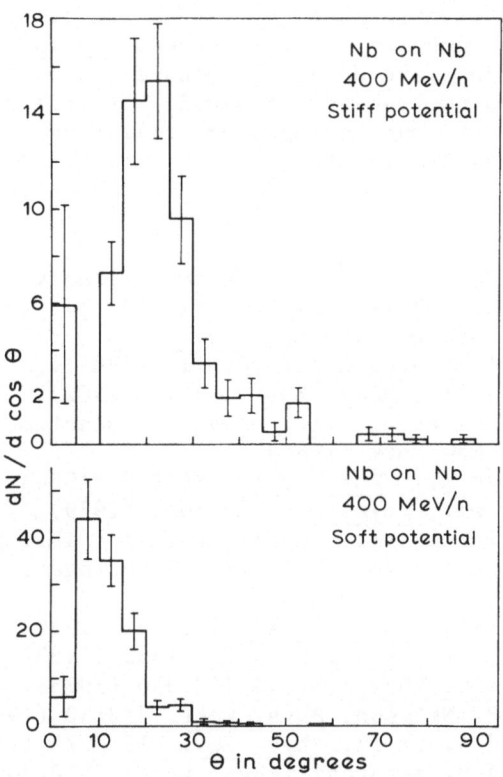

Fig. 14. Predicted flow angles for Nb+Nb collisions at 400 MeV/n, from Ref. 25.

sensitivity to the potential function, with the stiff potential producing more anisotropy. We thus have some hope of extracting the
high-density dynamics of heavy-ion collisions from this type of
measurement. It is interesting that the sensitivity seems to be
greatest at a beam energy of 400 MeV/n; at 200 and 800 MeV/n the flow
angle moves to forward direction.[24]

Finally, I would like to conclude with a suggestion that the
Uehling-Uhlenbeck equation be applied to the calculation of linear
momentum transfer. The data show divisions between regions of nearly
complete momentum transfer between target and projectile, and regions
where half of the momentum is transferred. At the higher energies
where the experiments are done, the mean-field contribution to the
longitudinal momentum transfer is small. It will be interesting to
see the effect of the nucleon-nucleon collisions in a well-defined
theory of the heavy-ion reaction.

References

1. A. Fetter and J. D. Walecka, "Quantum Theory of Many-Particle
    Systems," McGraw-Hill, New York (1971), p. 149.
2. H. S. Köhler and H. Flocard, Nucl. Phys. A323:189 (1979).
3. H. Tang et al., Phys. Lett. 101B:10 (1981).
4. J. W. Negele, Rev. Mod. Phys. 54:913 (1982).
5. K. T. R. Davies et al., in:"Heavy-Ion Reactions," D. A. Bromley,
    ed., Plenum, New York (1984).
6. A. Bonasera et al., Phys. Lett. 141B:9 (1984).
7. P. Bonche et al., Phys. Rev. C20:641 (1979).
8. K. T. R. Davies et al., Dynamical Fusion Thresholds, in:"Nuclear
    Physics with Heavy Ions," P. Braun-Munzinger, ed., Harwood, New
    York (1984).
9. E. Tomasi et al., Nucl. Phys. A373:341 (1982).
10. J. Barreto et al., Phys. Rev. C27:1335 (1983).
11. J. Randrup, Ann. Phys. (N.Y.) 112:356 (1978).
12. A. Dhar and B. Nilsson, Phys. Lett. 77B:50 (1978).
13. J. R. Nix and A. Sierk, Phys. Rev. C15:2075 (1977).
14. R. Balian and M. Veneroni, Phys. Lett. 136B:301 (1984).
15. C. Y. Wong and K. T. R. Davies, Phys. Rev. C28:240 (1983).
16. W. Nörenberg, Phys. Lett. 104B:107 (1981);
    H. Yadav and W. Nörenberg, Phys. Lett. 115B:179 (1982).
17. E. Uehling and G. Uhlenbeck, Phys. Rev. 43:552 (1933).
18. G. Bertsch, H. Kruse, and S. Das Gupta, Phys. Rev. C29:673
    (1984).
19. F. Harlow and A. Amsden, "Fluid Dynamics," Los Alamos Report
    LA-4700 (1971).
20. J. Aichelin and G. Bertsch, Phys. Lett. 138B:350 (1984).
21. J. Aichelin and G. Bertsch, to be published.
22. H. Kruse, B. Jacak, H. Stöcker, and G. Westfall, to be
    published.
23. B. Jacak et al., to be published.

24.   H. A. Gustafsson et al., Phys. Rev. Lett. 52:1590 (1984).
25.   C. Gale and S. Das Gupta, private communication.

SOME ASPECTS OF HEAVY ION MACROPHYSICS

Christian Ngô

Service de Physique Nucléaire - Métrologie Fondamental

91191 Gif-sur-Yvette Cedex, France

ABSTRACT

In these notes we review, in a schematic way, some aspect of the physics with heavy ions. In the first lecture we review how is possible to describe the dissipative phenomena observed above the Coulomb barrier, up to 10-15 MeV/u, using transport theories. The second lecture is devoted to the question of fusion and the appearance of a new mechanism : fast fission. It is shown that one can now have a global understanding of these phenomena within single picture. The third lecture presents, in a simplified way, some results obtained recently with heavy ions in the range of 30-50 MeV/u at GANIL and SARA.

Lecture I   : Dissipative phenomena in heavy ion physics.

Lecture II   : Fusion and fast fission.

Lecture III   : Some aspects of the physics between 20 and
                50 MeV/u.

Lecture 1 : DISSIPATIVE PHENOMENA IN HEAVY-ION PHYSICS

With nuclei we can investigate two sorts of things. First we can try to understand their structure using different probes like electrons, nucleons, nuclei, ... which we use to bombard target nuclei. From the experimental information obtained in these experiments, we try to understand how they are made and why they exist. For that it is necessary to develop models which explain their structure and could possibly do also predictions. The problem is really understood when the explanation, as well as the description of their nuclear structure, becomes simple but based on microscopic grounds.

It is also interesting to understand what happens when two nuclei interact with each other. This concerns the investigation of the reaction mechanisms and is a second thing which can be done with nuclei. Such studies become very interesting if what we observe is different from what is expected from the simple free nucleon- nucleon interaction when it is folded to take into account of the size of the nuclei. This is for instance the case when a coherent motion of nucleons can be observed, because it is really a new feature compared to a result consisting of only an interaction between single free nucleons.

The domain of heavy-ion reactions corresponding to bombarding energies ⩽ 10 MeV/u has shown the existence of collective dissipative motions.[1] This fact is rather unique in nuclear physics and the domain dealing with these cooperative phenomena has been called macrophysics. This field allows to study microsystems in strong interaction at, and out, of equilibrium. This is a rather unique opportunity for statistical physics. Furthermore quantum effects are also present in some cases, and the possibility of providing the system with an excitation energy going from zero to large values, allows to study the transition between a quantum system and a nearly classical one. The time scales which characterize the coherent motions as well as the individual ones are not so different. This situation is in constrast with the one of macrosystems which we are used to. This rather unique feature has called for extensions of usual statistical theories. In addition to that there will be also the possibility to see non Markovian effects and this might be a direction of interest in the near future.

The richness of macrophysics has been such over the last decade, that it now occupies a large part, both financially and in manpower, of the nuclear physics efforts. In this first lecture, I would like to give a feeling of how it is possible to explain collective dissipative motions in heavy ions reactions at bombarding energies ⩽ 10 MeV/u. We shall restrict ourselves to the physical ideas and therefore stay a simple as possible.

1. DISSIPATIVE PHENOMENA IN HEAVY ION PHYSICS

In heavy ion reactions at bombarding energies < 10 MeV/u, it is possible to observe products with a total kinetic energy which is much smaller than the incident one. This means that a large part of the kinetic energy in the relative motion can be converted into intrinsic excitation (heat). The reactions where this surprising phenomenon can be observed are usually called deep inelastic collisions.[1] Since heavy ion reactions can be described, to a good approximation, by classical mechanics, one may try to describe the dynamical evolution of the system by introducing friction forces in the interaction region. These forces will pump the energy which is in the relative motion and transform it into intrinsic excitation energy. In other words they will convert organized energy into desorganized one. The simplest form for the friction force we may think of, is a one which is proportional to the relative velocity with a form factor which lets it to occur only in the interaction region. This prescription was used in the past with great success[1,2]. It allowed to qualitatively understand a large body of experimental data concerning not only deep inelastic collisions but also fusion which is the most dissipative mechanism which can be observed in heavy ion reactions.

In order to illustrate the method, let us consider that the two heavy ions can be described by two classical particles which interact by means of the Coulomb and nuclear interaction potentials. Then the natural coordinates of the problem are the distance r separating their center and the polar angle θ. For a conservative system the equations for motion would be

$$\mu \ddot{r} = - \frac{\partial V_N(r)}{\partial r} + \frac{Z_1 Z_2 e^2}{r^2} + \frac{\ell(\ell+1)\hbar^2}{\mu r^3} \qquad (1)$$

$$\frac{d}{dt} (\mu r^2 \dot{\theta}) = 0 \qquad (2)$$

where μ is the reduced mass, Z_1 and Z_2 the atomic numbers of the colliding ions, ℓ the initial orbital angular momentum and $V_N(r)$ the nuclear interaction potential. A dot above a variable denotes a time derivative.

These equations of motion can be deduced from the following Lagrangian :

$$\mathcal{L} = \frac{1}{2} \mu \dot{r}^2 + \frac{1}{2} \mu r^2 \dot{\theta}^2 - V_N(r) - \frac{Z_1 Z_2 e^2}{r} \qquad (3)$$

The above dynamics only leads to an elastic scattering of the two nuclei. Dissipation can be obtained by adding in the right hand side of equations (1) and (2) friction forces of the form :
$- C_r f(r)\dot{r}$ for the radial motion, and $- C_\theta r^2 f(r) \dot{\theta}$ for the

Tangential one, where C_r and C_θ are the radial and Tangential friction constants, and f(r) is a form factor which is usually taken in such a way that it fits the data for a large number of systems.

By modifying eqs.(1) and (2) as indicated above, the colliding system becomes dissipative. The energy loss per unit of time, dE/dt, is equal to

$$\frac{dE}{dt} = - C_r \ f(\dot{r}) \ r^2 - C_\theta \ f(r) \ r^2\dot{\theta}^2 \qquad (4)$$

The above model can be generalized to take into account of a larger number of macroscopic variables like for instance the deformations of the two nuclei or their internal rotations. If these macroscopic variables are denoted by q^i the equation of motion can be deduced from the following Lagrangian :

$$= \frac{1}{2} \ m_{ij} \ \dot{q}^i \dot{q}^j - V(\{q^k\}) \qquad (5)$$

where m_{ij} is the inertia tensor (in the above example we had $m_{rr} = \mu$, $m_{\theta\theta} = \mu r^2$). Dissipation is taken care of by introducing the Raleigh dissipation function, J, which can be expressed as :

$$J = \frac{1}{2} \ \gamma_{ij} \ q^i q^j \qquad (6)$$

the equations of motion have then the following form :

$$\frac{d}{dt} \frac{\partial}{\partial \dot{q}^i} - \frac{\partial}{\partial q^i} = - \frac{\partial J}{\partial \dot{q}^i} \qquad (7)$$

And the energy loss per unit of time is just

$$\frac{dE}{dt} = - 2J \qquad (8)$$

2. NATURE OF THE DISSIPATION

Introducing friction forces is a convenient way to reproduce the data. However a theoretical justification for doing so is needed. Up to now there are several microscopic theories which have shown that friction forces proportional to the collective velocities can occur in heavy ion collisions [refs.1,3,4]. These theories are based on the fact that it is possible to divide the degrees of freedom of the system in two categories : the macroscopic, or collective degrees, which have a rather slow time evolution towards equilibration (a few 10^{-22}s to $\sim 10^{-21}$-10^{-20}s), and the intrinsic degrees which relax quickly to equilibrium ($\sim 10^{-22}$s or smaller). Under these assumptions it is possible to derive equations of motion for the macroscopic variables. This was done for instance by Hofmann and Siemens[3] using linear response theory. Let us briefly sketch the principle of such a derivation.

We shall divide our total system in two parts S and B. S will consist of the macroscopic variables and B will consist of the intrinsic degrees. The total Hamiltonian of the system is assumed to be split in three parts :

$$\hat{H} = \hat{H}_S + \hat{H}_B + \hat{H}_{SB} \qquad (9)$$

$\hat{H}_S$ is the Hamiltonian operator associated to the subsystem S, $\hat{H}_B$ the one connected to subsystem B, and $\hat{H}_{SB}$ represents the interaction between both subsystems. It is this part which allows energy to be transferred from one subsystem to the other.

The evolution of the total system is governed by the total density operator $\hat{W}(t)$ which depends explicitly upon all the variables of the system. However, in many cases we are not interested in the properties of the system B. Rather, we want to follow the time evolution of the subsystem S. The only thing we need to know on B is how it influences the evolution of S. Therefore the important quantity is the reduced density operator $\hat{\rho}$ which is obtained by averaging over the B-degrees of freedom :

$$\hat{\rho}(t) = Tr_B \, \hat{W}(t) \qquad (10)$$

where Tr_B means that the trace of is taken with respect to the B subsystem only. To get the equation of motion for $\hat{\rho}$, one starts from the Von Neumann equation :

$$i\hbar \, \dot{\hat{W}} = \left[\hat{H}, \hat{W}\right] \qquad (11)$$

and projects out this equation in the subspace S. This can be done by the projection method developped by Nakajima and Zwanzig which is fully described in the paper of Haake.[5] Let us just quote the result of such manipulations which lead to the so called Nakajima-Zwanzig equation :

$$\dot{\hat{\rho}} = - i \, \hat{L}_S^{eff} \, \hat{\rho} + \int_0^t ds \, \hat{K}(s) \, \hat{\rho}(t-s) + \hat{I}(t) \qquad (12)$$

where $\hat{L}_S^{eff}$ is some effective liouvillian operator, $\hat{K}(s)$ a kernel and $\hat{I}(t)$ corresponds to an inhomogeneity term. This equation is still reversible and the interesting thing is that it is non local in time (see second term of the right hand side). Of course such an equation is useless in this form for practical purposes and approximations, taking care of the physical situation under interest, have to be done. In the case of the Hofmann and Siemens theory they are the following :

1. It is assumed that at the beginning of the reaction the total density matrix factorises in two terms : one for subsystem B, the other for subsystem S. This amounts to say that, before they touch, the two subsystems do not interact. This is certainly true in the case of heavy ion collisions. As a consequence $\hat{I}(t)=0$.

2. Since the relaxation time for the intrinsic degrees is smaller than the one associated to the macroscopic variables it is possible to assume that at each stage of the reaction the intrinsic degrees are in statistical equilibrium and play the role of a heat bath for the macroscopic degrees. Furthermore, one may assume that this statistical equilibrium can be described by a temperature (canonical ensemble). Since we have to deal with a microsystem, this last approximation is by no means obvious but it can be shown[6] that it leads to a reasonable error compared to the case where, as it should be done, a microcanonical distribution is used.

The time scale difference between the internal and the collective degrees allows to do the Markov approximation as a further simplification. It means that the equation of motion for $\rho(t)$ will become local in time.

At this stage a remark should be done : the collective motions have a relaxation time which is larger, but not so much, than the one corresponding to the intrinsic degrees. Consequently the above approximations are only valid to a certain extent. In particular one might think that non Markovian effects could be present and it is a challenge for the experimentalists to find them.

3. Hofmann and Siemens have further assumed that the coupling between S and B is small. This allowed them to apply perturbation theory to the kernel $K(t)$. Assuming that this coupling takes the form

$$\hat{H}_{SB} = \sum_j \hat{Q}^j \, F_j(\hat{X}^i) \tag{13}$$

where $\hat{Q}$ is the operator associated to the collective degree Q^j and $\hat{F}_j$ the corresponding field operator for the intrinsic degrees (X_i are the intrinsic variables), they could use the methods of linear response theory.

It is interesting to note that the field $\hat{F}_j(\hat{X}^i)$ is a one-body operator. This is equivalent to say that the dissipation process is governed by the mean field. This is a reasonable approximation since the nucleons have a long mean free path in the nucleus due to the Pauli exclusion principle. Consequently most of the collisions that the nucleons will suffer occur with the mean field. That is the reason why such kind of friction has been called one-body dissipation in contrast to the two-body dissipation which is usual in our macroscopic world.

Let us call τ^* and τ^{coll} respectively the relaxation time associated to the intrinsic and collective motions. If we consider the system during a time Δt around t , where Δt is microscopically large ($\Delta t \gg \tau^*$) but macroscopically small ($\Delta t \ll \tau_{coll}$), it is possible to show that the dynamical evolution of the macroscopic system can be described by a quantum master equation. It is necessary to choose Δt as indicated above because of the following reasons : Δt should be sufficiently large to allow the intrinsic degrees to relax to equilibrium. Consequently they play the role of a heat bath for the collective degrees. However Δt should be sufficiently small for the collective variables to remain almost constant during this amount of time.

If we consider, for the sake of simplicity, only one macroscopic variable Q, then during Δt it will not change very much around its mean value $Q_0 = <Q>_{t_0}$. The coupling between S and B is known to be strong (indeed we know that in a deep inelastic reaction several tens MeV can be easily lost in the relative motion) and it is not possible to apply to it perturbation theory for which this coupling has to be small. However the difference :

$$\hat{H}_{SB}(\hat{Q},\hat{X}^i) - \hat{H}_{SB}(Q_0,\hat{X}^i) \approx (Q-Q_0) \left.\frac{\partial \hat{H}_{SB}}{\partial Q}\right|_{Q_0} \qquad (14)$$

is small and can be treated by perturbation theory. One has then to renormalize the Hamiltonian of the heat bath :

$$H_B^{ren} = \hat{H}_B + \hat{H}_{SB}(Q_0,\hat{X}^i) \qquad (15)$$

the renormalization of the intrinsic Hamiltonian at each stage of the reaction is the essence of the Hofmann and Siemens approach. It allows to remove a large past of the coupling but one needs, except for a pure harmonic motion, that the fluctuations around the mean values remain small.

The above prescription can be generalized to the whole trajectory followed by the two heavy ions, and a quantum equation of motion for the macroscopic system can be obtained. Since, in most of the cases, the collective motions show a classical behaviour, it is possible to go to the classical limit. A transport equation (Fokker Planck equation in the phase space of the collective degrees) is then obtained. This equation has a similar form as the one used to describe transport phenomena in macroscopic systems (see section 4). It contains, in addition to conservative terms, friction and diffusion coefficients. Such an equation allows a large amplitude collective motion to be described, if it is slow. This is the case for most of the col-lective motions observed in deep inelastic reactions.

If one has to deal with fast collective motions (we will see an example in section 5), they can exhibit quantum features and it is not possible to apply to them a classical transport equation. However it is possible, only in the case of a harmonic motion, to derive a transport equation for the Wigner transform of the reduced density matrix. We will see later (section 4) that this equation has the same mathematical structure as the classical one.

To obtain these transport equations an average over the intrinsic system has been done in order to get the reduced density operator $\hat{\rho}$. Since the collective motions are coupled to the intrinsic degrees, it is necessary to know something about the intrinsic system. It appears that the two only informations which are needed on the heat bath, are the response, and the correlation functions. If we know them it is possible to describe the dynamical evolution of the macroscopic system. Considerable effort have been and are made to calculate these quantities microscopically.[7] Indeed their knowledge allows the calculation of the friction and diffusion coefficients which are the key quantities entering the equations of motion of the macroscopic system (see section 4).

Several microscopic theories have been developed in the literature.[1,2,3] They differ in the way how to treat the coupling between the macroscopic and the microscopic system. However they all end up with a transport equation for the macroscopic variables in which enter friction and diffusion terms. This transport equation is similar to the one describing the evolution of a Brownian particle.

3. SIMPLIFIED APPROACHES TO DISSIPATION

In the above approach, friction is strongly connected to the mean field in which the nucleons evolve. The existence of a coupling between the macroscopic and the microscopic systems, together with the difference of time scales associated with them, leads to an irreversible flow of energy from the collective motions to the intrinsic degrees. During the process, a lot of 1p-1h excitations are created in the intrinsic system. Due to the residual interactions these excitations decay into more complicated states (heat). It is the existence of these residual interactions which make the relaxation time for the internal variables, τ^*, to be so small. Indeed microscopic calculations,[6,7] as well as measurements,[8] seem to indicate that $\tau^* < 10^{-22}$ s).

In a schematic classical picture, dissipation can be visualized in the following way :

Nucleons are assumed to be a free Fermi gas enclosed in a container which simulates their mean field. Since the mean free path of the nucleons is large compared to the dimension of the

system, most of the collisions that the particles suffer will be
with the walls of the container. Such a gas is usually called a
Knudsen gas. It is assumed that after a collision with the wall,
the particle velocity is randomly distributed. This hypothesis is
nothing more than saying that the internal degrees of freedom have
a relaxation time shorter than the one of the macroscopic variables.

A moving wall will be a source of dissipation. If the wall
velocity is much smaller than the Fermi velocity of the nucleons,
a friction force proportional to the wall velocity will appear,
and the motion of the wall will be slowed down. This has been
demonstrated first by Gross[9] (piston model) and worked out in
details by Blocki et al.[10] (wall formula).

When two nuclei are connected to each other by a window,
nucleons can be transferred from one container to the other. With
the same hypothesis of randomization as above and assuming that a
transferred nucleon never comes back, one can also show that a
friction force, proportional to the relative velocity of the two
nuclei, appears (window formula).[10] An interesting result of such
a description is that the friction coefficient in the radial directions is twice as large as the one in the tangential motion.[10]

With the simple above picture in mind, it is possible to
calculate the friction coefficients. Several calculations have
been done in this direction.[11-16] The most important physical
effects to be taken into account in these approaches are the following : tunneling of the single particle potential barrier separating the two nuclei, influence of the temperature on the occupation probabilities of the nucleons and on single particle potential, Pauli blocking and window velocity with respect to each
nuclei (see ref.[17] for a review).

Before we close this section, it should be noted that there
can also exist a coupling between the macroscopic variables. This
allows an energy transfer between the different collective motions. This is the way for instance how surface vibrations or
giant resonances can be excited.[18] Since these collective motions
can be coupled to the intrinsic system, they can be damped. The
coupling between collective modes can play an important role in
the sense that energy can be transferred from the relative motion
to some other collective mode (surface vibration for instance)
which can be in turn strongly damped. In this case one may have a
strong dissipation in the relative motion because the second collective mode plays the role of a doorway state (or of a cataly-
ser).

4. ILLUSTRATION OF A TRANSPORT EQUATION

We shall now illustrate the physics which is contained in a
transport equation. For that we shall consider only one collective

degree Q (P will the conjugate momentum), and assume that the associated motion is harmonic. Consequently the collective Hamiltonian reads :

$$H_{coll} = \frac{P^2}{2B} + \frac{1}{2} C Q^2 \tag{16}$$

where B and C are respectively the inertia and the stiffness coefficient associated to the harmonic oscillator. The collective motion will be coupled to a heat bath at temperature T. Hofmann and Siemens[3] have shown that the dynamical evolution of the oscillator is governed by a transport equation which has the following form :

$$\frac{\partial f}{\partial t} = \{H_{coll}, f\} + \gamma \frac{\partial}{\partial P} \frac{P}{B} f + D \frac{\partial^2 f}{\partial P^2} \tag{17}$$

In this equation f is the distribution function in collective phase space. In the case of a classical oscillator f is the classical distribution, whereas for a quantum oscillator f has to be understood as the Wigner transform of the reduced density matrix.[19] γ and D are respectively the friction and the diffusion coefficients. The first term in the right hand side is a Poisson bracket :

$$\{H_{coll}, f\} = \frac{\partial H_{coll}}{\partial Q} \frac{\partial f}{\partial P} - \frac{\partial H_{coll}}{\partial P} \frac{\partial f}{\partial Q} \tag{18}$$

Using eq.(16) it can be rewritten as :

$$\{H_{coll}, f\} = - \frac{P}{B} \frac{\partial f}{\partial Q} + C Q \frac{\partial f}{\partial P} \tag{19}$$

In transport theory, friction and diffusion are intimately related by means of the so called fluctuation dissipation theorem which can be expressed here by means of the Einstein relation

$$D = \gamma T \tag{20}$$

where T is the temperature. The above relation holds only if we have a classical harmonic oscillator. In the case of a quantum one, the connection between D and γ reads[19] :

$$D = \gamma T^* \tag{21}$$

where

$$T^* = \frac{\hbar\Omega}{2} \text{Cotgh} \frac{\hbar\Omega}{2T} \tag{22}$$

It should be noted that in the case of high temperature (T >> $\hbar\Omega$) the generalized Einstein relation (21) goes to the classical one (eq.(20)).

It is interesting to write down the equations of motion for the mean macroscopic variables $<Q>$ and $<P>$. They can be obtained from the transport eq.(17) by multiplying it by Q or P and integrating the phase space :

$$\frac{d<Q>}{dt} = \frac{<P>}{B} \qquad (23)$$

$$\frac{d<P>}{dt} = C<Q> - \gamma \frac{<P>}{B} \qquad (24)$$

or

$$B \frac{d^2<Q>}{dt^2} + - \Omega^2<Q> - \gamma \frac{d}{dt} <Q> \qquad (25)$$

we recognize in eq.(25) the classical equation of motion of a harmonic oscillator subject to a friction force proportional to the velocity. Therefore the transport equation, which is derived on microscopic grounds, provides a justification of the friction force introduced in the classical models for heavy ion collisions. In addition it gives the fluctuations around the mean values. In the case of a classical oscillator these fluctuations are of statistical nature, whereas, in the case of a quantum harmonic oscillator, they have a quantum origin (zero point motion).

In the case of the harmonic oscillator, the solution of the transport equation is a Gaussian which is therefore completely determined by the first and the second moments of Q and P. We have seen that the first moments obey eqs.(23,24). As far as the second moments are concerned, they obey the following set of linear coupled differential equations :

$$\frac{d\Sigma_{PP}}{dt} = - 2 C \Sigma_{PQ} - \frac{2\gamma}{B} \Sigma_{PP} + \gamma T^* \qquad (26)$$

$$\frac{d\Sigma_{QQ}}{dt} = \frac{2}{B} \Sigma_{PQ} \qquad (27)$$

$$\frac{d\Sigma_{PQ}}{dt} = \frac{\Sigma_{PQ}}{B} - C \Sigma_{QQ} - \frac{\gamma}{B} \Sigma_{PQ} \qquad (28)$$

where

$$\Sigma_{QQ} = \frac{1}{2} \int dQ \ dP \ f(Q,P,t) \ (Q - <Q>)^2 \qquad (29)$$

$$\Sigma_{PQ} = \frac{1}{2} \int dQ \ dP \ f(Q,P,t) \ (Q - <Q>) \ (P - <P>) \qquad (30)$$

$$\Sigma_{PP} = \frac{1}{2} \int dQ \ dP \ f(Q,P,t) \ (P - <P>)^2 \qquad (31)$$

are the second moments. Note that in the case of a classical harmonic oscillator $T^* = T$.

We can check that as $t \to \infty$ we get the equilibrium distribution. Indeed, at equilibrium

$$0 = 0 \; , \; \Sigma_{PQ} = 0 \; , \; \Sigma_{PP} = \frac{BT^*}{2} \; \text{ and } \; \Sigma_{QQ} = \frac{T^*}{2C} \qquad (32)$$

as it should be.

It is interesting to illustrate the difference between statistical and quantum fluctuations. This is shown in Fig. 1. In both cases considered there, we have an oscillator with the same stiffness coefficient coupled to a heat bath at temperature T=2 Mev. The inertia is chosen in such a way that, on the left hand side, the phonon energy is equal to 8 MeV, whereas on the right hand side, it is equal to 0.5 MeV. It means that the levels are separated by 8 MeV on the left, and by 0.t MeV on the right. The oscillator which is on the left is called a quantum oscillator whereas the one on the right is a classical one. Since the levels of the quantum oscillator are spaced by 8 MeV, which is much larger than T = 2 MeV, the heat bath does not excite the oscillator which remains in its ground state. We observe the zero point motion and the equilibrium distribution is represented by the dashed line. On the contrary, in the case of the classical oscillator, $\hbar\Omega \ll T$, consequently it will be excited by the heat bath. The equilibrium distribution is the Boltzmann distribution represented by the dashed line.

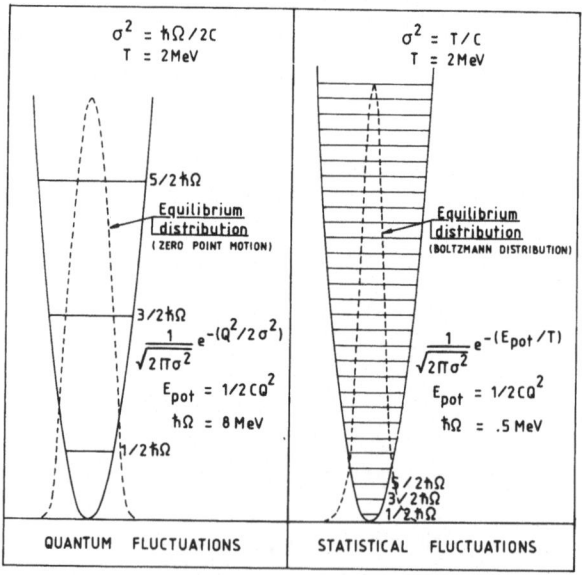

Fig. 1 - Schematic presentation of the difference between statistical and quantum fluctuations in the case of a harmonic oscillator (see text).

   The above considerations concerned the equilibrium stage
which is reached after a time larger than the relaxation time
associated to the harmonic oscillator. Now, we shall illustrate
the role of the different terms entering the transport equation.
This is schematically shown in Fig. 2. If we would only consider
the first term in the right hand side of eq.(17), we would
just have a Liouville equation in phase space. This would correspond to no friction and the trajectory followed by the system in
phase space is shown on top of the left hand side of Fig. 2 (the
units have been chosen in such a way that it is a circle). The
second term on the left hand side of eq.(17), proportional to γ,
leads to dissipation. It will be responsible for a drift of the
mean value towards equilibrium. The trajectory will have the form
schematically shown on top of the right side of Fig. 2 (in the
figure we show the trajectory for a quantum harmonic oscillator).
In this case equilibrium corresponds to a circle. If the oscillator becomes classical the radius of the circle will go to zero.
Each time we have dissipation, we know that we have also fluctuations. The effect of the fluctuations is given by the last term
in eq.(17). The physical effect of this term is to spread the
density distribution in phase space like a diffusion process. The
distribution will broaden when the time increases until it reaches
equilibrium. This is illustrated in the lower part of Fig. 2 where
the dashed area represents, say, some percentage of the density
distribution.

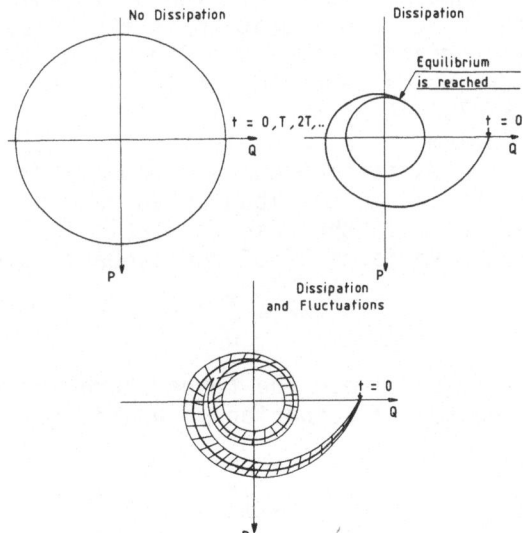

Fig. 2 - Schematic presentation of different type of trajectories
in phase space when the system is conservative (top left), where
there is friction (top right) and when there is friction and fluctuations (bottom). See text.

5. APPLICATIONS OF TRANSPORT THEORY

The method which has been described schematically above can of course be generalized to more complicated situations where several collective degrees are taken into account explicitly. A lot of calculations have been done and compared with experimental data.[1] They give an overall understanding of the deep inelastic process. We just would like to show two typical examples of such calculations.

The first one corresponds to the case where the collective motion behave classically. Then only statistical fluctuations can be observed. These models are able to calculate multidifferential cross section with respect to various macroscopic variables.[20] The higher is the order of the differential cross section, the more difficult it is to reproduce. Therefore we show in Fig. 3 the calculated Wilczynski plot corresponding to several Z values ($d^3\sigma/dEdZd\theta$) for the 280 MeV Ar + Ni system measured in ref. 22. The experimental data are displayed in Fig. 4. We see that the overall experimental picture is reproduced but still not the details.

The second example will concern charge equilibration which is known to be a fast collective mode exhibiting, in some cases, quantum features.[23] This is in particular the case for the 430 MeV ^{86}Kr + ^{92}Mo investigated in ref. [23]. The observable which is directly related to charge equilibration is the isobaric distribution (atomic number distribution for fixed value of the mass). The FWHM of this distribution turns out to be, at equilibrium, much too large to be explained by statistical fluctuations. Furthermore, when equilibrium is reached, it remains practically constant as a function of the excitation energy (temperature). Indeed, charge equilibration can be decribed by an harmonic oscillator coupled to a heat bath. In the case of statistical fluctuations we would expect for the variance σ^2 of the isobaric distribution :

$$\sigma^2 = \frac{T}{C} \tag{33}$$

where T anc C are respectively the temperature and the stiffness coefficient. For quantum fluctuation σ^2 would be larger and given by :

$$\sigma^2 = \frac{T^*}{C} \tag{34}$$

where T^* is defined by eq.(22). The experiment indicates that, for the Kr + Mo system, we are likely to observe quantum fluctuations. In Fig. 5 we show a comparison between the model based on a transport equation, and the data. We see that they agree quite well. In fact TDHF calculations have shown that the mode associated to

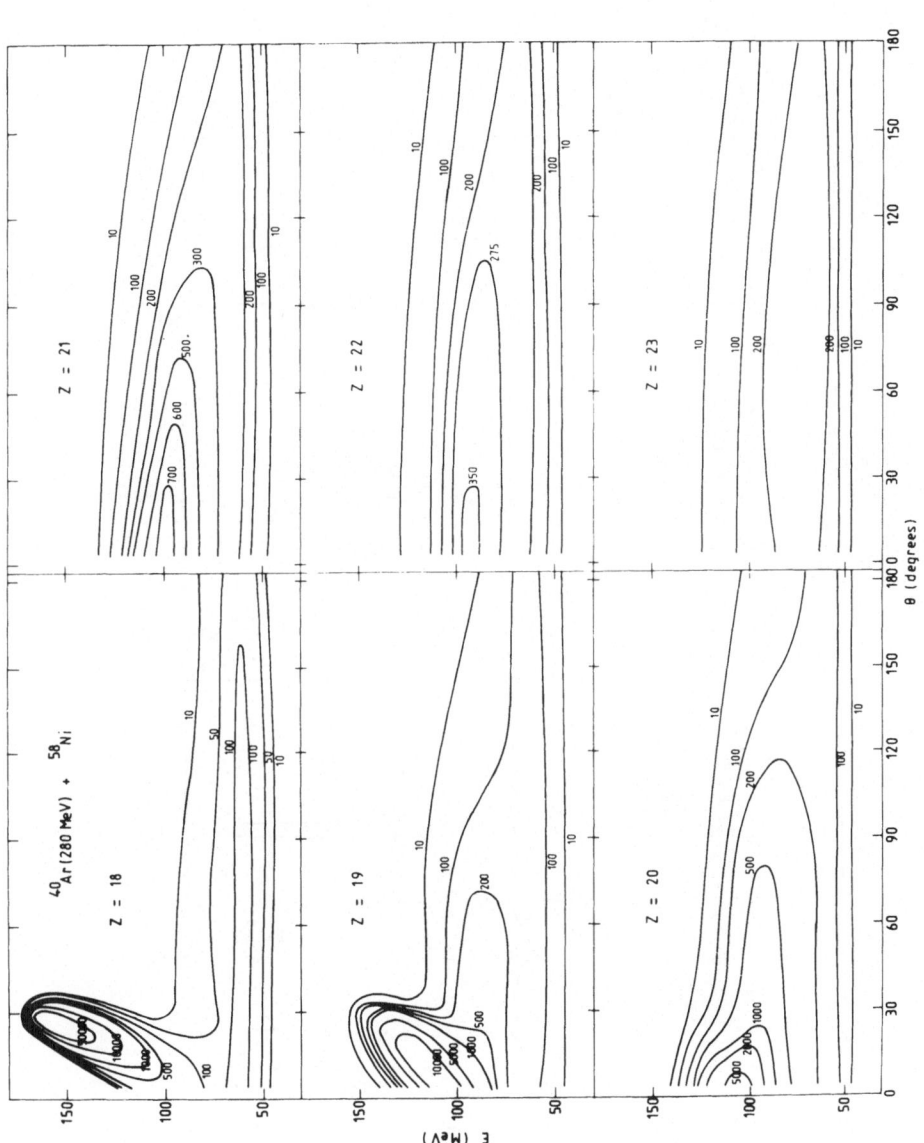

Fig. 3 – Calculated differential cross section $d^3\sigma/dEdZ\theta$ in $\mu b/MeV/rad/(charge\ unit)$ for the Ar + Ni system at 280 MeV. Calculation done in ref.[21].

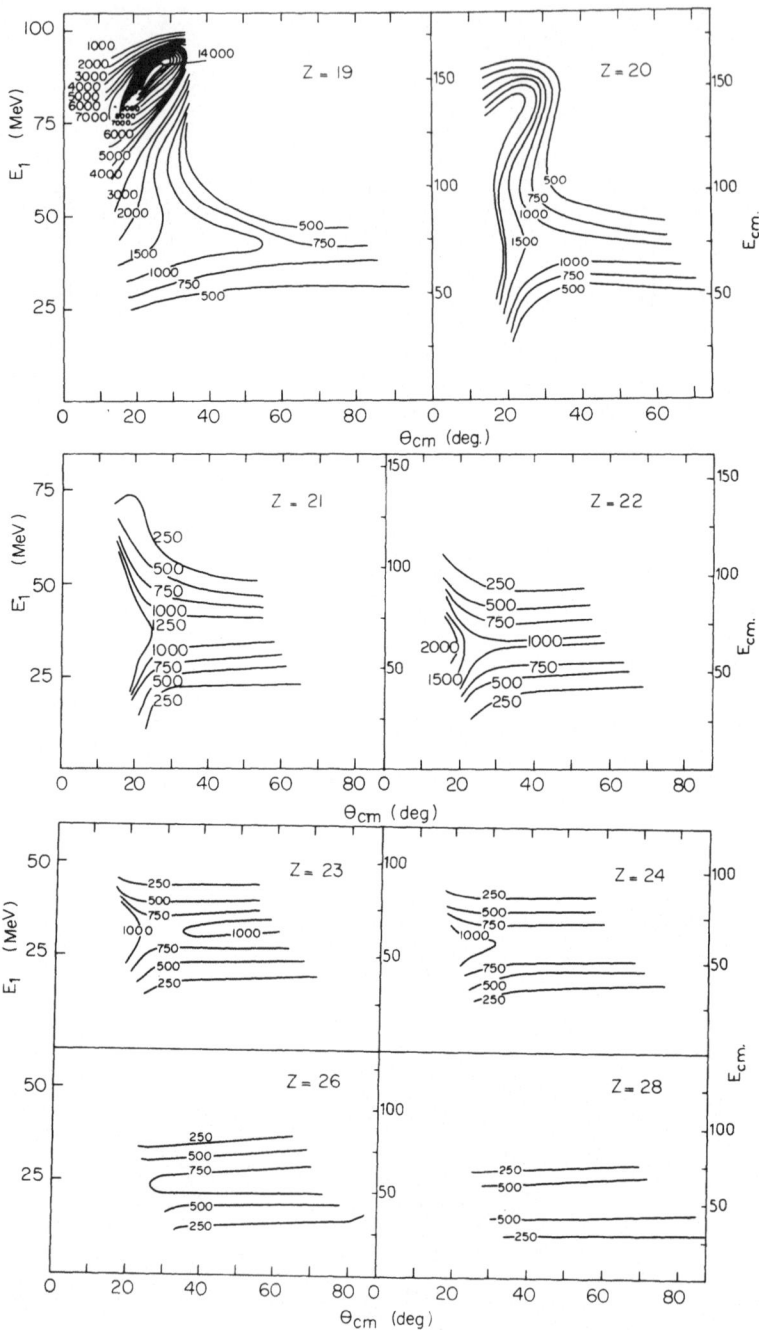

Fig. 4 - Experimental curves corresponding to Fig. 3 measured in ref.[22].

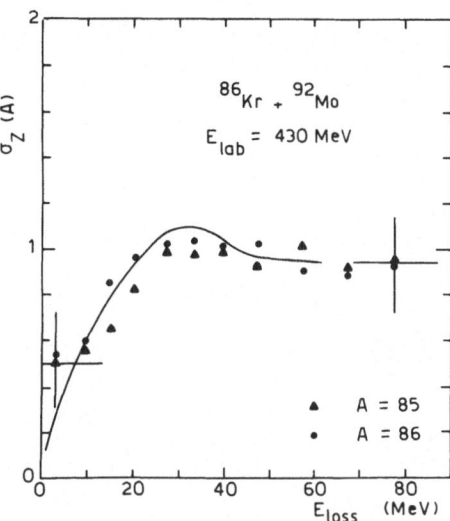

Fig. 5 - Calculation of the standard deviation of the isobaric distribution of A = 86 for the 430 MeV Kr + Mo system (full line) done in ref. 19 compared to the experimental data of ref.[23].

charge equilibration seems to be related to the longitudinal component of the giant dipole resonance of the composite system.[24]

These two short examples have shown us the power of transport theory for explaining dissipative phenomena in heavy ion physics. We are now at a stage where these calculations are refined in order to describe finer details of heavy ion collisions. A lot of work remains to be done in order to obtain extensively microscopic transport coefficients. On the experimental side, it is probably interesting to try to study the possible existence of non Markovian effects. For this, high resolution is needed and this is along the line of the new accelerators.

REFERENCES

1. For reviews see for instance M. Lefort and C. Ngô (1978), Ann. Phys., 3, 5 ; C. Ngô (1983), Note CEA-N-2354.
   Proc. of the International Conference on selected aspects of heavy ion physics, Saclay (1982), Nucl. Phys. A387.
   Proc. of the International Conference on theoretical approaches to heavy ion reaction mechanisms, Paris (1984), to appear in Nucl. Phys.
2. D.M.E. Gross and M. Kalinowski (1974)͑, Phys. Lett. 48B, 302.
3. M. Hofmann and P.J. Siemens (1976), Nucl. Phys. A257, 165 ; (1977) Nucl.Phys. A275, 464.

4. D. Agassi, C. Ko and H. Weidenmüller (1977), Ann. Phys. 107, 140.
5. F. Haake (1983), Springer tracts in modern physics : quantum statistics in optics and solid state physics, n°66, 98.
6. M. Hofmann (1977), Fizika 9, suppl. 4, 441.
7. P. Johansen, P.J. Siemens, A. Jensen and H. Hofmann (1977), Nucl. Phys. A288, 152.
   K. Sato, A. Iwamoto, K. Harada, S. Yamaji and S. Yoshida (1978), Z. Phys. A288, 383.
   S. Yamaji, A. Iwamoto, K. Havada and S. Yoshida (1981), Phys. Lett. 106B, 433 ;
   A.S. Jensen, J. Leffers, K. Reese, H. Hofmann and P.J. Siemens (1982) Phys. Lett. 117B, 5 ;
   A.S. Jenson, L. Leffers, K. Reese and P.J. Siemens (1982), Phys. Lett. 117B, 2157.
8. B. Tamain, R. Chechik, H. Fuchs, F. Hanappe, M. Morjean, C. Ngô, J. Peter, M. Dakowski, B. Lucas, C. Mazur, M. Ribrag and C. Signarbieux (1979), Nucl. Phys. A330, 253.
9. D.H.E. Gross (1975), Nucl. Phys. A240, 472.
10. J. Blocki, Y. Boneh, J.R. Nix, J. Randrup, M. Robel, A.J. Sierk and W.J. Swiatecki (1968), Ann. Phys. 113, 330.
11. J. Randrup (1978), Nucl. Phys. A307, 319 and (1978), Ann. Phys. 112, 356.
12. C.M. Ko, G.F. Bertsch and D. Cha (1978), Phjys. Lett. 77B, 174.
13. G.J. Ball and S.E. Koonin (1982), Nucl. Phys. A388, 125.
14. F. Stancu and D. Brink (1982), Phys. Rev. 25C, 2450.
15. M. Pi, M. Barranco, X. Viñas, C. Ngô and E. Tomasi (1983), Nucl. Phys. A406, 325.
16. M. Pi, M. Barranco, X. Vinas, G. La Rana, S. Leray, C. Ngô and E. Tomasi Nucl. Phys. in press.
17. M. Barranco, M. Pi, X. Vinas, G. La Rana, S. Leray, C. Ngô and E. Tomasi (1984), International Conference on the theoretical approaches to heavy ion reaction mechanismes, Paris, to appear in Nucl. Phys.
18. R.A. Broglia, C. Dasso and A. Winther (1976), Phys. Lett. 61B, 113.
19. H. Hofmann, C. Grégoire, R. Lucas and C. Ngô (1979), Z. Phys. A293, 229.
20. C. Ngô and H. Hofmann (1977), Z. Phys. A282, 83 and Phys. Lett. 65B, 97.
21. C. Grégoire, C. Ngô and B. Remaujd (1982), Nucl. Phys. A383, 392.
22. J. Galin, B. Gatty; D. Guerreau, M. Lefort, X. Tarrago, R. Babinet, B. Cauvin, J. Girard and H. Nifenecker (1976), Z. Phys. A278, 347.
23. M. Berlanger, A. Gobbi, F. Hanappe, U. Lynen, C. Ngô, A. Olmi, H. Sann, H. Stelzer, H. Richel and M.F. Rivet (1979), Z. Phys. A291, 133.
24. P. Bonche and C. Ngô (1984), Phys. Lett. 105B, 17.

Lecture II : FUSION AND FAST FISSION

I. FUSION

a) General considerations

Fusion is the most dissipative phenomenon observed in heavy ion reactions. Indeed all the nucleons are involved, all the kinetic energy in the relative motion is transformed into intrinsic excitation energy of the compound system and all the initial orbital angular momentum is transformed into spin of the fused nucleus. It is of fundamental importance to know under which conditions two heavy ions fuse together and to calculate the probability of such a process. It is also interesting to know what happens to the fused system.

Experimentally it turns out that two heavy nuclei with a product of the atomic numbers $Z_1 Z_2 > 2500$-3000 cannot fuse.[1] For this reason, even if the superheavy element would exist, it would not be possible to synthesize it by the fusion of two heavy nuclei. The reason why there is no fusion, in the case of very heavy systems, comes from the disappearance of the pocket in the total interaction potential.[2] This is due to the Coulomb force which becomes so attractive that it cannot be counteracted by the nuclear force at any separation distance. It is possible to calculate, in a static way, the condition for which the pocket disappears in the case of a head-on collision.[3] One finds that when :

$$\frac{Z_1 Z_2}{C_1 C_2 (C_1 + C_2)} > 8.7 \tag{1}$$

the pocket disappears and fusion is no longer possible. In eq.(1) C_1 and C_2 are the central radii of the projectile and of the target :

$$C_i = R_i - 1/R_i \tag{2}$$

and

$$R_i = 1.16 \, A_i^{1/3} \tag{3}$$

However dynamical effects decrease a bit this limit of 8.7 and we shall see the reason below.

For a given system, where fusion is possible, one observes that, if the bombarding energy is not too far above the Coulomb barrier, the fusion cross section goes almost linearly as a function of the inverse of the center of mass bombarding energy. However at higher energies, when larger values of the orbital angular

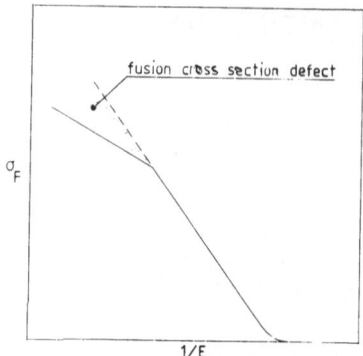

Fig. 1 - Schematic presentation of the fusion excitation function as a function of 1/E, the inverse of the center of mass bombarding energy. At high bombarding energies a fusion cross section defect is observed compared to the behaviour followed just above the fusion threshold.

momentum are involved, the measured fusion cross section becomes smaller than what can be expected by an extrapolation the preceding straight line (see Fig. 1). Such a fusion cross section defect is also observed for systems in the region where fusion just disappears[4] (heavy systems). This can be understood as due to an increase of the fusion threshold (see Fig. 2).

The preceding experimental facts have to be understood in a single picture. In this lecture, I would like to present a simple dynamical model which is able to do that.[5] We will see that it is only based on very simple physical arguments.

Up to now there is no simple picture which is able to describe the fusion cross sections for all the heavy ion combinations that we can imagine and with bombarding energies in a range going from the Coulomb barrier to about ~ 10-15 MeV/u. The reason

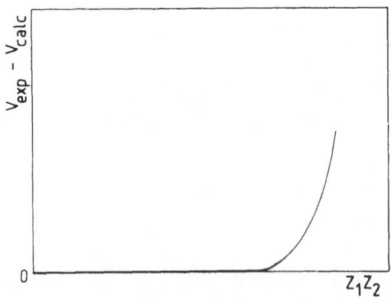

Fig. 2 - Schematic plot of the difference between the experimental and the theoretical fusion threshold as a function of the product Z_1Z_2 of the atomic numbers of the two ions.

~ 10-15 MeV/u. The reason for that can be found in the fact that the dynamics plays a very important role in heavy ion collisions. We have seen, in the preceding lecture, that dissipation occurs as soon as the two heavy nuclei start to strongly interact.   Therefore friction will also play an important role in the fusion process. In a pure static picture, fusion can be obtained, for a given value of the initial orbital angular moment ℓ, if the bombarding energy is larger than the corresponding static fusion barrier associated to this particular collision (see Fig. 3). However friction may act before the system reaches this barrier and some energy loss in the relative motion will result. This is illustrated in Fig. 4 for a head-on and a non central collision. We see that for the system to reach the barrier, it will be necessary to provide it with a certain amount of extra energy above the static fusion barrier. This supplement of kinetic energy is necessary to compensate the friction forces which are acting before the system reaches the barrier. Let us write down the condition for fusion. For that we need to introduce the critical angular momentum, ℓ_{CR}, which is the largest ℓ value leading to fusion. The experimental fusion cross section, σ_F, is usually expressed in terms of ℓ_{cr} by means of the following relation :

$$\sigma_F = \frac{\pi}{k^2} (\ell_{CR} + 1)^2 \tag{4}$$

where k is the wave number. Eq.(4) assumes that they are the lowest ℓ values which contribute to fusion, up to the largest one ℓ_{CR}, and that the sharp cut off approximation is valid.

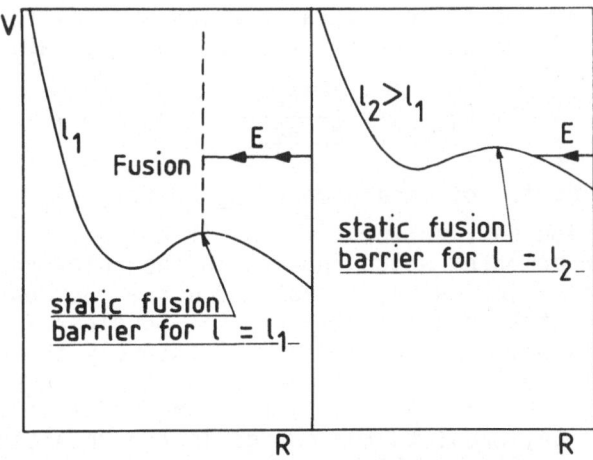

Fig. 3 - In a static picture we can have fusion if the bombarding energy is larger than the static fusion barrier. On the left this is possible, but not on the right.

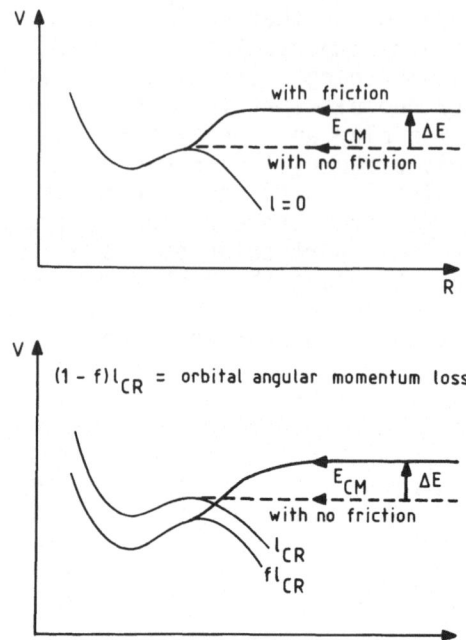

Fig. 4 - Schematic illustration of the fact that some extra kinetic energy is needed to overcome the static fusion barrier : on top is the case of a head-on collision. In the bottom is the case when the orbital angular momentum is equal to ℓ_{CR}. There the total interaction potential including centrifugal energy changes due to angular momentum loss.

Since ℓ_{CR} is the largest ℓ value which is able to pass the fusion barrier, it should satisfy the following equation:

$$E = V(R_{f\ell_{CR}}) + \frac{f^2 \ell_{CR} \hbar^2}{2\mu \, R^2_{f\ell_{CR}}} + \Delta E_R + \Delta E_t \qquad (5)$$

where E is the center of mass bombarding energy, $R_{f\ell_{CR}}$ the position of the fusion barrier for $\ell = f\ell_{CR}$ (f being the fraction of orbital angular momentum which remains in the relative motion after tangential friction has acted) and μ the reduced mass. V(R) represents the total interaction potential (nuclear + Coulomb) for a head-on collision. ΔE_R and ΔE_t are respectively the energy losses in the radial and in the tangential motions. Since in dynamical calculations the energy loss in the tangential motion is, to a good approximation, equal to the change in the rotational energy,[6] we can rewrite eq.(5) as follows :

$$E = V(R_{f\ell_{CR}}) + \frac{\ell_{CR}^2 \hbar^2}{2\mu \; R_{f\ell_{CR}}^2} + \Delta E \qquad (6)$$

where ΔE will be the dynamical energy surplus. It is the extra energy which one has to provide the system with, in order to compensate for friction forces. Therefore for a given ℓ value, we can define a dynamical fusion barrier which equals the static one plus the dynamical energy surplus :

Dynamical fusion barrier = static fusion barrier + dynamical energy surplus

---

Since it is possible to reproduce the fusion cross section for many systems at not too high bombarding energy using static barriers,[3] it means that the dynamical energy surplus is zero for them. Let us now try to estimate ΔE from experiment.

For that we shall use equation (6). For a given system we shall take the bombarding energy E. Then $R_{f\ell_{CR}}$ and $V(R_{f\ell_{CR}})$ will be calculated using the energy density potential[3,7] which has proved to be very successful in reproducing the fusion thresholds for a large body of systems. From eq.(6) we can deduce a value for the dynamical energy surplus, and try to plot this quantity as a function of a variable which should be strongly correlated to it. A natural one we may think of is the following[8] :

$$X_{eff} = \frac{1}{4\pi\gamma} \left(\frac{Z_1 Z_2 e^2}{C_1 C_2 (C_1 + C_2)} + \frac{f^2 \ell^2 \hbar^2}{m} \frac{A_1 + A_2}{A_1 A_2} \frac{1}{C_1 C_2 (C_1 + C_2)^2} \right) (7)$$

where γ is the surface tension coefficient of nuclear matter ($\gamma \approx$ 1 MeV/fm²) and m the nucleon mass. X_{eff} represents the ratio between the Coulomb plus centrifugal forces, over the nuclear force, at distance $C_1 + C_2$. We see that in the definition of X_{eff} it enters the factor f, which is the proportion of orbital angular momentum remaining in the relative motion after tangential friction has acted. The choice of a value for f is not clear[3] : indeed should we take the sticking or the rolling limit?, or something else? This is illustrated in Fig. 5 where ΔE has been plotted as a function of X_{eff} for the heavy systems investigated at GSI [ref.[4]] (circles) and for the Ar + Ho system measured in ref.[9] (crosses). In the former case f was taken to be equal to 5/7 (rolling) whereas in the later case f was taken for the sticking limit. It is only with these choices that ΔE could exhibit a similar mean behaviour and this tells us that a static approach of the problem will be difficult since, for a given system, we do not know how to choose this factor f. Therefore we should now really consider the dynamics of the problem.

b) Dynamical approach to fusion

In order to check the above qualitative ideas about fusion, a simple dynamical model has been developed in ref.[5] to describe the fusion process. It describes the collision of the two heavy ions by means of two macroscopic variables : the distance R separating the center of mass of the two nuclei, and the polar angle θ. The dynamical evolution of the system will be treated in a similar way as the one described in the first section of lecture I. The nuclear potential between the two heavy ions, which was deduced using the energy density formalism, is taken from ref.[3]. Friction forces proportional to the velocities are introduced in the radial

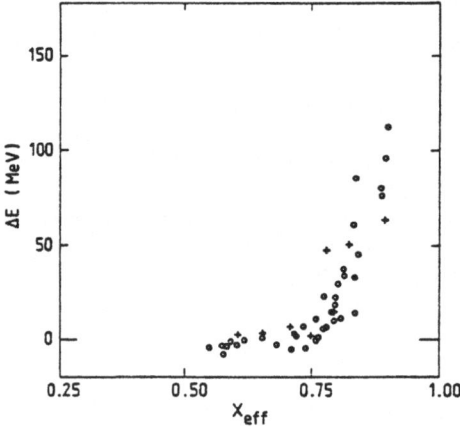

Fig. 5 - The dynamical energy surplus is plotted as a function of X_{eff} defined in eq.(7). The circles are associated to the systems investigated in ref.[4] with f = 5/7 (Rolling). The crosses correspond to the Ar + Ho system of ref.[9] with f given by the sticking condition. From ref.[3].

and tangential motions. The equations of motion which have been used are the following :

$$\mu \, \frac{d^2 R}{dt^2} = - \, \frac{\partial V_\ell(R)}{\partial R} - C_R \, g(R) \, \dot{R} \qquad (9)$$

$$\frac{d\ell}{dt} = - \, \frac{C_t}{\mu} \, g(R) \, (\ell - \ell_{st}) \qquad (10)$$

C_R and C_t are respectively the radial and the tangential friction coefficients. According to the-one body picture [10] they are such that $C_R = 2 \, C_t$. In eq.(9) a limiting value for the orbital angular momentum has been introduced : it is the sticking limit which should not be overcomed if we consider the two nuclei as two solid objects. $V_\ell(R)$ is the total interaction potential (nuclear + Coulomb + centrifugal energy) and g(R) the form factor which was chosen as follows :

$$g(R) = \frac{1}{1 + \exp(\frac{s-0.75}{0.2})} \qquad (11)$$

where

$$s = R - C_1 - C_2 \qquad (12)$$

represents the separation distance between the surfaces of the two nuclei at half density. The fact that the form factor depends explicitly upon s rather than upon R is from our belief that the friction between the two heavy-ions is dominatedby a surface-surface effect. It turns out that if we take C_R = 31 000 MeV fm^{-2} $(10^{-23}s)^2$ it is possible, with this value kept fixed, to reproduce the fusion cross section for a large number of systems. We refer the readers to ref.[5] where a lot of comparisons between experimental and theoretical fusion excitation functions are shown. In Fig. 6 we just illustrate the model with a few systems going from light to heavy ones. We observe a rather good agreement between experiment and theory which encourages us to believe that the dynamics is of great importance in the fusion of two heavy ions.

From the dynamical calculation it is interesting to see if the dynamical energy surplus has a strong correlation with some quantity which is easily accessible. This turns out to be the case and is shown in Fig. 7 where ΔE is plotted as a function of, $s_{f\ell_{CR}}$, the position of the static fusion barrier corresponding to $\ell = f~\ell_{CR}$. We observe that all the points are focussed around a mean behaviour which can be parametrized as follows :

$$\Delta E = 1100 \left(1.57 - s_{f\ell_{CR}}\right)^3 \qquad \text{when} \qquad s < 1.57 \text{ fm}$$

and   (13)

$$\Delta E = 0 \qquad\qquad \text{when} \qquad s > 1.57 \text{ fm}$$

We observe in Fig. 7 that ΔE increases a lot when the interdistance between the two surfaces becomes smaller than 1.4 fm. It is necessary for the system to reach a distance smaller than $\sim$ 1 fm, ΔE will become so large that fusion will no longer be possible in practice. This shows the existence of a saturation distance beyond which fusion will not be possible.

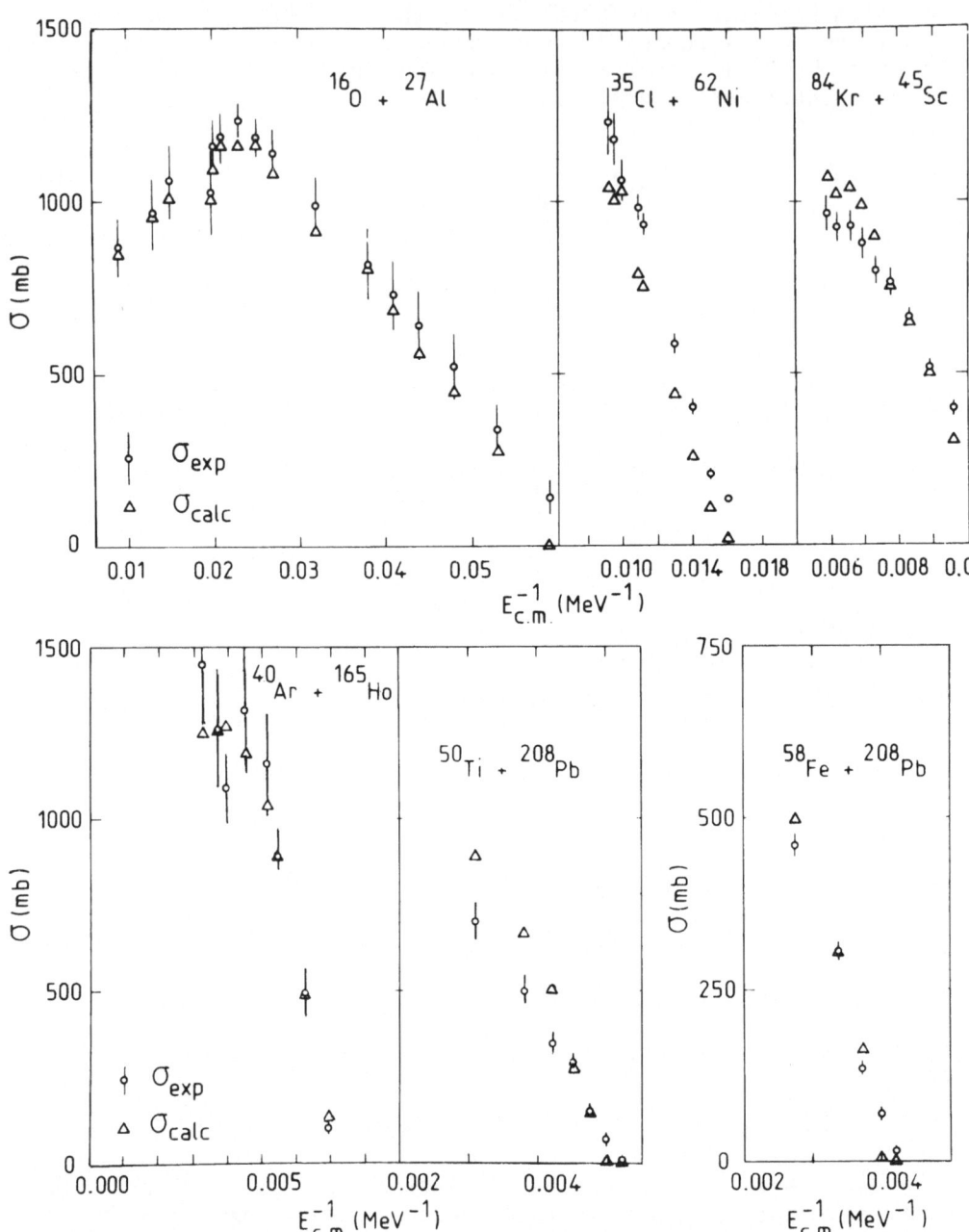

Fig. 6 - Comparaison of the experimental cross sections with those computed using the dynamical model of ref.[5]. References corresponding to experimental points can be found also in this reference. This figure has been extracted from ref.[11].

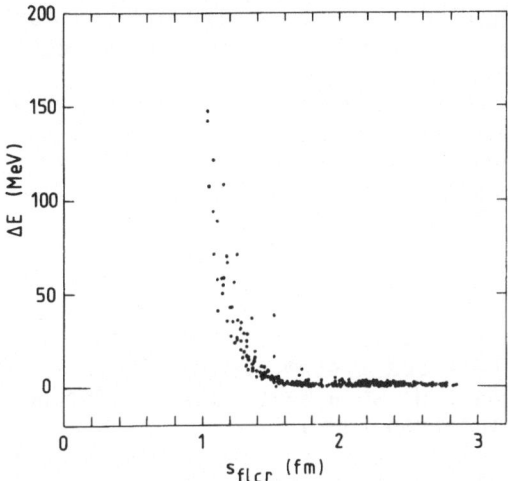

Fig. 7 - Dynamical energy surplus needed to pass the fusion barrier as a function of $s_{f\ell_{CR}}$ the position of the fusion barrier for $\ell = f\ell_{CR}$. From ref.[5].

Let us now summarize the results of the model in simple physical terms : from e. (6) we see that the fusion cross section is given by the following expression :

$$\sigma_F = \pi \ R^2_{f\ell_{CR}} \ (1 \ - \ \frac{V(R_{f\ell_{CR}}) + \Delta E}{E})$$ (14)

For a given system where fusion is possible, when we increase the bombarding energy, we increase the critical orbital angular momenta but at the same time the position of the static barrier decreases. At the beginning (in the region just above the fusion threshold) $s_{f\ell_{CR}}$ is in general large enough for ΔE to be zero. But as ℓ_{CR} increases, $s_{f\ell_{CR}}$ is decreasing and can reach the region where $\Delta E = 0$. Then the dynamical fusion barrier becomes different from the static one and a fusion cross section defect $\approx R^2_{f\ell_{CR}} \ \Delta E/E$ is observed. At very high bombarding energies, either ΔE becomes too large, or the pocket in the total interaction potential has disappeared due to the centrifugal energy, and the critical angular momentum saturates.

For a given value of the initial orbital angular momentum, say $\ell=0$ for instance, the position of the static fusion barrier will decrease when the size of the nuclei increases due to the Coulomb field. In the region where fusion first disappears (systems investigated in ref.[4]) $s_{f\ell_{CR}}$ will reach the region where

ΔE ≠ 0 and the dynamical fusion barrier will be larger than the static one lead in this way to a fusion cross section defect.

Therefore dynamical effects allow to understand, in a single picture, the fusion process. Furthermore the success of this simple model in reproducing the experimental data indicates that fusion is mainly determined in the entrance channel, before the minimum distance of approach is reached.

2. FAST FISSION

We have seen how fusion occurs and we shall now investigate what happens to the fused system. In particular we shall try to answer the question : do we always form a compound nucleus when two heavy nuclei fuse together?

a) Compound nucleus formation and fusion

A compound nucleus is an entity which has completely forgotten about its formation except for some macroscopic parameters which should satisfy conservation laws. Such a nucleus is characterized by an excitation energy and an angular momentum. It will be unstable with respect to particle evaporation (leading to residual nuclei) but it may also undergo fission. The fission probability depends of course on the height of fission barrier which in turns, decreases strongly with increasing angular momentum.[11] For a certain value of ℓ, which we shall denote by ℓ_{B_f} the fission barrier even vanishes. Since some amount of time is needed to form a compound nucleus in the real sense, it is reasonable to think that it is not possible to form it if it has no fission barrier $(\ell > \ell_{B_f})$. In the case where fusion would be identical to compound nucleus formation ℓ_{CR} should always be smaller than ℓ_{B_f}. However a compilation of the existing data shows that this is not true and several measurements indicate that ℓ_{CR} can considerably exceed ℓ_{B_f}[ref.[12]]. A possibility to explain the data would be to say that σ_F does not only contain complete fusion but that there is a contribution of incomplete fusion. This last mechanism corresponds to the case where fast particles are emitted before the two remaining fragments fuse together. However, measurements of light particles associated to this process show a too small multiplicity[13] and cannot account for the difference between ℓ_{CR} and ℓ_{B_f}. Therefore we are led to conclude that fusion cannot be identified with compound nucleus formation and a new question qppears : what happens when $\ell_{B_f} \leqslant \ell \leqslant \ell_{CR}$ for which there is fusion but not compound nucleus formation?

A series of experiments has given some insight into this problem. In ref.[9] the evolution of the full width half maximum (FWHM) of the fission-like products has been investigated as a function of the excitation energy of the "compound nucleus", for the Ar + Ho system. It was observed that the FWHM increases with the excitation energy E^* (see Fig. 8). Because the temperature T also increases with E^*, we expect the mass distribution to broaden due to statistical fluctuations. However, an estimation of this

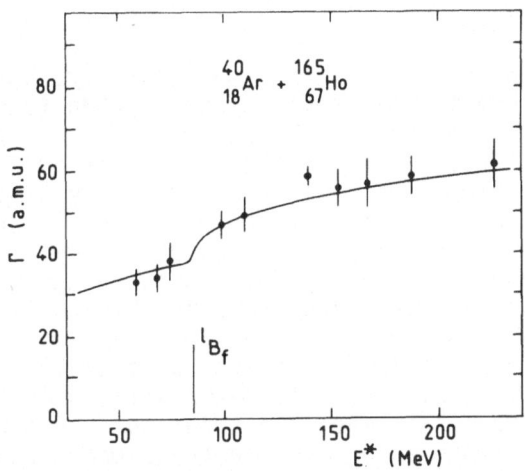

Fig. 8 - **Full width half maximum, Γ, of the fission like mass distribution, as a function of the excitation energy of the fused system, for Ar + Ho. The dots are the experimental points in ref.[9]. The full curve is the result of the calculation of ref.[12].**

effect gives a too small increase compared to what is observed experimentally. This is likely due to a change of the stiffness coefficient of the potential energy surface along the mass asymmetry coordinate which decreases with increasing ℓ value. If we notice, in Fig. 8, that the FWHM changes a lot in the region which is close to ~ 80 MeV, we are tempted to say that a mechanism different from ordinary fission contributes at high bombarding energies. It is precisely in this region that ℓ values larger than ℓ_{B_f} start to contribute to the fusion process. Therefore one might suggest the existence of a mechanism which is different from fission following compound nucleus formation and which occurs when $\ell_{B_f} \leqslant \ell \leqslant \ell_{CR}$. This preliminary conclusion is supported by investigations on heavier systems (Cl + Au in ref.[13] and Cl + U in ref.[14]). Indeed, as the system becomes more massive, ℓ_{B_f} decreases and for the Cl + U system, for example most of the ℓ values are greater than ℓ_{B_f}. In this case the FWHM is almost constant over investigated bombarding energy range (240-350 MeV) whereas it varies in the case of the Cl + Au system (204-31 MeV) but to a smaller extend than for the Ar + Ho combination.

The fact that there exists a new mechanism when $\ell_{B_f} < \ell < \ell_{CR}$ which we will call later on fast fission is, from the experimental point of view, just a guess. We shall see below that there exists theoretical calculations which predict this mechanism in a natural way.

b) Fast fission

In heavy ion collisions, the two nuclei are assumed to be spherical, or close to this configuration, when they are far apart from each other. When they reach the interaction region various shape degrees of freedom are excited. For instance a neck appears between the two heavy ions creating in this way a single composite system with two centers. If we want follow the future evolution of the fused system to be followed, we need to have a good description of these changes of shape. In fact these excitations will transform a potential landscape where the two nuclei are spherical (sudden potential) in one where some the shape degrees of freedom have relaxed to equilibrium (adiabatic potential). Instead of trying to describe explicitly the deformation degrees of freedom, which is an enormous task when one tries to describe also the fluctuations which are associated to these macroscopic variables, it is tempting to simulate this transition between the sudden and the adiabatic potential, in a phenomenological way. This has been done in ref.[15] where a dynamical transition between a sudden interaction potential in the entrance channel, and an adiabatic one in the exit channel has been done. The degree of completeness of the transition depends, of course, upon the overlap between the two ions.

The physics which is behind this simulation is to see whether a transition between the fusion valley (that the system follows in the entrance channel) and the fission valley, is possible under certain circumstances. This notion of two valleys has been pointed out by Swiatecki[16] and confirmed recently by microscopic calculations.[17]

In the model of ref.[15] the collision between the two nuclei is described by means of four macroscopic variables ; the distance separating the center of mass of the two nuclei, the polar angle, the mass asymmetry of the system and the neutron excess of one of the fragments. The deformations are simulated as it is indicated above. The dynamical evolution of the collision is followed by means of a transport equation which was derived by Hofmann and Siemens and which we have discussed in the first lecture.

When some conditions are fulfilled, the model reveals the existence of a mechanism which is intermediate between deep ine-

lastic reactions and compound nucleus formation. This is illustrated in Fig. 9 where typical mean trajectories are shown as a function of the mean mass asymmetry, and of the mean radial distance,
for the 340 MeV Ar + Ho system.

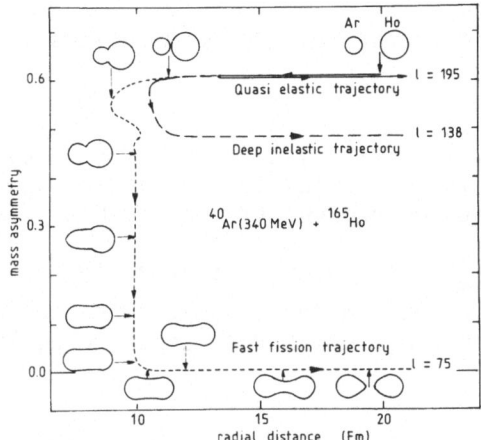

Fig. 9 - Few mean trajectories for various initial values of the
orbital angular momentum, ℓ, plotted in the plane radial distance-
mass asymmetry. Three kinds of mechanisms are illustrated in this
plot : 1) quasi-elastic process for ℓ=195, 2) deep inelastic collision for ℓ=138 and 3) fast fission phenomenon for ℓ=75. For
$\ell < \ell_{B_f}$ = 72, a compound nucleus is formed. This figure has been
extracted from ref.[15].

- ℓ=195 is a quasi-elastic trajectory with little mass and energy
exchanged between the two nuclei.

- ℓ=138 is typical of a deep inelastic collision : some mass
transfer occurs between the two ions and a large damping of the
initial kinetic energy is observed.

- ℓ=75 shows a new kind of phenomenon. The system is trapped in
the pocket of the entrance potential. Mass asymmetry relaxes to
equilibrium and, at the same time, the sudden potential switches
to the adiabatic one. However with this value of the angular momentum the compound nucleus has no fission barrier. Consequently
there exists no pocket in the adiabatic potential to keep the
system for a while : it will decay in almost two equal fragments.
This mechanism has been called fast fission since it proceeds
faster than ordinary fission where the stage of forming a compound
nucleus is needed. In contrast to ordinary fission which corresponds to a decay of a one-center system, fast fission results from
a decay of a two-center composite nucleus (see Fig. 10). The interaction time turns out to be of the order of $\sim 10^{-20}$s which is
intermediate between a deep inelastic collision and a compound

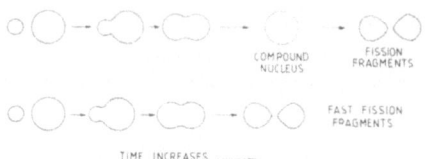

Fig. 10 - Schematic picture of compound nucleus fission and fast fission.

nucleus formation. The fast fission mechanism provides a way to go directly from the fusion valley to the fission valley without passing by the compound nucleus stage (see Fig. 11).

- When $\ell < 72 = \ell_{B_f}$, the system is trapped in the entrance potential but remains trapped in the adiabatic one because the fission barrier is non zero : we form a real compound nucleus.

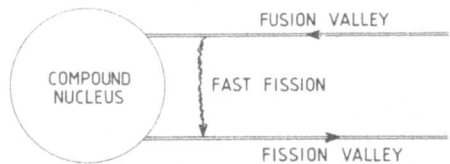

Fig. 11 - Schematic representation of fast fission as a mechanism which allows to go directly from the fusion to the fission valley.

To summarize, for a system like Ar + Ho, we can observe fast fission only if $\ell_{B_f} < \ell < \ell_{CR}$.

When the fissility parameter Z^2/A of the compound system increases, the saddle configuration becomes more and more compact. For big nuclei it can become less elongated than the pocket configuration. For a symmetric system this occurs when :

$$\frac{Z^2}{A} > 38.5 \qquad (14)$$

In this case, even if $\ell < \ell_{B_f}$, the system which is trapped in the pocket of the entrance potential cannot remain trapped in the adiabatic one, although there exists a fission barrier, because it is located already outside the saddle configuration. We have again fast fission but this special case has been called quasi-fission by Swiatecki[8] who was the first to point out such a possibility.

The macroscopic model described above predicts that fast fission has properties which are very similar to those of fission following compound nucleus formation. This makes difficult to get an unambiguous experimental evidence of this mechanism. At this

stage the model explains, in a simple way, the data which were not
understood before but that is all. The most advanced experimental
proof that might be found is the measurements of Ho et al. who
found evidence of light particles evaporated by a two-center system similar to the one leading to fast fission.

It is interesting to have a look at the fusion excitation
function of the Ar + Ho system (Fig. 12). The fusion cross section
is the sum of the compound nucleus and of the fast fission cross
sections. We observe that just above the fusion threshold we have,
for this particular system, only compound nucleus formation. Fast
fission starts to contribute only when ℓ values larger than ℓ_{B_f}
fuse. Since a transport equation is used to describe the dynamics,
one can calculate the statistical fluctuations in the mass asymmetry coordinate, construct the fast fission mass distribution at
a given bombarding and add to it the one corresponding to fission
following compound nucleus formation (taken from ref.[19]). In
Fig. 8 the calculated FWHM of the fission-like mass distribution
(full line) is compared with the data and the agreement appears to
be rather good.

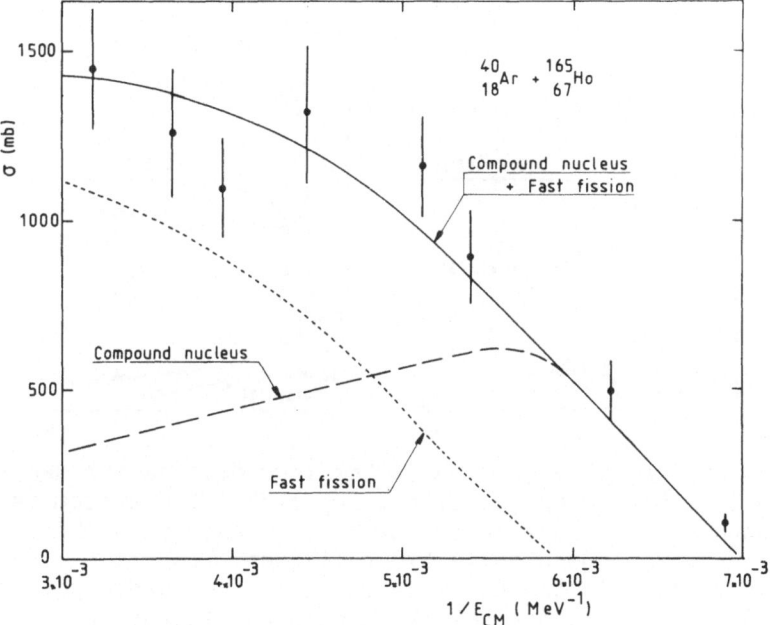

Fig. 12 - Experimental fusion cross section (dots) from ref.[9]
plotted as a function of $1/E_{CM}$, the inverse of the center of mass
bombarding energy. It is compared with the calculated fusion cross
section of ref.[15] (full curve). The fusion cross section is the
sum of the compound nucleus and of the fast fission cross sections. Their corresponding excitation functions are also shown in
the figure. This figure is extracted from ref.[15].

Four classes of dissipative heavy ion collision can be predicted by the macroscopic model of ref.[15]. This is schematically
illustrated in Fig. 13 where the sudden and the adiabatic potentials are represented as a function the interdistance R separating
the two nuclei. This one dimensional plot is just to get a feeling
of the collision which in fact occurs in multidimensional space.

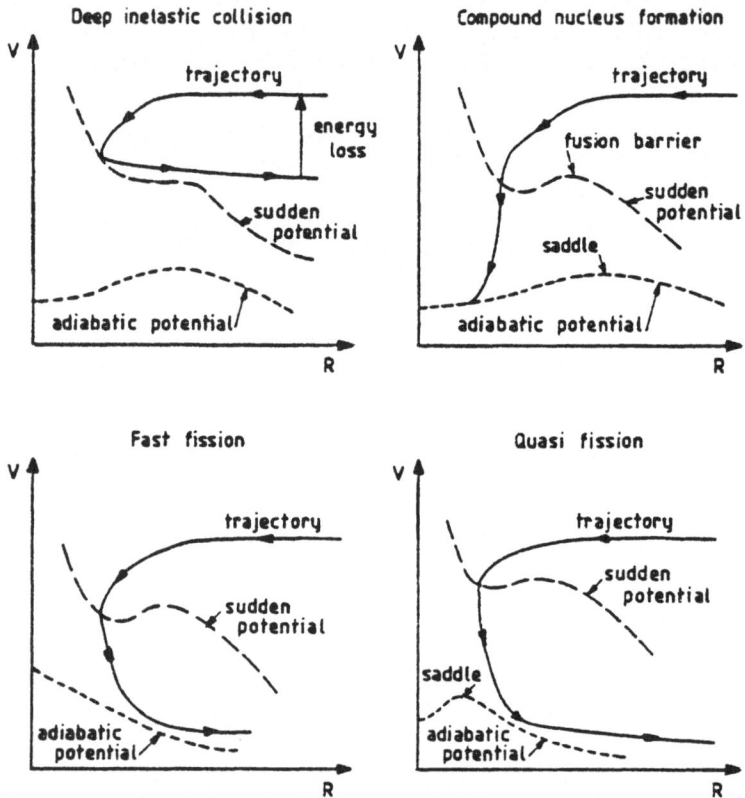

**Fig. 13 - Typical illustration of the four dissipative mechanisms
occuring in a heavy ion reaction : Top left :** the system is
not trapped but it looses a lot of kinetic energy in the relative
motion : we have a deep inelastic collision. **Top right :** the
system is trapped in the entrance channel. The sudden potential
goes to the adiabatic one but the saddle configuration is elongated enough to keep the system trapped : we have compound nucleus
formation. **Bottom left :** the system is trapped but the fission
barrier of the compound nucleus has vanished due to angular momentum. Therefore it desintegrates in two almost equal fragments
because mass asymmetry had time to reach equilibrium : we have
fast fission. **Bottom right :** the compound nucleus has a fission barrier but the saddle configuration is too compact to keep
the trapped system : we have also fast fission or quasi-fission.

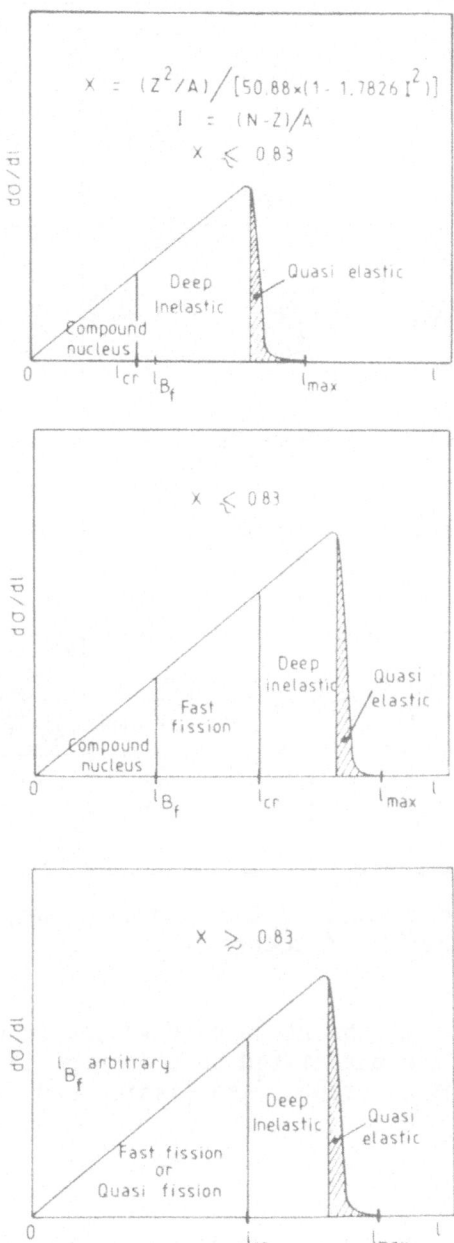

Fig. 14 - Schematic representation of the different ranges of ℓ values associated to the four dissipative mechanisms which can be observed in heavy ion reactions.

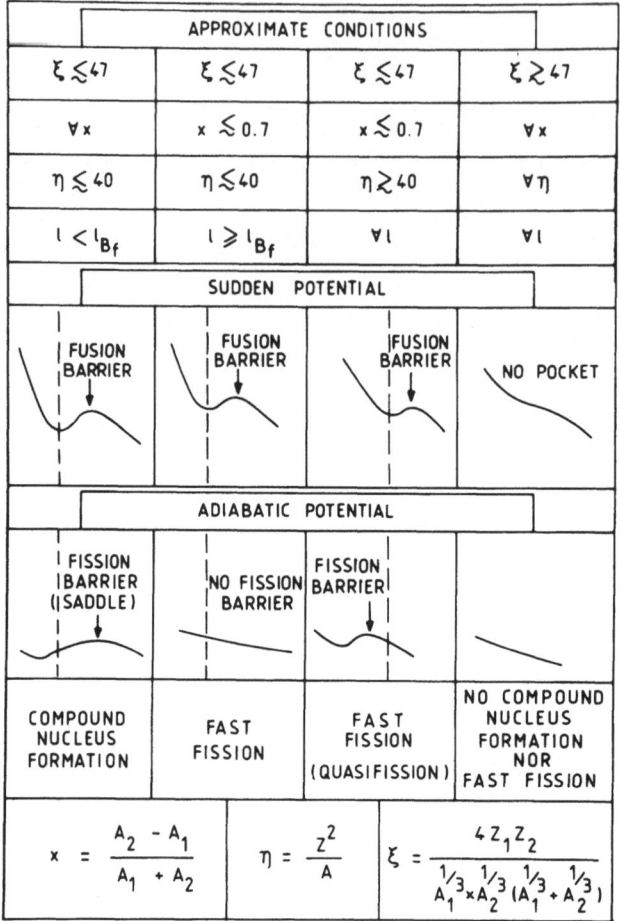

Fig. 15 - Schematic summary of the different mechanisms following fusion and their domains of occurence.

In Fig. 14 we show the range of ℓ values to which these dissipative phenomena are associated and, in Fig. 15 we summarize the conditions under which fusion, fast fission and quasi-fission occur.

CONCLUSION

During the recent years a great progress has been done in the understanding of fusion. New ideas have been introduced and we have now a rather good description of this phenomenon. Other dynamical descriptions have also been proposed for fusion[8] and they reach conclusions which are similar in the average (see ref.[3] for a comparison between the different approaches). It seems that at

present the models are beyond the experiment and are able to predict new phenomena like fast fission which still need to be really
confirmed experimentally. Therefore many clever experiments at
high resolution are needed in order to agree or to contradict the
models.

REFERENCES

1. M. Lefort, C. Ngō, J. Peter and B. Tamain (1973), Nucl. Phys.
   A216, 166.   2. C. Ngō, B. Tamain, J. Galin, M. Beïner and
   R.J. Lombard (1975), Nucl. Phys. A240, 353.
3. C. Ngo, Proc. of International Conference on nuclear physics,
   Florence, (1983), p. 321.
4. R. Bock, Y.T. Chu, M. Dakowski, A. Gobbi, E. Gross, A. Olmi,
   H. Sann, D. Schwalm, U. Lynen, W. Muller, S. Bjornholm, H.
   Esbensen, W. Wolfi and E. Morenzoni (1982), Nucl. Phys.
   A388, 334.
5. T. Suomijarvi, R. Lucas, C. Ngo, E. Tomasi, D. Dalili and
   J. Matuszek, to appear in Nuovo Cimento.
6. U. Mosel, 2nd Europhysics study Conf. on the dynamics of heavy
   ion collisions, Hvard (1981) (North-Holland), p. 1.
7. H. Ngo and C. Ngo (1980), Nucl. Phys. A348, 140.
8. W. Swiatecki (1981),Phys. Scripta, 24, 113 and (1982), Nucl.
   Phys. A376, 2/5.
9. C. Lebrun, F. Hanappe, J.F. Lecolley, F. Lefebvres, C. Ngö,
   J. Peter and B. Tamain (1979), Nucl. Phys. A321, 207.
   B. Borderie, M. Berlanger, D. Gardes, F. Hanappe, L. Nowicki,
   J. Peter, B. Tamain, S. Agarwal, J. Girard, C. Grégoire,
   J. Matuszek and C. Ngo (1981), Z. Phys. A299, 263.
10. J. Blocki, Y. Noneh, J.R. Nix, J. Randrup, M. Robel,
    A.J. Sierk and W.J. Swiatecki (1978), Ann. Phys. 113, 330.
11. S. Cohen, F. Plasil and W.J. Swiatecki (1974), Ann. of. Phys.
    82, 557.
12. C. Grégoire, C. Ngo, E. Tomasi, B. Remaud and F. Scheuter
    (1982), Nucl. Phys. A387, 37.
13. V. Bernard, C. Grégoire, C. Mazur, C. Ngö, M. Ribrag,
    G.Y. Fan, P. Gonthier, H. Ho, W. Kühn and J.P. Wurm (1982),
    Nucl. Phys. A385, 319.
14. S. Leray, X.S. Chen, C. Grégoire, C. Mazur, C. Ngö, M. Ribrag ,
    E. Tomasi G.Y. Fan, H. Ho, A. Pfoh, L. Schad and J.P. Wurm,
    Nucl. Phys. in press.
15. C. Grégoire, C. Ngo and B. Remaud (1981), Phys. Lett. 99B, 17
    and (1982), Nucl. Phys. A383, 392.
16. W.J. Swiatecki (1972), J. Phys. 33, 45.
17. J.F. Berger, M. Girod and D. Gogny, International Conference
    on the theoretical approaches to heavy ion reaction mechanism ,
    Paris (1984), to appear in Nucl. Phys.
18. L. Schad, H. HO, G.Y. Fan, B. Lindl, A. Pfoh, R. Wolski and
    J.P. Wurm, (1984) preprint.
19. C. Grégoire aned F. Scheuter (1981), Z. Phys. A303, 337.

Lecture III : SOME ASPECTS OF THE PHYSICS BETWEEN 20 AND 50
MeV/u

The heavy ion mechanisms at bombarding energies smaller than
about 10-20 MeV/u are dominated by the mean field.[1] As a consequence it is possible to observe dissipative phenomena for the
collective motions which are excited during the collision. The
situation changes a lot at very high bombarding energies
(> 100 MeV/u) where it is found that the physics is dominated by
the nucleon-nucleon interaction and where a collective behaviour
seems to be very scarce.[1]

The intermediate bombarding energy region, which is located
between 20 and 100 MeV/u is now being available. It is interesting
to see how the transition between the regime dominated by collective motions and the one dominated by the nucleon-nucleon interaction takes place. This region is interesting to investigate
because the incident kinetic energy of the nucleons can be of the
same order than the Fermi energy. As the bombarding energy will be
raised from 20 to 100 MeV/u, we expect the Pauli principle, which
prevents, for a large part, the collisions between nucleons, to be
less and less effective. Consequently one might expect, as a
simple guess (and if dissipation is still present) some kind of a
transition between the one-body and the two-body friction. We
would then become closer to an hydrodynamic picture. However increasing the incident velocity of the projectile means also that
the interaction time will decrease. Consequently the collective
motions which have a relaxation time which is much larger than
this interaction time will no longer be excited. This means that,
if collective motions of the same nature as those seen at smaller
bombarding energy can still be observed, it will only be those who
have a short relaxation time (charge equilibration for instance).
We have seen, in the first lecture, that the relaxation time for
the intrinsic degrees was very short : of the order of $\sim 10^{-22}$s.
However if the collision time becomes of that order, the problem
of equilibration and statistical equilibrium of the internal degrees becomes questionable and connected to it, the possible existence of dissipation. If there is no global statistical equilibrium, memory effects can be present (non Markovian process) and
the reaction might also proceed by other mechanisms different from
the ones observed at low bombarding energy.

The differences between the reaction mechanisms in the three
energy domains discussed above, might probably find their origin
in the relative balance between the intrinsic velocity of the
nucleons and the relative velocity of the two ions. This is
schematically represented in Fig. 1. At low bombarding energy the
relative velocity of the two ions is much smaller than the mean
velocity of the nucleons. Therefore we understand why there are
collective motions which are slow compared to the time evolution
of the intrinsic system. The interaction between the two nuclei
will create a perturbation of the mean field but not a complete

destruction. At very high bombarding energies it is the contrary. The relative velocity of the two nuclei is much larger than the instrinsic velocity of the nucleons and their binding is negligeable compared to the incident kinetic energy. The mean field represents a very small perturbation in the system which will be dominated by the nucleon-nucleon interaction. At medium bombarding energies both velocities become of the same order and it will not be possible to neglect either one compared to the other. We are in a transition region which will not be easy to treat theoretically and experiments will also not be easy to do since, in addition to heavy fragments, a non negligible amount of light particles, which are not all of evaporative nature, will be present.

In this energy domain there are several questions which we would like to answer. Let us briefly quote some of them in an arbitrary way :

- Can we still observe collective motions of nucleons in this energy range?

- Do we have dissipative phenomena and what will be the nature of friction? Will it be of one-body type (1p-1h excitations followed by a decay to more complicated states) or will we observe two-body friction as it is the case for macroscopic objects?

- Can we observe a gas-liquid transition[2]?

- How much energy, linear and angular momenta can we deposit in the fragments? Such a question is strongly connected to the fusion process. Indeed if we can deposit a large amount of excitation energy in the fused system, we can reach the boiling of nuclear matter which should occur at temperatures of the order of ~ 8 MeV.

In heavy-ion reactions which are induced in this energy range, it is interesting to know whether the available incident energy will be converted into temperature, or into compressional energy. In fact we very likely will observe a mixture of both kinds of energy but one needs to know the relative proportion of them. The compressed system will expand, afterwards probably at constant entropy.[3] This will produce a cooling which can possibly lead to condensation of nuclear matter into a liquid and a gas phase. If such a process is also dissipative there will be nevertheless entropy production in the expansion phase.

In this lecture I would like to give a feeling, in a simplified manner, of some aspects of the results which have been obtained between 20 and 50 MeV/u. The first aspect will concern linear momentum transfer and the second one the possible fracture of nuclei when they collide at such high velocities. This last problem is related to the question of whether nuclei can behave, under some circumstances, as a crystal.

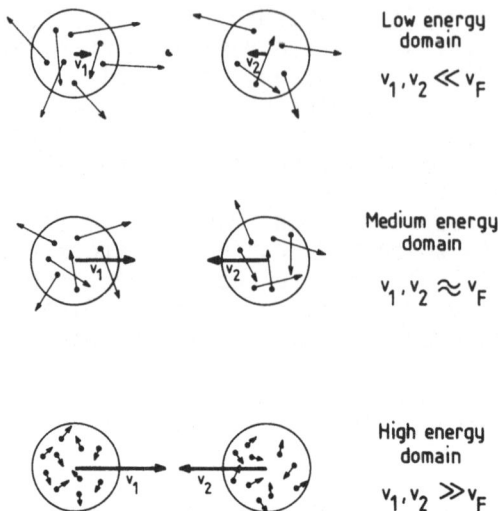

Fig. - 1 Schematic presentation of the three energy domains. The wide arrows indicate the velocity of the ions and the small ones, distributed randomly the velocity of the nucleons in the frame of reference where the nucleus is at rest.

1. LINEAR MOMENTUM TRANSFER

When the bombarding energy is smaller than 7-8 MeV/u two heavy ions can fuse together. If the fused system is heavy enough, or has a lot of angular momentum, it will have a large probability to fission. For a given mass ratio and a total kinetic energy of the products, the most probable correlation angle between the two fragments can be easily calculated in the laboratory system by just applying simple kinematics. Indeed the two fission fragments are emitted at 180° from each other in the frame of the fissioning nucleus and one has, to go to the laboratory system, just to add, to their velocity the recoiling velocity of the fused nucleus (see Fig. 2). Nevertheless there will be a distribution around the most probable value of the correlation angle because the fission frag-ments are excited. Consequently they will de-excite by emitting light particles which will change the orientation, as well as the value of the initial velocity. This distribution will extend both in and out of the reaction plane. If we now remove the condition of a fixed mass ratio, and of a fixed kinetic energy for the fra-gments, we will create an additional broadening of the correlation angle between the two fragments, but only in the angle oriented along the reaction plane. At low bombarding energies the recoiling velocity of the fused system is equal to the velocity, V, of the center of mass :

$$V = v_1 \frac{A_1}{A_1 + A_2} \qquad (1)$$

where v_1 is the velocity of the projectile, A_1 its mass and A_2 the mass of the target. In this case we say that there is full linear momentum transfer.

When the bombarding energy is raised, complete fusion is not the only possible mechanism producing a fissioning nucleus. We can have incomplete fusion where fast light particles are emitted before the remaining nuclei fuse together. The fused system will not recoil with the velocity of the center of mass because these light particles have removed a part of the initial linear momentum of the projectile. Consequently the correlation angle between the two fragments will be larger than in the case discussed above.

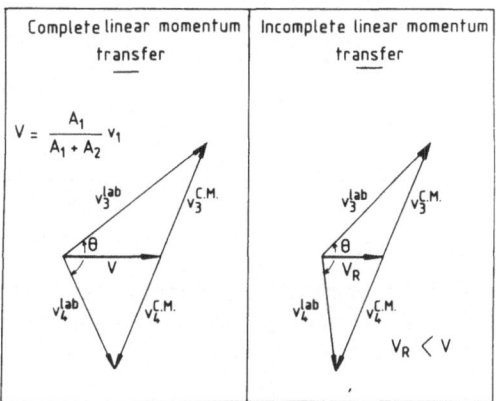

Fig. 2 - Principle of the measurement of linear momentum transfer using the correlation angle technique: when less and less linear momentum is transferred from the projectile to the fused nucleus the opening angle between the two fission fragments, θ, increases.

In such a situation we say that we have incomplete linear momentum transfer. The fraction of linear momentum transferred, ρ, can be defined as

$$\rho = \frac{p_3 + p_4}{p_1} \qquad (2)$$

where $p_{1_{//}}$, $p_{3_{//}}$ and $p_{4_{//}}$ are respectively the parallel linear momentum components (along the beam axis) of the projectile and of the fission fragments. For complete linear momentum transfer $\rho=1$, whereas, it is zero for non-linear momentum transfer.

When projectiles up to ^{12}C, ^{16}O are accelerated at bombarding energies ranging between 10 and 90 MeV/u, it turns out that the most probable amount of linear momentum transferred from the projectile to the fissioning nucleus represents a larg part of the incident one. This is illustrated in Fig. 3 which shows ρ as a

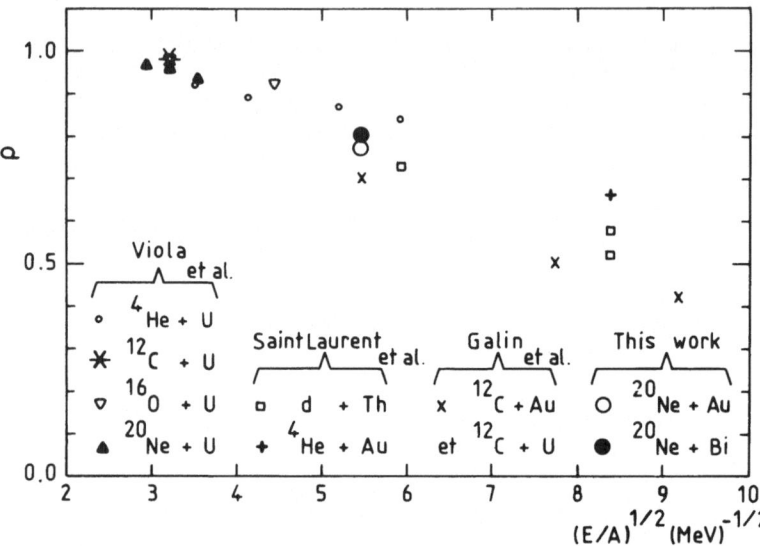

Fig. 3 - Fraction of full momentum transfer as a function of the square root of the incide t energy per nucleon for different systems [data from refs.[4-7]]. From ref.[7].

function of the incident velocity for different systems investigated in the literature. We observe that ρ decreases as $\sqrt{E/A}$ increases. This behaviour can be represented, to a good approximation, by the following expression[7] :

$$\rho = -0.092 \sqrt{E/A} + 1.273 \quad \text{for} \quad \sqrt{E/A} \geqslant 3.2 \text{ [MeV/u]}^{1/2}$$

and   (3)

$$\rho = 1 \quad \text{for} \quad \sqrt{E/A} \leqslant 3.2 \text{ [MeV/u]}^{1/2}$$

In Fig. 3 we have also plotted the points corresponding to the Ne + Au and Ne + Bi systems which have been investigated at SARA using the 30 MeV/u Ne beam.[7] We however do not know whether these systems will still follow the behaviour given by eq.(3) when the bombarding energy is increased above the Fermi energy (see below).

In order to illustrate how the correlation function of the two fission fragments looks like, we show, in Fig. 4, the result of such a measurement for the Ne + Au system at 30 MeV/u [ref.[7]]. The arrow indicates the position where the most probable value of the correlation angle, θ, should be, assuming full linear momentum transfer. In fact we see that the maximum of the distribution is

shifted towards larger values of θ and it corresponds to ρ ≈ 0.8.
Around 160° we have also a contribution of another mechanism : the
fission of the quasi Au target resulting from a very peripheral
collision where little linear momentum and energy is transferred
from the projectile to the target. Such a mechanism is usually
called sequential fission. It is easy to deduce a physical information from the most probable values which are observed in the
experiment. However for one particular event, one has to be very
careful before drawing any conclusion. Indeed the distribution
function displayed in Fig. 4 corresponds mostly to the fission
products following incomplete fusion plus, for the part close to
160°, a little bit of sequential fission. The large width of the
distribution around its most probable value comes mainly from the
existence of a mass and kinetic energy distribution for the fission products, and from the evaporation of the excited fission
fragments. All these effects induce a broadening around the most
probable value. On top of this, we also have a distribution in the
linear momentum distribution of the incomplete fusion nucleus.
This explains why we observe, for instance, fission fragments with
a θ smaller than the one corresponding to full linear momentum

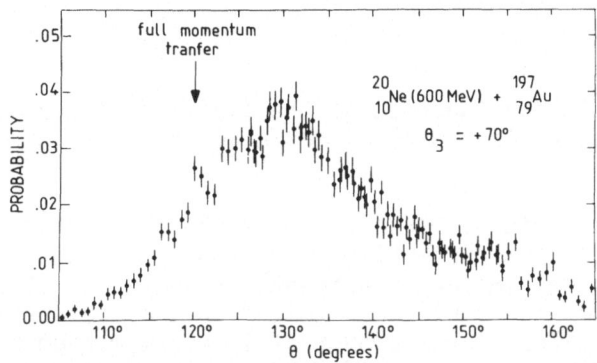

Fig. 4 - Correlation function of the two fission fragments observed in the reaction Ne + Au at 30 MeV/u. From ref.[7].

transfer. Of course these products do not correspond to a linear
momentum transfer larger than 100 %, but have this value of θ just
because of the preceding effects. Since a direct deconvolution of
the results is extremely difficult, and perhaps impossible, it
seems safer to perform a Monte-Carlo simulation of the data as it
has been done in ref.[7].
      Eq.(3) seems to give a good representation of the experimen-
tal data concerning linear momentum transfer. However, the use of
heavier projectiles has shown that this is not the case and ρ
depends not only on the bombarding energy but also on the size of
the projectile. This has been demonstrated using the 44 MeV/u Ar
beam accelerated at Ganil.[8,9] If we apply eq.(3) to the Ar + U

system we would get for the most probables values : $\rho \approx 0.66$ and θ
≈ 105 for the particular arrangement of the detectors. In Fig. 5
we show the measured correlation function for the Ar + U system[8],[9] (similar results have been obtained with Ar + Au).[8] Compared with Fig. 4 we see a very different picture. Indeed the most
probable θ lies close to 170˘ and corresponds to a sequential
fission of the quasi target. The probability of detecting two
fission fragments decreases as θ decreases from the preceding
region indicating that there are a small number of events coming
from a fission of a fused system. The fact that the Ar + Au, U
systems do not follow the systematic given by eq.(3) is very

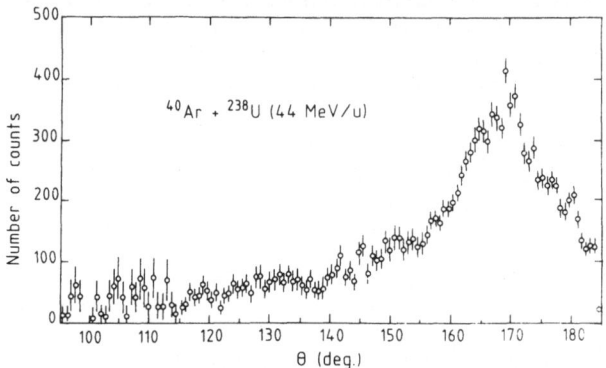

Fig. 5 - Correlation function of the two fission fragments observed in the reaction Ar + U at 44 MeV/u. From ref.[9].

surprising. It indicates that not only the bombarding energy, but
also the mass of the projectile is an important parameter to be
considered. Since similar experiments,[10] performed at 27 MeV/u
with an Ar beam show a picture similar to the one observed with
lighter projectiles (large amount of linear momentum transfer),
would be interesting to know if a Ne beam at $\geqslant 40$ MeV would give
something similar to Ar or to ^{12}C. Indeed, the results obtained
with the Ar beam seem to indicate that the Fermi energy ($\varepsilon_F \sim 40$
MeV) plays an important role in the process when the size of the
projectile is large enough : for bombarding energies below ε_F/u
the reaction mechanisms would be close to the one observed at low
bombarding energies and, above ε_F/u, the situation would change.
Therefore more experiments in this energy region are needed to
extract the physical content of what happens.

2. VERY INELASTIC FRAGMENTS IN 35 MeV/u Kr INDUCED REACTIONS

In the preceding section we have seen that the amount of
linear momentum which can be transferred from the projectile to
the fused nucleus, depends not only on the bombarding energy, but
also on the mass of the projectile. We may also wonder whether
some other mechanism may also depend upon the sizes of the projectile.

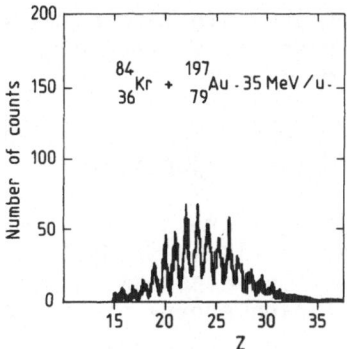

Fig. 6 - Atomic number distribution of the very inelastic events.

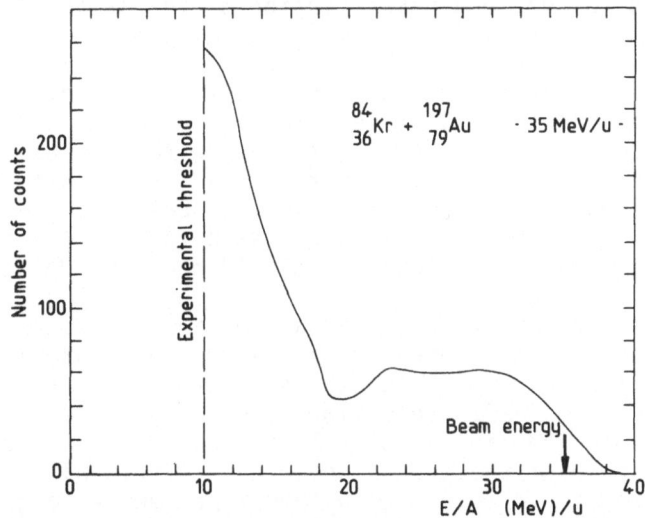

Fig. 7 - Kinetic energy per nucleon of all the products with a bombarding energy > 10 MeV/u. We see the presence of a lot of events with a kinetic energy < 18 MeV/u. Their atomic number distribution is displayed in Fig. 6.

which are detected in ref.[13,14] would come from the fusion between the participants and the projectile spectators followed by a fission of this system. Of course it is only speculations about the mechanism and more exclusive experiments are needed to precise the nature of these reactions.

From the two aspects discussed in this lecture it seems that the energy domain between 20 and 100 MeV/u is really a transition region between the two physics which have been extensively investigated at low and high bombarding energies. If it would remain so, the physics would not be so rich as it is just above the Coulomb barrier where many collective phenomena take place.

    With Ar projectiles the main mechanisms which are observed are
the following (when we restrict to fragments with a mass smaller
than half of the total mass) :

- Fragmentation, where we detect products with a mass and a velocity close to the one of the projectile.

- Light products which have a cross section which decreases when
the fragment mass increases. These products might be explained by
some liquid phase transition mechanism,[2,3] or by a cold fragmentation as in the case of high energy proton induced reactions.[12]

- Depending upon the bombarding energy, and on the target mass,
one might also see fission products following incomplete fusion,
or associated to the sequential fission of the target-like products.

    In the same way as an ^{40}Ar projectile has revealed different
features, compared to ^{12}C projectile, as far as linear momentum
transfer is concerned, it turns out that, with a Kr beam, new
features have been observed.[13] In refs.[13,14] the first 35 MeV/u
Kr beam provided by the Ganil facility has been used to bombard a
^{197}Au target. Experimental set-up was such that it was possible to
detect several heavy fragments in coincidence. In addition to the
usual reactions which are expected in this bombarding region and
briefly quoted above, it was found a new kind of events which has
not been observed with lighter projectiles. Their atomic number
distribution (see Fig. 6) lies between 20 and 28 with a most probable value around Z=24. Their kinetic energy is much smaller than
the incident one and lies below ~ 18 MeV/u. In Fig. 7 we show a
velocity spectrum of the events detected a little bit backwards
the grazing angle. In this region they represent a large part of
the differential cross section. Nevertheless they are also present
forwards the grazing angle, but they represent only a small part
(~ 1 %) of the fragmentation products.

    These very inelastic events cannot be understood in terms of
the usual pictures used in this energy domain : fragmentation or
liquid-gas phase transition. We may think at several origin for
them. For example, they could come from a could break-up of the
projectile. This would mean that the nuclei behave, under certain
circonstances, like a cristal. We can also imagine, and that might
be more likely, that we have a kind of participant spectator picture similar to the one used at high energies. However instead of
having in the exit channel three components corresponding to the
participants, the projectile and target spectators, we would have,
due to a reminiscence of the mean field, an interaction between
the participants and one of the spectators. This interaction could
lead to a fusion between the participants and the spectators of
the projectile or of the target. The fused system being highly
excited would fission. Therefore we would have a sequential two-
step mechanism giving three fragments in the exit channel. Those

ACKNOWLEDGEMENTS

The materials presented in these lectures have been obtained over the last ten years by the collaboration between many institutes : Saclay, Orsay, Munich, GSI, Caen, Lyon, Bruyères-le-Châtel, Strasbourg and Barcelona. I would like to express my gratitude to all my colleagues for the nice discussions we have had together and for the pleasure I had in participating to these collaborations.

I would also like to thank Mesdames F. Lepage and E. Thureau for the material preparation of the manuscript and Monsieur J. Matuszek for drawing a lot of the figures presented here.

REFERENCES

1. For a review see for instance the Proceedings of the International Conference on theoretical approaches to heavy ion reaction mechanisms, (Paris, 1984) to appear in Nucl. Phys.
2. M.W. Curtin, H. Toki and D.K. Scott (1983), Phys. Lett. 123B, 289.
3. G. Bertsch and P.J. Siemens (1983), Phys. Lett. 126B, 9.
4. V.E. Viola, B.B. Back, K.L. Wolf, T.C. Awes, C.K. Gelbke and H. Breuer (1982), Phys. Rev. 26C, 178.
5. F. Saint-Laurent, M. Conjeaud, R. Dayras, S. Harar, H. Oeschler and C. Volant (1982), Phys. Lett. 110B, 372.
6. J. Galin, H. Oeschler, S. Song, B. Borderie, M.F. Rivet, I. Forest, R. Bimbot, D. Gardes, B. Gatty, H. Guillemot, M. Lefort, B. Tamain and X. Tarrago (1982), Phys. Rev. Lett. 48, 1787.
7. G. La Rana, G. Nebbia, E. Tomasi, C. Ngo, X.S. Chen, S. Leray, P. Lhénoret, R. Lucas, C. Mazur, M. Ribrag, C. Cerruti, S. Chiodelli, A. Demeyer, G. Guinet, J.L. Charvet, M. Morjean, A. Péghaire, Y. Pranal, L. Sinopoli and J. Uzureau (1983) Nucl. Phys. A407, 233.
8. S. Leray, G. Nebbia, C. Grégoire, G. La Rana, P. Lhénoret, C. Mazur, C. Ngo, M. Ribrag, E. Tomasi, S. Chiodelli, J.L. Charvet and C. Lebrun (1984), Nucl. Phys. in press.
9. J.L. Charvet, S. Chiodelli, C. Grégoire, C. Humeau, G. La Rana, S. Leray, P. Lhénoret, J.P. Lochard, R. Lucas, C. Mazur, M. Morjean, C. Ngö, Y. Patin, A. Péghaire, M. Ribrag, S. Seguin, L. Sinopoli, T. Suomijarvi, E. Tomasi and J. Uzureau, International Conference on theoretical approaches to heavy ion reaction mechanisms (Paris, 1984) p. 82.
10. J. Galin, private communication.
11. J. Lopez and P. Siemens (1984), Preprint.
12. J. Aichelin, J. Hüfner and R. Ibarra (1983), Preprint MPI-H-1983-V32.

13. D. Dalili, P. Lhénoret, R. Lucas, C. Mazur, C. Ngo,
    M. Ribrag, T. Suomijarvi, E. Tomasi, B. Boishu; A Genoux-
    Lubin, C. Lebrun, J.F. Lecolley, F. Lefebvres, M. Louvel,
    R.R. Regimbart, J.C. Adloff, A. Kamili, G. Rudolf and
    F. Scheibling (1974), Z. Phys. A316, 371.
14. J.C. Adloff, B. Boishu, D. Dalili, A. Genoux-Lubin,
    A. Kamili, C. Lebrun, J.F. Lecolley, F. Lefebvres,
    P. Lhénoret, M. Louvel, R. Lucas, C. Mazur, C. Ngo,
    R. Regimbart, M. Ribrag, G. Rudolf, F. Scheibling,
    T. Suomijarvi and E. Tomasi, International Conference on
    theoretical approaches to heavy ion reaction mechanisms
    (Paris, 1984), p. 68.

PERSPECTIVES OF ULTRARELATIVISTIC NUCLEUS-NUCLEUS COLLISIONS

Hans J. Specht

Physikalisches Institut der Universität Heidelberg
Heidelberg, Germany

Comment: The lectures delivered at this school survey the expectations for the new field of ultrarelativistic nucleus-nucleus collisions. Nearly all the material covered is available in the literature. Therefore, only an extended abstract is reproduced in the following, containing the primary motivation for the field, the organization of the lectures, a short description of a major forthcoming experiment, and a list of review-type references.

INTRODUCTION

In the past decades, low energy nuclear physics has traditionally studied the properties of bulk nuclear matter under conditions, in terms of baryon density and temperature, which have remained relatively close to the ground state. Particle physics, on the other hand, has been concerned in parallel with the properties of the strong (and electroweak) interaction of isolated particles at higher and higher energies. It is the most exciting prospect of ultrarelativistic nucleus-nucleus collisions that the completely unexplored field of the properties of extended hadronic matter in the limit of high energy density, reachable so far only in the hot early universe or in certain other astrophysical events, may become accessible to terrestrial experiment.

In the framework of Quantum Chromodynamics (QCD), one of the fundamental issues is connected with the understanding of confinement. Plausibility arguments as well as lattice results predict the deconfinement of hadronic matter into a plasma of quarks and gluons at high energy density, due to the property of asymptotic freedom.

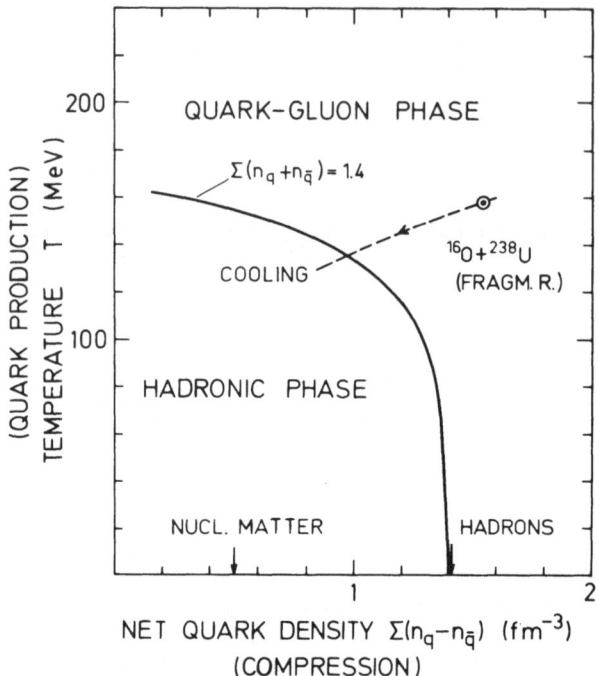

Fig. 1. Phase diagram of strongly interacting matter in the temperature versus baryon density plane. The borderline separating the two phases has been obtained numerically on the basis of simple relativistic Fermi gas expressions, assuming the critical density to be that of normal nucleons (1.4 quarks/fm^3).

In a phase diagram (Fig.1) of temperature versus net quark density, the possibility of a phase transition between normal and deconfined matter is readily understood in a pedestrians way. When the total density of quarks plus antiquarks $\Sigma(n_q + n_{\bar{q}})$ in a system is low, the quarks are confined inside individual hadrons, surrounded by the normal vacuum. However, as the energy density is raised (either by increasing the baryon-, i.e. net quark density $\Sigma(n_q - n_{\bar{q}})$ via compression, or by producing hadrons, i.e. additional quark-anti-quark pairs via temperature, or by both), the hadrons begin to overlap and the system finally undergoes a transition to a state in which the quarks and gluons are no longer locally confined, but free to move within the total volume. Due to the much larger number of degrees of freedom in this quark-gluon phase, the energy density (for a given temperature) is larger by at least one order of magnitude than that of the hadronic phase.

Ultrarelativistic nucleus-nucleus collisions appear to be the only laboratory tool able to reach sufficiently high energy densities over sufficiently extended space-time volumes to study the consequences of this picture.

ORGANIZATION OF THE LECTURES

(i)   A remainder: The elementary constituents of matter, quarks and leptons. The structure of hadrons. Forces and field quanta. A comparison of QED and QCD. References: Modern textbooks on particle physics[1,2].

        Quark-gluon plasma as the high energy density limit of extended hadronic matter. Analogies to QED phenomena. Simple thermodynamics of massless quarks and gluons. Results from QCD lattice calculations. References: Proceedings of the Quark Matter Conferences in Bielefeld[3], Brookhaven[4], Helsinki[5]; a collection of monographs[6].

(ii)   The time-sequence of a nucleus-nucleus collision. Particle production in hadron-hadron collisions. Rapidity-, transverse momentum- and transverse energy dependences. Event structure; soft processes and jets. Particle production in hadron-nucleus collisions. Energy deposition and energy densities. Cosmic ray data for nucleus-nucleus collisions. References: particle production in general[7,8]; pA and AA collisions in [3-6].

(iii)   Plans for future experiments at the CERN SPS. The details of the NA34 experiment. Possible signatures for the deconfinement transition and the appropriate experimental techniques. The role of transverse energy measurements; calorimetry. Target techniques. Particle spectra; strangeness production; $<p_T>$ versus E_T; the production of exotica. Particle identification techniques. References: the discussion of signatures in [3-6]; the proposals of the NA34 collaboration[9,10].

(iv)   Lecture (iii) continued. The production of real photons and lepton pairs in hadron collisions and the expectations for nuclear collisions. High-p_T processes as probes of single parton interactions. Low-p_T processes as probes of non-perturbative QCD phenomena; the need for better understanding. Experimental techniques for the measurement of photons, muon pairs and electron pairs. References: the discussion of signatures in [3-6]; a specific review of photon and lepton pair production[11]; the proposals of the NA34 collaboration[9,10].

THE NA34 EXPERIMENT

The forthcoming experiment of the NA34 collaboration consists of two parts, one[9] devoted to hadron-nucleon collisions, the other[10] to nucleus-nucleus collisions; the case of hadron-nucleus collisions is common to both. The overall lay-out, shown in Figure 2, contains five major components: a target plus vertex detector system, a 4π-calorimeter with separate sections for electromagnetic and hadronic energy, a compact forward electron spectrometer (for [9]), a forward muon spectrometer, and an external spectrometer at sideward angles. Smaller components exist, in addition, for the detection of real photons. A sixth major component, an electron spectrometer at larger angles (for[10], not shown in Fig.2), may be added in the future. The nuclear experiment aims at (i) measuring as many observables as possible for a given event, and (ii) comparing with high accuracy to hadron-nucleus collisions under identical instrumental conditions.

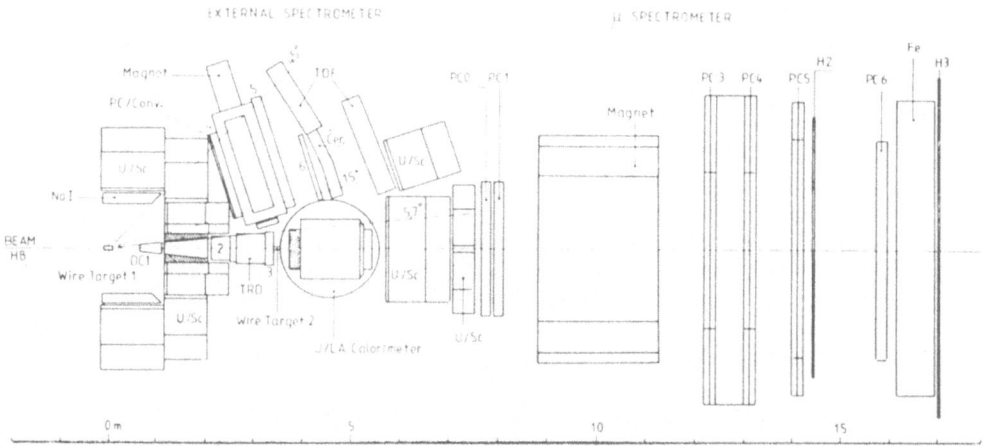

Fig. 2. Top view of the experimental configuration of the NA34 collaboration[9,10].

REFERENCES

1. D.H. Perkins, Introduction to High Energy Physics (1982)
2. F.E. Close, An Introduction to Quarks and Partons, Academic, New York (1979)
3. Quark Matter Formation and Heavy Ion Collisions, Proc. of the Bielefeld Workshop, Editors M. Jacob and H. Satz, World Scient. Publ., Singapore (1982)
4. Quark Matter 1983, Proc. of the Third Int. Conf. on Ultrarelativistic Nucleus-Nucleus Collisions, Brookhaven 1983, Editors T.W. Ludlam and H.E. Wegner, Nucl. Phys. A418 (1984)
5. Quark Matter 1984, Proc. of the Fourth Int. Conf. on Ultrarelativistic Nucleus-Nucleus Collisions, Helsinki 1984, to appear in Lecture Notes in Physics, Springer, Heidelberg
6. M. Jacobs and J. Tran Thanh Van (editors), Quark Matter Formation and Heavy Ion Collisions, Phys. Rep. 88 (1982) No.5
7. G. Giacomelli and M. Jacob, Phys. Rep. 55C (1979) 1
8. P.D.B. Collins and A.D. Martin, Hadron Interactions, Hilger, Bristol, and Univ. of Sussex Press (1984)
9. H. Gordon et al., Proposal P189 to the SPSC, CERN-SPSC/83-51 (1983); accepted as NA34
10. H. Gordon et al., Proposal P203 to the SPSC, CERN-SPSC/84-43 (1984), accepted as NA34/2
    The present list of names: H. Gordon, T. Ludlam, V. Polychronakos, D.C. Rahm, I. Stumer, C. Woody, T. Åkesson, H. Atherton, H. Breuker, C.W. Fabjan, U. Goerlach, S. Katsanevas, U. Mjornmark, J. Schukraft, W.J. Willis, P. Glässel, A. Pfeiffer, J. Soltani, H.J. Specht, N.J. DiGiacomo, P.L. McGaughey, W.E. Sondheim, J.W. Sunier, S. Almehed, G. Jarlskog, B. Lörstad, L.A. Hamel, C. Leroy, Y. Sirois, G. Beaudoin, J.M. Beaulieu, P. Depommier, H. Jeremie, L. Lessard, A. Lounis, S. Mayburov, A. Shmeleva, V. Cherniatin, B. Dolgoshein, Yu. Golubkov, A. Kalinovsky, V. Kantserov, P. Nevsky, A. Sumarakov, V. Sidorov, W. Cleland, J. Thompson, A. Gaidot, F. Gibrat, G.W. London, J.P. Pansart, D. Bettoni, M. Goldberg, N. Horwitz, G.C. Moneti, L. Olsen, O. Bernary, S. Dagan, D. Lissauer, Y. Oren (Brookhaven, CERN, Heidelberg, Los Alamos, Lund, Montreal, Moscow, Novosibirsk, Pittsburgh, Saclay, Syracuse, Tel Aviv)
11. H.J. Specht, Direct Photon and Lepton Production in High Energy Collisions, in Ref.[5].

CRITICAL PHENOMENA IN NUCLEAR COLLISIONS

David K. Scott

National Superconducting Cyclotron Laboratory
and
Departments of Physics and Astronomy and of Chemistry
Michigan State University
East Lansing, Michigan 48824, U.S.A.

INTRODUCTION

The possibility of creating states of matter resembling those prevailing in the early Universe constitutes a major incentive for the study of high energy heavy ion collisions.[1-3] If energy densities of 2-4 GeV/fm^3 can be achieved, then according to quantum chromodynamics, a phase transition to a quark-gluon plasma may occur. Such a transition represents an extreme example of a critical phenomenon which may exist at high temperatures and densities. Many questions on reaching such a state can be raised--whether, for example, there is sufficient time for the phase transition to manifest itself in the brief time during a nuclear collision when high compressions are thought to be reached, and whether phase equilibrium can be sustained by a finite system consisting of a small number of particles. It is of interest, therefore, to explore the consequences of other critical phenomena, which may set in at lower temperatures and densities and which may be attainable with existing accelerators.

The variety of realized and anticipated states of nuclear matter is indicated[4] in Fig. 1. At high density cold nuclear matter forms a gas of quarks interacting through gluon exchange. As the density decreases to a few times normal nuclear density, the quarks cluster into hadrons. At still lower density the hadrons are ordered by standing waves of pions into a liquid crystal of neutrons and protons. Just above nuclear matter density the well known superfluid state is possible due to pairing correlations. The diagram also illustrates that matter at low density but at high temperature again forms a quark plasma, containing quarks, antiquarks, and gluons.

As the temperature drops, neutrons, protons, and pions condense into
an hadronic gas, in a similar fashion to the evolution of the early
Universe following the Big Bang. At lower temperatures larger clusters of deuterons, alpha particles emerge once the temperature
becomes commensurate with nuclear binding energies. Once the entropy
per baryon falls below unity, the deuterons are expected to disappear
due to the Mott transition. It is possible that alpha particles may
undergo Bose condensation, a state which appears at the bottom left
hand corner of the phase diagram.

Another state of matter lies in the middle of the diagram
between the gas and liquid phases. The liquid-gas phase transition
has certain similarities to the physics of the quark plasma transition but lies at much lower temperatures and densities. In this
paper some of the low temperature critical phenomena are explored,
viz a fast mechanical instability leading to the break-up of a
nuclear system into fragments, and a slower chemical instability
involving a transition between gas and liquid phases. Both phenomena
rely for their development on a hydrodynamical description of nuclear
collisions, and raise issues which may be germane to phase instabilities of more exotic kinds, such as the quark-gluon plasma and pion
condensation.

The next section of the paper describes the underlying physics
of the instabilities, and is followed with a discussion of the relevant time scales for the processes to be established. Some of the
present experimental evidence is then presented.

CRITICAL CONDITIONS AT LOW TEMPERATURE

An accepted picture of high energy heavy ion collisions is
shown on the right of Fig. 2 and illustrates the division of a reaction into participants and spectators. This behavior is established in nuclear collisions at high energies and is to be compared
with the slower, more gentle, evolution associated with the TDHF
description shown at the left. We shall assume that the participant-
spectator description is valid at energies of 50-100 MeV/nucleon.
(Some justification for this assumption will be presented later.)
The participant zone is initially compressed and heated; during the
subsequent expansion, when the density and temperature drop, it is
possible that the system passes through conditions favorable to a
division into liquid and gaseous phases, thereby influencing the
production of complex fragments. Although frequent discussions of
this phenomenon have appeared in the literature (see, for example,
Refs. 6-8) pertaining to the evolution of neutron stars and supernovae, its possible manifestation during the dynamical evolution of
a heavy ion collision has only recently been raised.[9-14] In the
following, a simplified analytical treatment is followed to

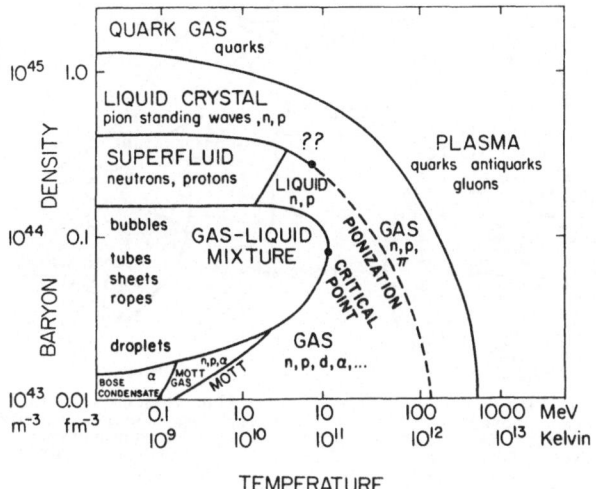

Fig. 1. Phase diagram for nuclear and hadronic matter.

illustrate the basic ideas; more formal treatments can be found in the above references.

Consider the following parameterization of the energy per particle, E, in a nuclear system as a function of temperature, T, and density, ρ;

$$E = E_0 + \frac{K}{18}\left(\frac{\rho - \rho_0}{\rho_0}\right)^2 + \frac{\pi^2}{4\varepsilon_F}T^2\left(\frac{\rho_0}{\rho}\right)^{2/3}$$

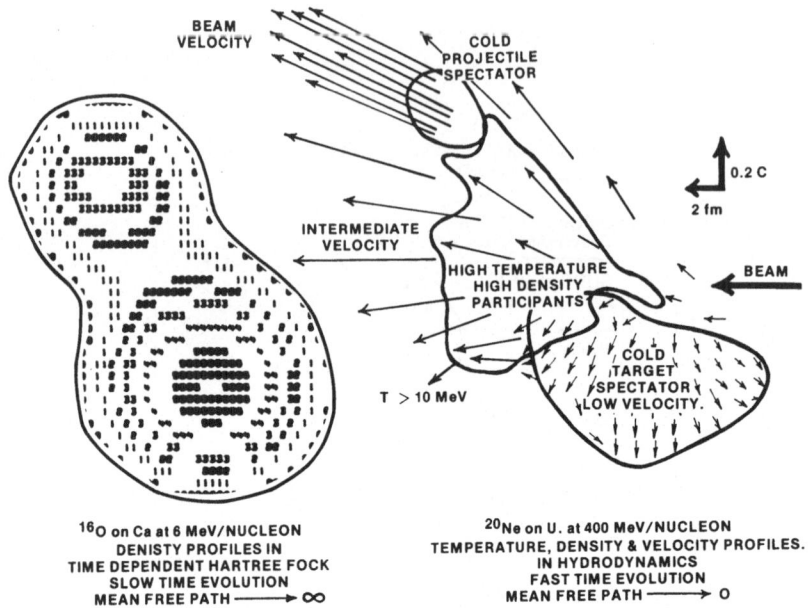

Fig. 2. The participant-spectator picture of high energy heavy ion
collisions is illustrated at the right by temperature and
density profiles calculated with a hydrodynamical model.
The participant zone initially contains high density and
high temperature nuclear matter. This view of a reaction
is contrasted with the picture obtained from a TDHF calculation at low energies shown on the left.

which is comprised of the ground state binding energy per particle
E_0, the compressional term containing the incompressibility K, and
a thermal contribution derived from the low temperature approximation for a Fermi gas; ε_F is the Fermi energy and ρ_0 the normal
nuclear matter density of approximately 0.16 nucleons/fm^3. From the
free energy F = E - TS, with the specific entropy

$$S = \frac{\pi^2}{2\varepsilon_F} T \left(\frac{\rho_0}{\rho}\right)^{2/3} ,$$

we obtain

$$F = E_0 + \frac{K}{18} \left(\frac{\rho-\rho_0}{\rho_0}\right)^2 - \frac{\pi^2}{4\varepsilon_F} T^2 \left(\frac{\rho_0}{\rho}\right)^{2/3} .$$

The pressure as a function of ρ and T can be calculated from

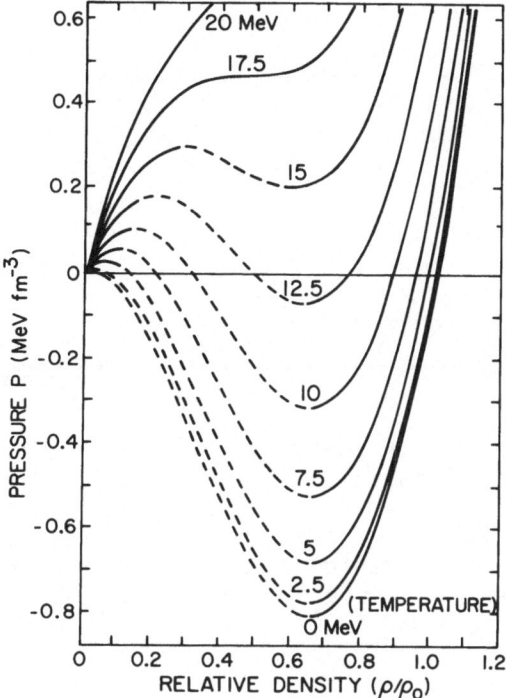

Fig. 3.   The equation of state for a nuclear system is illustrated
         by isotherms for pressure versus density. The isotherms
         separate into two regions which can be identified with
         liquid and gas phases in a Van der Waals system, connected
         by a region of negative incompressibility (shown dashed).

$$P = \left(\frac{\partial F}{\partial V}\right)_T = \rho^2 \left(\frac{\partial F}{\partial \rho}\right)_T$$

Results for a more complete calculation[6] are shown in Fig. 3, where
it can be seen that the equation of state has the form of a Van der
Waals system, for the simple reason that the nuclear and molecular
systems are analogous; both are subject to short range attractive
forces and very short range repulsions.

     As in the Van der Waals system, there exist liquid and gaseous
phases. For the unphysical region (shown dashed in Fig. 3) where
the slope of P versus ρ is negative (implying a negative incompressibility) a Maxwellian construction is employed, along which the
liquid and gas phases coexist. This region of coexistence is illustrated[12] more clearly in Fig. 4, which also shows that as the temperature increases, the apex of the coexistence region coincides

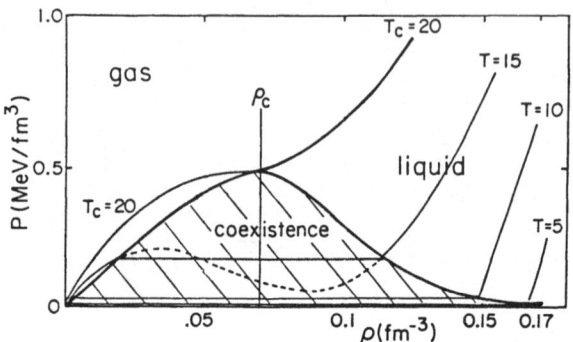

Fig. 4. Pressure-density isotherms in which the region of negative
         compressibility is suppressed (compare Fig. 3) by the Max-
         well construction describing a phase transition between
         liquid and gas. The region of phase coexistence is shown
         by the cross hatching. Above T = 20 MeV only a single gas
         phase exists irrespective of the density. The critical
         density is 0.065 fm^{-3}.

with the inflection point of the critical temperature. This point
corresponds to the condition,

$$\frac{\partial P}{\partial \rho} = \frac{\partial^2 P}{\partial \rho^2} = 0$$

A solution of the above analytical expression with K $\cong$ 210 MeV, consistent with measurements of the monopole excitation, give a result
close to that obtained from a more detailed analysis, viz $T_c \cong$ 18 MeV
and $\rho_c \cong$ 0.07 nucleons fm^{-3}. For all higher temperatures only a
gaseous phase exists.

    Most studies of heavy ion collisions at relativistic energies
have been conducted in regions where the temperature exceeds 20 MeV.
In the later discussion on observable consequences we shall show
that temperatures for the participant zone of 20 MeV and lower are
generated in collisions below 100 MeV/nucleon, implying that intermediate energy heavy ion studies will be most important for the
investigation of these critical phenomena.

Another type of instability, which would occur on a faster time scale than the liquid-gas transition, has also been discussed recently.[15] This concerns a mechanical instability where the compressibility of the nuclear system $K = \rho \frac{\partial P}{\partial \rho}$ becomes negative. Such a region is easily identified in Fig. 3 by the dotted lines, but is shown more clearly in Fig. 5 where K is plotted as a function of density for different temperatures. The region of zero or negative compressibility falls below the horizontal axis. The corresponding boundary is transposed onto the plot of internal energy as a function of density in Fig. 6. Here the region labelled "unstable zone" defines where nuclear matter becomes dynamically unstable. Expressed alternatively the boundary traces out the locus of the tensile stress of nuclear matter as a function of temperature. The authors of Ref. 15 argue that the occurrence of nuclear fragmentation as a dominant reaction process depends on whether the system enters this unstable region. Since the zone has a boundary of lower than normal nuclear matter density, it is necessary to consider how it can be reached, given that a nuclear system is prepared in a reaction at normal or greater than normal density.

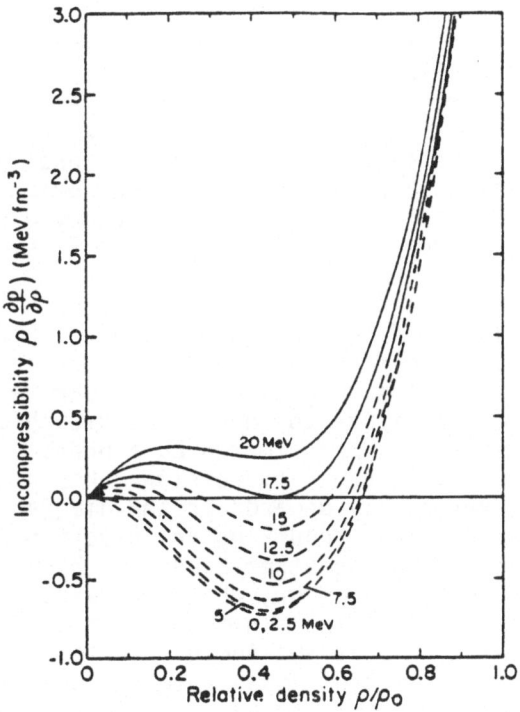

Fig. 5.   The rigidity $\rho \frac{\partial P}{\partial \rho}$ plotted as a function of density for various tempeartures. The region of negative incompressibility, shown dashed in Fig. 2, corresponds to the part of the diagram below the horizontal axis. The rigidity of the liquid is much larger than that of the gas.

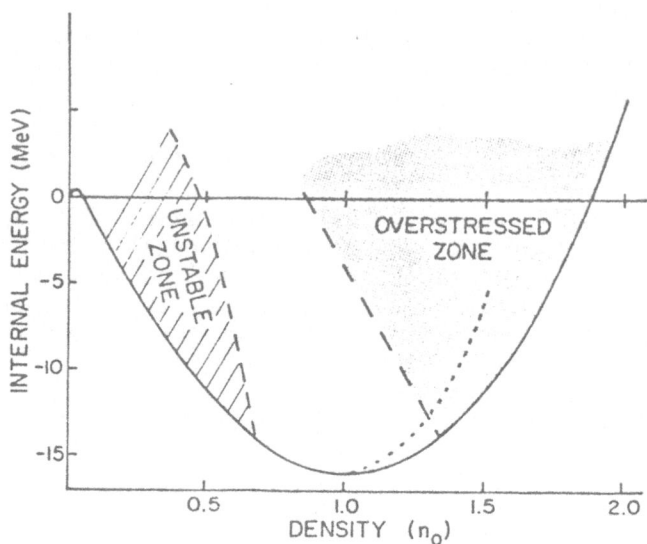

Fig. 6.   The internal excitation energy per nucleon is shown as a
          function of relative density. The shaded portion on the
          left, labelled "unstable zone," defines the region of nega-
          tive incompressibility. The shaded area on the right
          defines the overstressed zone from which a nuclear system
          will be able to reach the unstable zone through an isen-
          tropic expansion. The dotted locus extending into the over-
          stressed zone indicates the trajectory expected from theo-
          retical calculations of heavy ion collisions.

     The region labelled "overstressed zone" on Fig. 6 defines a
boundary which will under certain conditions allow the system to
reach the unstable region. Initially a nuclear reaction carries the
system from the ground state to some point with higher internal
energy. If the energy is transferred by a proton, it is plausible
to assume that no significant compression takes place. On Fig. 6
the system moves vertically upwards on the diagram from the minimum
at normal nuclear density, requiring an injection of approximately
10 MeV per particle. On the other hand, for heavy ion collisions we
expect some compression to occur, which prepares the system initially
at a point to the right of the minimum of normal density. From this
condition it is assumed that the nuclear system will expand along
an isentrope until a point of equal internal energy is reached on
the left. The justification for an expansion at constant entropy
is based on cascade calculations,[16] which indicate that little dissipation takes place; other evidence comes from our knowledge of the
monopole vibration, which has a damping width much smaller than its
excitation.[17] In Fig. 6 the region of appropriate initial conditions
which will provide access to the fragmentation zone is defined by

the dashed boundary of the shaded region labelled "overstressed zone." For example, a compression of 1.4 over normal density is predicted[18] in TDHF calculations at 10 MeV/nucleon (see the dashed line to the left of the S = 0 isentrope in Fig. 6), and if it is assumed that all of this excitation energy is thermalised, then the threshold for fragmentation would be lowered to 3 or 4 MeV/nucleon. The corresponding incident energy in the laboratory, assuming again equal participation from target and projectile, would be in the region of 12 MeV/nucleon. Above this threshold the system will always come apart in fragments. An observation of the onset of fragmentation might therefore provide a means of inferring the density at which thermalisation takes place; the energy threshold for fragmentation should be an increasing function of the initial density.

TIME SCALES

The mechanical and chemical instabilities we have discussed are quite different. One is a first order transition applicable to processes that occur slowly enough for an equilibrium to be established across the phase boundary. According to the picture of expansion and rarefaction of the initial compressed zone on a time scale commensurate with the frequency of the monopole vibration, this time can be estimated from the typical excitation energy in a medium weight nucleus,[16] $E \cong \hbar\omega \cong 15$ MeV, resulting in an expansion time of the order 10^{-22} sec. This time scale is the relevant one for a mechanical instability. It is not clear, however, if a liquid-gas chemical instability can be established on this short time scale, since there must exist sufficient time for equilibration to be established across the phase boundary. The time required for this equilibration to occur is of the same order as the evaporation time, an estimate of which can be obtained from the theory of thermionic emission.[19] Thus the current density can be expressed as

$$J = \frac{em}{2\pi^2\hbar^3} T^2(1 - r) e^{-W/T}$$

where r is the quantum mechanical reflection coefficient (taken as 0), W is the work function (taken as 8 MeV), and T is the temperature. By definition

$$J = \frac{\Delta q}{\Delta t} \frac{1}{A}$$

where A is the surface area of the emitting source. If we set q = e (equivalent to the emission of one nucleon) then $\Delta t = \tau_{evap}$.

Assuming a spherical geometry, so that $A = 4\pi R^2$ where $R \cong 3.5$ fm. as determined from the participant-spectator model for intermediate impact parameters[20] and consistent with determinations from pion interferometry measurements,[21] the evaporation time is found to be

$$\tau_{evap} \cong 3.5 \times 10^{-21} \cdot \frac{1}{T^2} \cdot e^{8/T} \text{ sec.}$$

The resulting values given in Table 1 are in good agreement with results deduced from an empirical fit to the measured widths of compound nuclei for $A = 20 - 100$.[22] Comparing the evaporation time with the time required for disassembly it appears that for $T \gtrsim 8.1$ MeV (henceforth referred to as the breakeven temperature) the liquid-gas phase instability may develop. A more detailed study of time scales is given in Ref. 23.

Since a liquid-gas phase instability exists only for temperatures below the critical temperature and above the breakeven temperature, it is obvious that if the breakeven temperature were higher than the critical temperature, the liquid-gas instability would not develop. The critical temperature of approximately 20 MeV predicted in Refs. 10 and 12 was deduced on the assumption that the binding energy per nucleon in nuclear matter is 16 MeV/u compared to the phenomenological binding energy per nucleon of 8 MeV for finite nuclei. A more thorough treatment of this question and of effective mass considerations is given in Ref. 13, where it is shown that in finite nuclei the predicted critical temperature lies between 13.4 MeV and 8.1 MeV depending on the choice of effective mass. Thus for temperatures above 8 MeV the liquid-gas instability may develop and there is sufficient time for it to do so.

Collision damping has been neglected throughout this discussion. A simple approach[24] to the problem begins with the equation for a damped non-driven oscillator:

$$\ddot{x} + \gamma \dot{x} + \omega_0^2 x = 0$$

$$x = \rho - \rho_{min} \text{ (S)}$$

Table 1. Nucleon evaporation times as a function of temperature.

T (MeV)	5	10	15	20
t (10^{-22} sec)	6.9	0.77	0.27	0.13

$$\Gamma = \frac{\gamma}{\omega_0^2 - \frac{1}{4}\gamma^2}$$

where γ is the damping coefficient (assumed constant) and ω_0 is the undamped harmonic oscillator frequency. The dimensionless damping coefficient can be deduced from the experimental measurements of the monopole oscillation characteristics. The variable ρ_{min} is the value of the density when the excitation energy is a minimum for a given value of the entropy. The damping constant determines the rate at which energy is transferred to thermal energy from the collective motion, thus determining the temperature. For a given value of the density, the entropy can be calculated and ρ_{min} can be determined. Although the oscillation in the density coordinate is not a true harmonic motion we shall use this approximation for small excursions from the equilibrium density. A solution to the above equation is of the form,

$$x \cong Ae^{-\gamma t/2}\cos(\omega t)$$

where

$$\omega^2 = \omega_0^2 - \frac{1}{4}\gamma^2$$

The time required for disassembly is then

$$t \cong \frac{\pi}{\sqrt{\omega_0^2 - \frac{1}{4}\gamma^2}}$$

The empirical full width at half maximum (FWHM) of the monopole excitation is typically $\cong 4$ MeV for an excitation energy of 15 MeV.[25] Table 2 illustrates how values of the damping constant influence the breakeven temperature for the onset of the liquid-gas instability. In Fig. 7 the overstressed region is redefined assuming that the damping constant remains fixed at $\Gamma = 0.27$ ((indicative of T = 0) damping) and is not a function of temperature. The minimum excitation energy of the overstressed region becomes 4.75 MeV above the

Table 2. Breakeven temperatures for various values of the dimensionless damping constant.

Γ/Γ (T $\cong$ 0)	0	5	10	20
$T_{breakeven}$ (MeV)	8.1	7.6	6.9	5.8

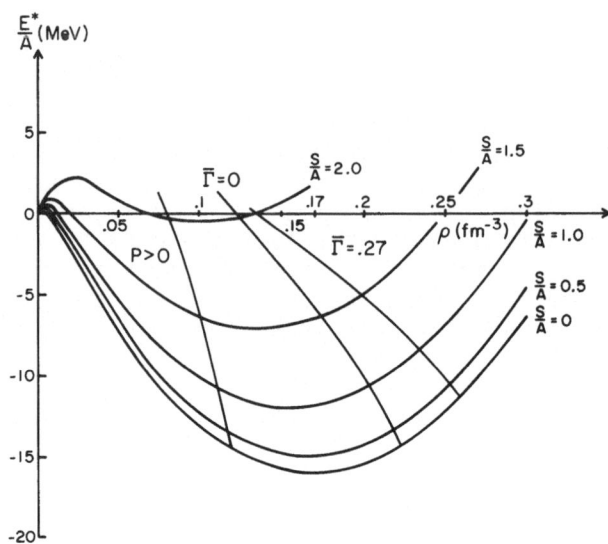

Fig. 7. Compared to Fig. 6, this figure shows two determinations of
         the boundary of the overstressed zone. One, labelled $\Gamma = 0$,
         is identical to that of Fig. 6, and the other, $\Gamma = 0.25$,
         takes into account damping effects during the expansion,
         increasing the threshold energy for reaching the unstable
         zone.

binding energy of normal nuclear matter (E = -16 MeV) with $\rho \cong 1.5$
times normal nuclear density. For an equal mass of projectile and
target the minimum incident energy required is 4 times the excitation energy, i.e. approximately 19 MeV per nucleon. If this incident energy is insufficient to generate a compression of 1.5 times
normal nuclear density, then the minimum required energy will be
even larger.

EXPERIMENTAL CONSEQUENCES

     So far there have been few experiments specifically directed at
observing the influence of low energy critical phenomena. In this
section we review some of the types of data which may be relevant.
It is well known that in high energy proton-induced spallation the
cross section for fragment production increases dramatically up to
energies of a few GeV, followed by a levelling off in the cross
section. An example[26] is illustrated in Fig. 8 for p + Ag leading
to ^{24}Na. The saturation is usually attributed to a limitation of
the energy deposition in the nucleus when it becomes transparent to
protons of a few GeV. On the other hand, the behavior may be related
to the onset of fragmentation when the system reaches the overstressed

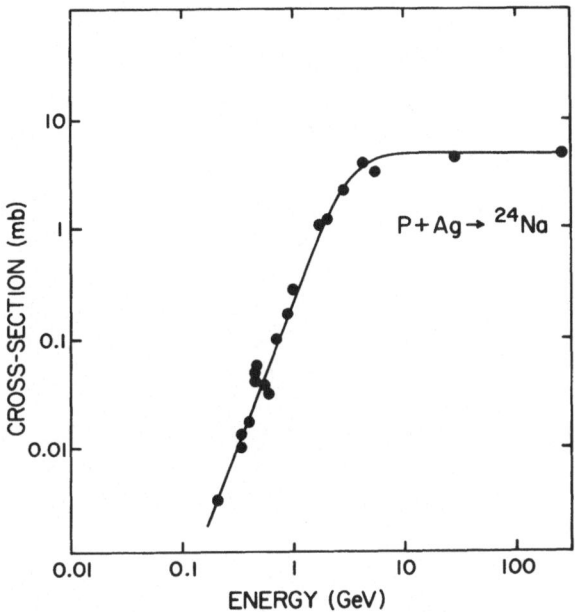

Fig. 8. The cross section for the production of ^{24}Na in spallation
         reactions induced by protons on Ag as a function of inci-
         dent energy. Following a rapid rise, the cross section
         saturates at energies above 2 GeV.

region.[15] For the system p + Ag with about 108 nucleons, our previous discussion would imply that about 1 GeV of energy is necessary.
Of clear interest here would be a comparison with heavy ion induced
fragmentation in order to discover if the saturation sets in at a
lower energy in the presence of some compression. A preliminary
comparison of this type if shown[27] in Fig. 9, from which there is
some indication that a lower threshold may indeed exist in heavy ion
proton induced reactions leading to the production of ^{24}Na and Sc
isotopes with a Au target.

    We should also point out that other approaches to high energy
proton induced break-up of a target have recently been discussed;
for example,[28] in the process of "cleavage" a high energy proton
drills a hole through the nucleus and when the hole expands bonds
are broken to release the observed nuclear fragments. A simple
argument can be used to show that the energy per particle released
in this process may be considerably smaller than the estimate of
the energy of 10 MeV/particle required to access the overstressed

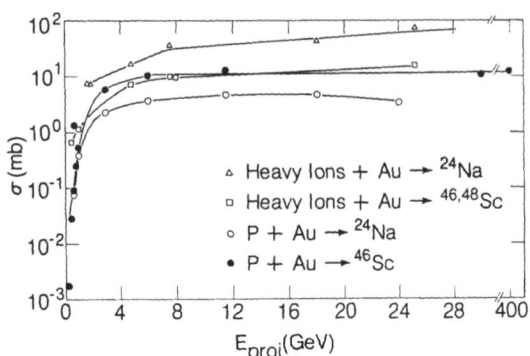

Fig. 9. A comparison of yields of ^{24}Na and ^{46,48}Sc in proton and
heavy ion induced reactions. There are some indications
for differences in the energy at which saturation sets in.

zone. The cleavage model does appear to give a satisfactory description of the absolute production cross sections and energy distribution of the fragments, although so far it has mainly been applied to
more massive fragments than have been explained by the models of
critical phenomena.

At very high proton energies of 30-350 GeV, the proton induced
data have also been interpreted in terms of critical phenomena associated with a liquid-gas phase transition.[29,30] The observation that
the mass yield of fragments obeys a power law in fragment mass
number, A_F, viz

$$Y(A_F) \propto A_F^{-\tau}$$

with $\tau \cong 2.6$ is taken as a signature for fragment formation near the
critical point. According to the theory of condensation in gases
and liquids developed by Fisher,[31] this power law describes the size
distribution of resulting droplets, and τ is predicted to have a
value in the interval $2 < \tau < 3$. An example of the distribution of
fragments fitted with this power law is shown in Fig. 10, using an
exponent of 2.34. Values of the critical density are found to be

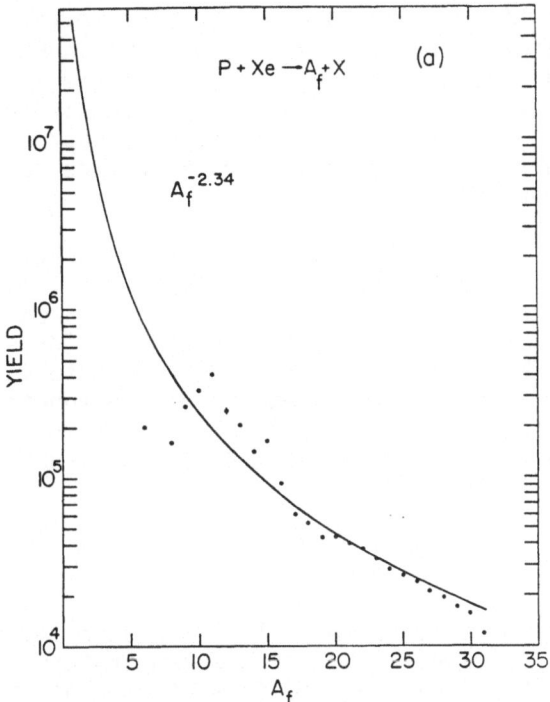

Fig. 10.   Yield of fragments from the reaction P on Xe, typical of
the incident energy range from 80 to 350 GeV. The line
describes a power law dependence $A_F^{-2.34}$ indicative of
critical phenomena of condensation.

$\cong \rho_0/2$ in this study; the extracted critical temperature, $T_C \cong 4$ MeV,
is, however, well below the value of 17.5 MeV mentioned earlier. We
recall that this value was appropriate for the case of infinite
nuclear matter. When a more complete treatment of effective mass
and binding energy corrections are included the value of T_C is lowered[13] to the region 8-13 MeV, but it is still much higher than
4 MeV. We must note, however, that there are difficulties in extracting temperatures from the experimental data, since the values obtained
from the distribution of isotopes and energies differ.

The pure power law form applies strictly only at the critical
temperature. Following the treatment of Siemens,[32] one can write in
general that the probability P_A of finding a cluster A in the fluid
is given by

$$P_A \propto A^{-\tau} \exp[-b(T)A^{2/3}]$$

where $b(T) = 4\pi r_0^2(T) \frac{\sigma(T)}{T}$ with $\sigma(T)$ the surface energy coefficient. At zero temperature $\sigma(T)$ takes the value of 1.14 MeV/fm^2, familiar from the liquid drop mass formula. As the temperature is increased the liquid density approaches the critical density of about 0.065 nucleons per fm^3; since the gas density also approaches the critical density, the density difference at the interface vanishes and the surface energy approaches zero.

In Fig. 11 a calculation of the surface energy as a function of temperature is shown[33] for three values of compressibility, where the nuclear potential energy is treated within Breuckner's energy-density formalism with appropriate corrections for surface and asymmetry effects. We see that for K ≅ 230, close to values of the compressibility derived from measurements of the monopole excitation, σ is predicted to vanish at T ≅ 12 MeV. A rather similar value is obtained by Sauer et al.[34] who parametrize the surface energy coefficient by

$$\sigma = 1.09(1 - 7.16 \times 10^{-3}T^2) \text{ MeV fm}^{-2}$$

which reaches zero at 11.82 MeV. It is encouraging that these derivations result in critical values of the temperature in rather close agreement with the values deduced from studies of the critical temperature of the liquid-gas phase instability in finite nuclei. The consistency implies that studies of fragment production as a function

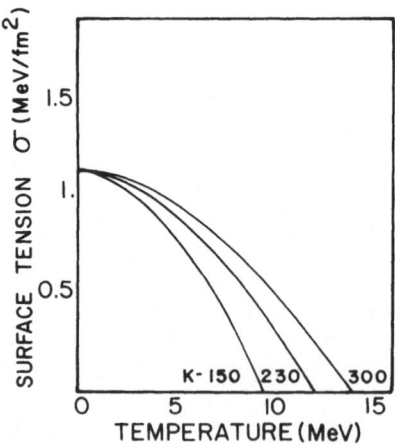

Fig. 11.   The surface tension σ as a function of temperature for several ground state compressibilities.

of energy could add to our knowledge of the temperature dependence of nuclear parameters such as the surface energy coefficient. The expected A dependence[32] of the cross section for values of the temperature below, near, and above the critical temperature are shown in Fig. 12. In this calculation a critical exponent of 2.0, a temperature of 8 MeV, and a surface coefficient of 10 MeV were used for $T < T_c$. The dotted line for $T > T_c$ gives a more rapid fall-off as would be predicted, for example, by the coalescence model.[35] In this model the production cross section of a fragment of mass A depends on the Ath power of the single nucleon cross section, giving rise to a steeper A dependence than in the expression for $T = T_c$.

In order to determine the incident energies required in a heavy ion collision to reach temperatures appropriate to the instabilities, we refer to Fig. 13 which gives values of temperature derived from a study of particle emission from the participant zone of local high temperature and high density nuclear matter. The temperatures were extracted[36] from the energy spectra of emitted light fragments ranging from protons to ^{12}C in reactions induced by incident projectiles from α particles to Argon. Spectra were fitted with a "moving source model," characterized typically by a velocity half the projectile velocity, i.e. a source of intermediate rapidity such as would be expected if projectile and target contribute roughly equal

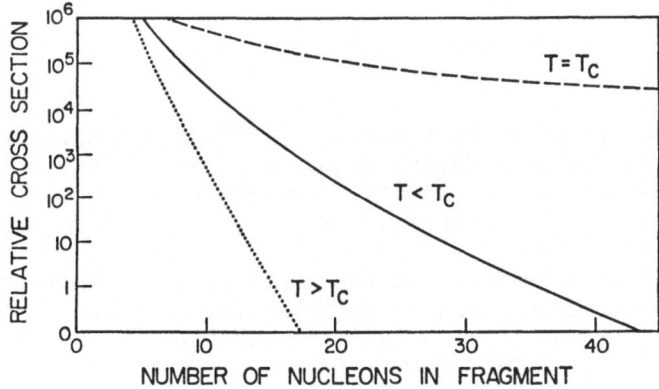

Fig. 12.   Expected dependence of cross sections on fragment mass A for temperatures above (dotted line), at (dashed line), and about half (full line) the critical temperature T_c. The critical exponent used was 2.0, the surface coefficient was 10 MeV, and the temperature was 8 MeV for the full line (see Ref. 32 for details).

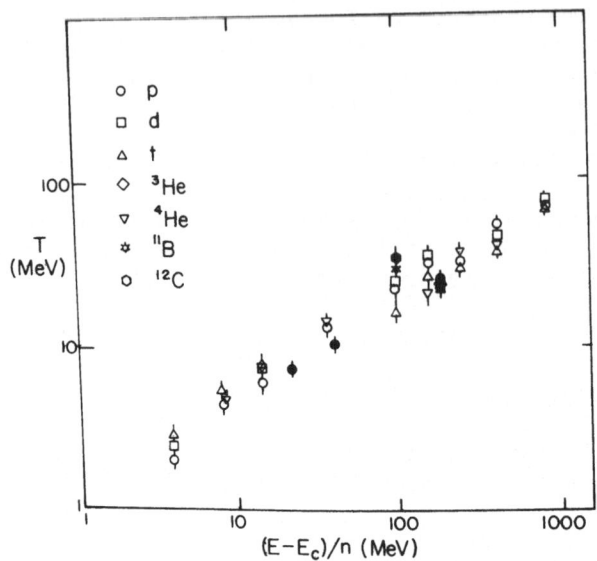

Fig. 13. Plot of temperatures as a function of incident energy per
          nucleon above the barrier, derived from a parameterization
          of the energy spectra of emitted light fragments with a
          localized moving source. Results are shown for different
          emitted fragments p, d, t, ^{3}He, ^{4}He, ^{11}B, and ^{12}C in reac-
          tions induced by α, ^{16}O, ^{20}Ne, and ^{40}Ar (see Ref. 36 and
          references therein for details.

numbers of nucleons to the formation of the hot zone. At high energies of several hundred MeV/nucleon and above, direct evidence for
the existence and size of such a localized zone comes from experiments on two-particle interferometry.[20,37]

Recently experiments[38] on two-particle interferometry have been
applied in much lower energy reactions; for the case of two protons.
The experimental correlation function is shown in Fig. 14 for the
case of ^{16}O on Au at 25 MeV/nucleon. The curves shown in the figure
are the results of model calculations for the case of incoherent
emission from a source of negligible lifetime and gaussian spatial
distribution of rms radius $\sqrt{3/2}$ r_0. The resulting value of r_0 is
approximately 4 fm.

Additional evidence for the formation of localized zones has
come recently from "sub-threshold" pion emission for reactions of

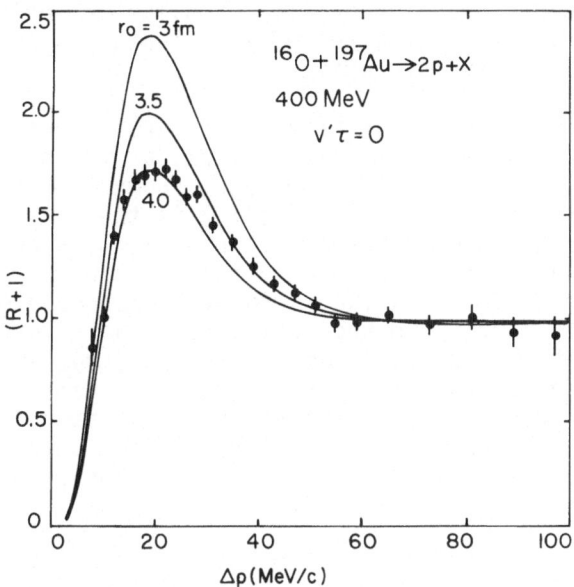

Fig. 14.   The experimental correlation function is plotted as a
function of proton relative momentum. The theoretical
curves correspond to different assumed radii, with a best
fit corresponding to approximately 4 fm.

^{12}C on a variety of targets in the energy region 35 to 80 MeV/
nucleon.[39] For ^{12}C on Au, the data are fitted under the assumption
of a localized source containing about 45 particles. At normal
matter density, the equivalent radius parameter is close to 4 fm.

In the remainder of this paper, on account of the evidence discussed above, we shall assume that localized depositions of energy
are a general feature of a variety of nuclear collisions, and that
the temperatures of these zones are also described by the trends of
Fig. 13.

We note that the trend of temperatures in Fig. 13 is roughly
that expected for a Fermi system composed of equal numbers of projectile and target nucleons. Thus, if we write the expression for
temperature in a Fermi system as $E^* = T^2/16$, where E^* is the excitation energy per particle and the factor 1/16 comes from the level
density parameter, then the result $E = 1/4 \, (E_L/A)$ follows, where
E_L/A is the incident laboratory energy per particle. Then $T = 2\sqrt{E_L/A}$,
and for $T \cong 20$ MeV we obtain $E_L/A = 100$ MeV/nucleon in agreement

with the experimental result. Temperatures below the critical value
of 18 MeV are therefore appropriate to collisions below 100 MeV/
nucleon, placing the observation of the liquid-gas phase instability
in the intermediate energy regime.

A power law distribution of fragment cross sections has been
found to apply in several other reactions, as for example[40] in the
reaction of ^{12}C on Ag at 30 MeV/nucleon, shown in Fig. 15. Here the
dashed line corresponds to $\sigma_Z \propto Z^{-2.6}$; Z is the charge of the fragment and is roughly proportional to the mass. The authors of this
work estimate an upper bound for the temperature reached in the
reaction of about 12 MeV, which is the value appropriate for equal
participation of target and projectile nucleons as shown in Fig. 13.
For reactions of this type, in which fragments much heavier than the
projectile are produced, it is quite possible that a much larger
volume of the target nucleus participates; the associated temperatures would then be commensurately lower. More detailed systematic
studies as a function of incident energy will be required to clarify
the situation.

Models which do not explicitly incorporate critical behavior
can also be used to explain some aspects of the data we have been
discussing. In Fig. 15, for example, the data are compared with a

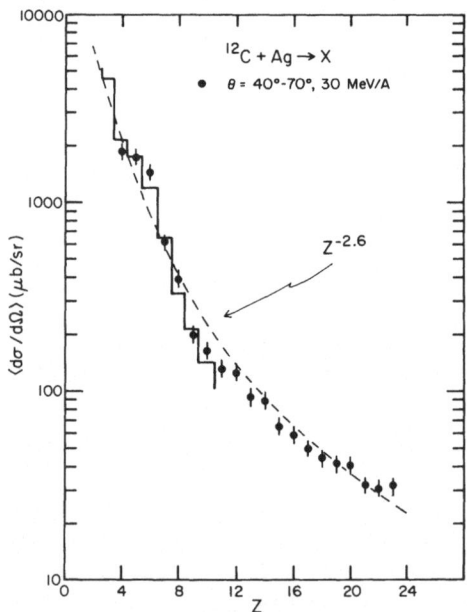

Fig. 15. Element production cross sections in the reaction ^{12}C + Ag
at incident energy of 30 MeV/nucleon. The dashed line corresponds to a power law dependence $Z^{-2.6}$, and the histogram
to a Hauser-Feshbach statistical emission model.

statistical model calculation[41] for the idealised case of emission
from a completely equilibrated composite system. Isotropic particle
emission was assumed, angular momentum effects were neglected, and
the level density was assumed to correspond to an ideal Fermi gas at
normal density. As indicated by the solid histogram the calculation
gives a tolerable account of the data. A similar calculation of the
fragment distribution in Fig. 10 for high energy proton induced
reactions met with comparable success.[42] It bears repeating that
only through detailed studies over a range of energies will it be
possible to distinguish the different theoretical models, and to
confirm the presence of critical phenomena. Other examples of power
law behavior are discussed in Refs. 43 and 44.

     In the discussion of mechanical instabilities we made use of
the cascade result that entropy stays fairly constant during the
expansion and cooling of the participant zone. Applying this con-
cept also to the case of a liquid-gas phase instability leads[14] to
Fig. 16, in which the regions of phase mixture and negative compressibility are defined on a pressure-density diagram with lines of
constant entropy. For entropies as high as S/A $\lesssim$ 3 the isentropic

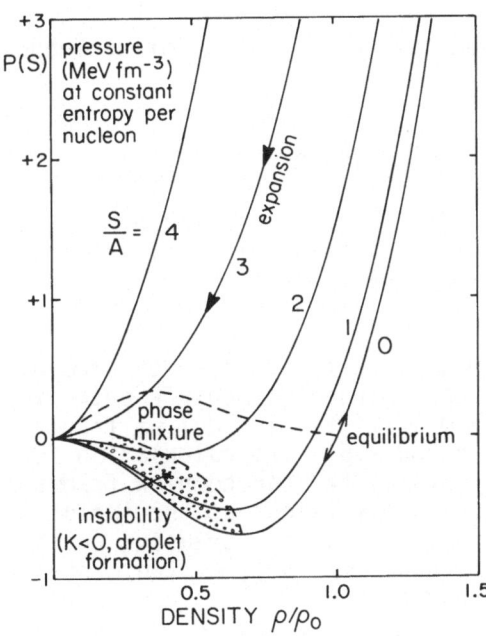

Fig. 16.   The pressure versus density for a nuclear system as a
           function of the entropy per particle in the system. The
           arrows indicate the expansion at constant entropy of the
           initial compressed and heated zone. The regions of liquid-
           gas phase admixture and of negative incompressibility dis-
           cussed for Figs. 3-7 are indicated.

lines intersect the region of liquid-gas phase mixture, implying
that effects of this instability may be present up to very high bombarding energies of several hundred MeV/nucleon. Now consider a
system prepared with initial compression and heating such that the
specific entropy S = 2. During the subsequent expansion the temperature and density will drop such that the entropy

$$S = \frac{\pi^2}{2\varepsilon_F} T \left(\frac{\rho_0}{\rho}\right)^{2/3}$$

remains constant. At the intersection with the phase boundary, if
a phase transition takes place, the entropy will no longer stay
constant but instead the volume will increase at constant temperature
along a locus parallel to the horizontal axis in Fig. 16. For a
system consisting of nucleons only the entropy increase can be derived
from the Sackur-Tetrode equation:

$$\frac{S}{A} = \frac{5}{2} + \ln \left(\frac{n_Q}{\rho}\right)$$

where n_Q is the quantum concentration of nucleons. The entropy
change is then[14]

$$\Delta S = \ln \rho_{GAS} - \ln \rho_{LIQUID} .$$

Taking values of ρ from Fig. 16 we find an entropy increase of $\Delta S \approx 2$
for the case in point. A measurement of entropy as a function of
bombarding energy may therefore be a means of detecting the onset of
critical phenomena.

Recently it was suggested that a measurement of the yield of
complex fragments can be used to measure entropy.[14] In Fig. 17 the
results of a quantum statistical model of fragment production are
shown for S/A = 1.5 and 2.5. The calculation for S/A = 1.5 is compared with experimental data[45] on complex fragment emission in reactions of ^{12}C in nuclear emulsions at the incident energies in the
range 50-110 MeV/nucleon. With increasing entropy the yield of
heavy fragments decreases rapidly, by two orders of magnitude in the
case of nitrogen fragments.

Until recently, emphasis on entropy measurement in heavy ion
reactions has focused on the ratio of deuterons to protons emitted
from the participant zone, resulting[46] in values which are always
in excess of the entropy at the critical point (S $\approx$ 3.3, see
Fig. 16). It may be that such light fragments are produced mainly
from a gaseous phase. The comparison with heavier fragments emitted

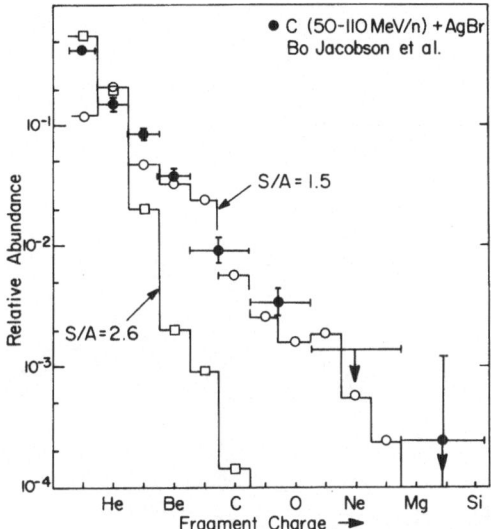

Fig. 17. The charge distribution Y(Z) observed in high multiplicity
events for the reaction of 50-110 MeV/nucleon carbon in
emulsions. The open symbols and histograms are the distributions predicted with a quantum statistical model for
two values of the entropy.

from reactions over a wide energy range will be important for estab-
lishing the onset of critical phenomena. Very recently the dynamics
of the first order liquid-gas phase transition have been investigated
in the final expansion stage of a high energy heavy ion collision.[47]
The high entropy values extracted from light particle abundances
are explained as a consequence of the liquid-gas phase coexistence
at the break-up of nuclear matter. The idea is illustrated in
Fig. 18, which shows the entropy of the system at the break-up
$n = n_{BU} = 0.015$ fm^{-3} for different beam energies. The full lines
represent the entropy of the liquid-gas phase in an isoergic and
adiabatic expansion. The hatched area defines the range where the
entropy of the whole system may vary, being defined by the adiabatic
and the maximum by the isoergic expansion. There is a suggestion
that the experimental data for light particle production A = 1-3
are associated with values of S in the gas phase, whereas the inclusion of a wider range of measured fragments A = 1-14 yields values
of S closer to the total value of S. This observation may support
the coexistence of two phases, but there are several problems associated with the determination of entropy, particularly for the heavy
fragments.

Perhaps the most tantalizing evidence for the existence of
critical phenomena to date is provided[48] by Fig. 19. According to

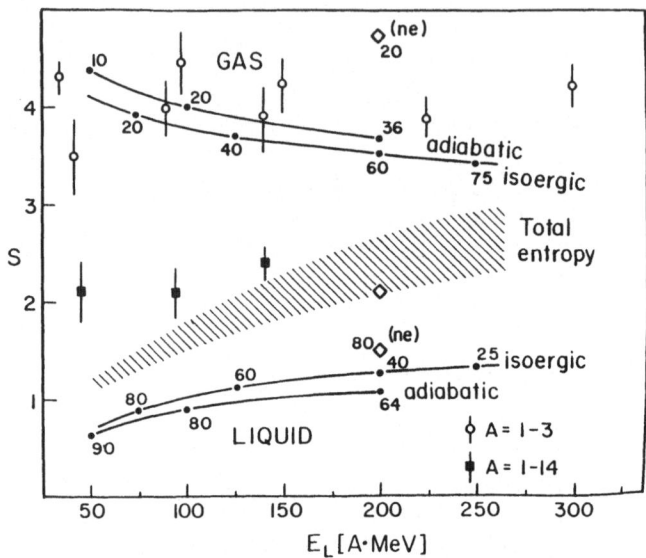

Fig. 18.   Entropy of nuclear systems at break-up for different beam
           energies. The theoretical values are compared with mea-
           sured entropies as discussed in the text.

the prescriptions outlined earlier in this section for determining
temperatures of emitting hot zones, the systematics of the production
of complex fragments have been studied as a function of temperature
for a variety of colliding systems, including proton and heavy ion
induced reactions. For each reaction the cross section as a function
of fragment mass is fitted with a power law distribution, as we discussed for Figs. 10 and 15. An example of the fit for p + U at
5.5 GeV is included in the insert. The apparent exponent appears
to reach a minimum and subsequently to increase. Such a trend is
consistent with the expectations from Fig. 12, if the critical temperature lies in the region of 10-12 MeV. We recall that such a
value is expected for finite nuclear systems.

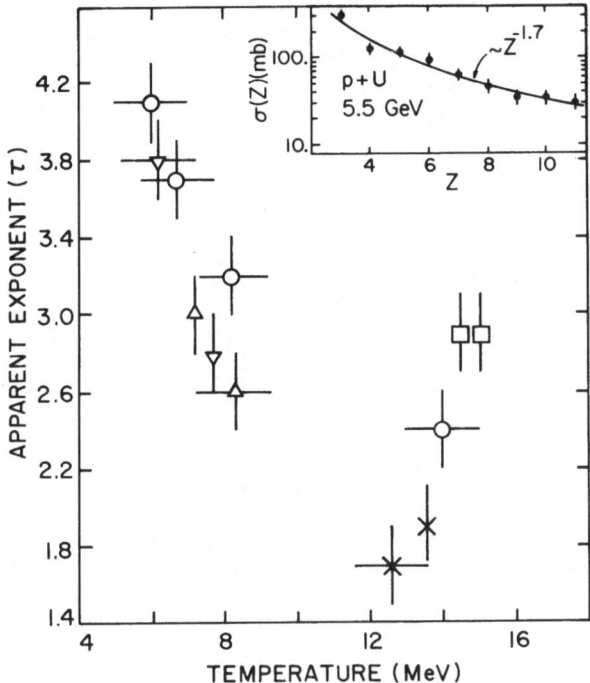

Fig. 19. The apparent exponent, τ, of the power-law fit to the
          fragment distributions as a function of the temperature, T.
          The systems are: circles, p + Ag (0.21-4.9 GeV); crosses,
          p + U (4.9, 5.5 GeV); squares, p + Xe, p + Kr (80, 350 GeV),
          triangles and inverted triangles, C + Ag, C + Au (180,
          360 MeV). The inset shows a typical power-law fit to a
          fragment distribution.

     There are many difficulties associated with the interpretation
of Fig. 19. In particular, due to a lack of a consistent set of
data for a single colliding system as a function of bombarding energy,
results from many different exponents are combined. Some of the difficulties are summarised in Refs. 49, 50. As Fig. 20 illustrates
when the data are restricted to proton induced reactions only, the
trend is less supportive of a minimum and may be explained by the
influence of the Coulomb barrier on the emission of the fragments
(dotted line).

     As we have alluded previously, there may also be alternative
explanations of the production systematics which are not explicitly
related to phase transitions and instabilities. For a review of
these approaches, see Ref. 51.

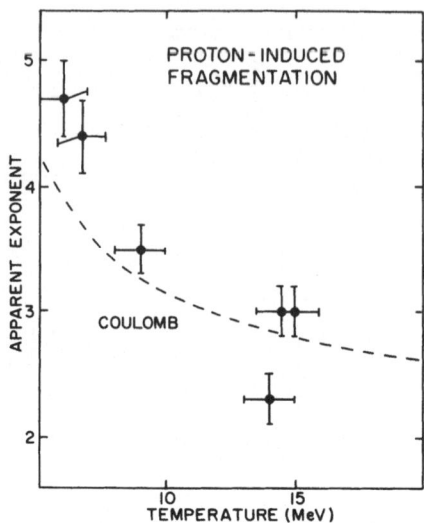

Fig. 20. Apparent behaviour of power law exponent τ as a function
          of temperature in proton induced fragmentation reactions.
          Shown for a comparison is an estimate of the Coulomb effect
          on τ.

CONCLUSION

    This paper has dealt with the possibility of investigating
critical phenomena in nuclear systems, with particular emphasis on
the liquid-gas and mechanical instabilities which can develop at a
critical temperature or at a critical incident energy. Although the
phenomena are frequently discussed and incorporated into calculations
of neutron stars and supernovae, the possibility of realising the
critical conditions during the evolution of a nuclear collision has
only recently been recognised. Many of the criteria required for
the creation of exotic phases of nuclear matter at high density and
temperature, such as the quark-gluon plasma, may be illustrated by
phase transitions of the liquid-gas type, the existence of which are
based on more accepted knowledge of the equation of state. In particular, questions of time scales and of hydrodynamical behavior
are quite pertinent in both cases.

    At present there is only sketchy evidence for the existence of
critical phenomena of the type we have discussed. Some of the new
accelerators will permit new studies of critical phenomena by providing beams of heavy ions over the critical region from 20 to
200 MeV/nucleon.

ACKNOWLEDGMENTS

I am grateful to many colleagues for discussion, ideas, and data pertaining to this paper. In particular I wish to thank S. Angius, N. Anantaraman, G.F. Bertsch, D. Boal, C.B. Chitwood, M.V. Curtin, G.M. Crawley, D.J. Fields, C.K. Gelbke, L. Harwood, B. Hasselquist, B.V. Jacak, J. Kupstas, W.G. Lynch, A.D. Panagiotou, P.J. Siemens, H. Stöcker, H. Toki, B.M. Tsang, H. Utsunomiya, and G.D. Westfall.

My thanks go to Carol Cole for rapid preparation of the manuscript, and to M. Blosser, M. Johnson, and O. McHarris for the illustrations.

This work was supported by the National Science Foundation under Grant No. PHY80-17605.

REFERENCES

1. M. Gyulassy in Proceedings of the International Conference on Nucleus-Nucleus Collisions (Michigan State University, 1982), eds. G.F. Bertsch, C.K. Gelbke, and D.K. Scott, Nucl. Phys. A400, 31C (1983).
2. H.J. Specht, ibid., p. 43C
3. H. Satz, ibid., p. 541C.
4. P.J. Siemens, Nucl. Phys. A428, 189C (1984).
5. H. Stöcker, J.A. Maruhn, and W. Greiner, Phys. Rev. Lett. 44, 725 (1980).
6. W.A. Kupper, G. Wegmann, and E.R. Hilf, Ann. of Phys. 88, 454 (1974).
7. B. Friedman and V.R. Phandaripande, Nucl. Phys. A361, 502 (1981).
8. D.Q. Lamb, J.M. Lattimer, C.J. Pethick, and D.G. Ravenhall, Phys. Rev. Lett. 41, 1625 (1978); Nucl. Phys. A360, 459 (1981).
9. P. Danielewicz, Nucl. Phys. A314, 465 (1979).
10. M. Barranco and J.R. Buchler, Phys. Rev. C22, 1729 (1980).
11. H. Schulz, M. Lunchow, G. Röpke, and H. Schmidt, Phys. Lett. 119B, 12 (1982).
12. M.W. Curtin, H. Toki, and D.K. Scott, Phys. Lett. 123B, 289 (1983).
13. H. Jacaman, A.J. Mekjian, and L. Zamick, Phys. Rev. C27, 2782 (1983).
14. H. Stöcker et al., Proc. of the Int. Conf. on Nucleus-Nucleus Collisions (Michigan State University, 1982), eds. G.F. Bertsch, C.K. Gelbke, and D.K. Scott, Nucl. Phys. A400, 63C (1983).
15. G.F. Bertsch and P.J. Siemens, Phys. Lett. 126B, 9 (1983).
16. G.F. Bertsch and J. Cugnon, Phys. Rev. C24, 2514 (1981).

17. F. Serr, G. Bertsch, and J.P. Blaizot, Phys. Rev. C22, 922 (1980).
18. P. Bonche, S. Koonin, and J. Negele, Phys. Rev. C13, 1226 (1976).
19. J.S. Blakemore in Solid State Physics (W.B. Saunders, Philadelphia, 1974), p. 110.
20. D.K. Scott in Dynamics of Heavy Ion Collisions, eds. N. Cindro, R.A. Ricci, and W. Greiner (North Holland, Amsterdam, 1981), p. 241 and references therein.
21. S. Nagamiya et al., Phys. Rev. C24, 971 (1981).
22. H. Bohning, Proc. of the Int. Conf. on Nuclear Reactions Induced by Heavy Ions, eds. R. Bock and W.R. Hering (North Holland, Amsterdam, 1970), p. 633.
23. J. Cugnon, Phys. Lett. 135B, 374 (1984).
24. M.W. Curtin, H. Toki, and D.K. Scott, Michigan State University Cyclotron Laboratory Preprint MSUCL-426.
25. D.H. Youngblood et al., Phys. Rev. C23, 1997 (1981).
26. R.E.L. Green and R.G. Korteling, Phys. Rev. C22, 1594 (1980) and references therein.
27. K. Aleklett, W. Loveland, P.L. McGaughey, K.J. Moody, R.M. McFarland, R.H. Kraus, Jr., and G.T. Seaborg, Preprint of Contributed Papers to the 6th High Energy Heavy Ion Study and 2nd Workshop on Anomalons (June 1983), Lawrence Berkeley Laboratory Publication LBL-16281, p. 38.
28. S. Bohrmann, J. Hüfner, and M.C. Nemes, Max Planck Institute, Heidelberg Preprint MPIH 1982-V16.
29. J.E. Finn, S. Agarwal, A. Bujak, J. Chuang, L.J. Gutay, A.S. Hirsch, R.W. Minich, N.T. Porile, R.P. Scharenberg, B.C. Stringfellow, and F. Turkot, Phys. Rev. Lett. 49, 1321 (1982).
30. R.W. Minich, S. Agarwal, A. Bujak, J. Chuang, J.E. Finn, L.J. Gutay, A.S. Hirsch, N.T. Porile, R.P. Scharenberg, B.C. Stringfellow, and F. Turkot, Phys. Lett. 118B, 458 (1982).
31. M. Fisher, Physics V3, 255 (1967).
32. P.J. Siemens, Nature, 305, 410 (1983).
33. H. Stocker and J. Burzlaff, Nucl. Phys. A202, 265 (1973).
34. G. Sauer, H. Chandra, and U. Mosel, Nucl. Phys. A264, 221 (1976).
35. S. Das Gupta and A.Z. Mekjian, Phys. Reports 72, 131 (1981) and references therein.
36. S. Angius, M.Sc. Thesis, Michigan State University 1982; S. Angius and D.K. Scott, Annual Report (Michigan State University Cyclotron Laboratory 1981-1982), p. 24.
37. S. Nagamiya, Proc. of Int. Conf. on Nucleus-Nucleus Collisions (Michigan State University, 1982), eds. G.F. Bertsch, C.K. Gelbke, and D.K. Scott, Nucl. Phys. A400, 399C (1983).
38. W.G. Lynch, C.B. Chitwood, M.B. Tsang, D.J. Fields, D.R. Klesch, C.K. Gelbke, A.D. Panagiotou, G.R. Young, T.C. Awes, R.L. Ferguson, F.E. Obenshain, F. Plasil, and R.L. Robinson, Phys. Rev. Lett. 51, 1850 (1984).
39. J. Aichelin, Phys. Rev. 52, 2340 (1984).

40. C.B. Chitwood, D.J. Fields, C.K. Gelbke, W.G. Lynch, A.D. Panagiotou, M.B. Tsang, H. Utsunomiya, and W.A. Friedman, Phys. Lett. 131B, 289 (1983).
41. W. Friedman and W.G. Lynch, Phys. Rev. C28, 16 (1983).
42. W. Friedman and W.G. Lynch, Phys. Rev. C28, 950 (1983).
43. H.H. Gutbrod, A.I. Warwick, and H. Weiman in Selected Aspects of Heavy Ion Collisions (Saclay, 1982), eds. M. Martinot, C. Ngô, and P. Gugenberger (North Holland-Amsterdam, 1982), p. 177C.
44. U. Lynen et al., ibid., p. 129C.
45. B. Jakobsson, G. Jöhnson, B. Linquist, and A. Oskarsson, Z. Phys. A307, 293 (1982).
46. J. Kapusta, Preprint, 1983.
47. L.P. Csernai, Preprint, University of Minnesota UMTNP-108, 1984.
48. A.D. Panagiotou, M.W. Curtin, H. Toki, D.K. Scott, and P.J. Siemens, Phys. Rev. 52, 496 (1984).
49. D.H. Boal, Phys. Rev. C30, 119 (1984).
50. D.H. Boal, Michigan State University Preprint MSUCL-467, to be published in Proceedings of the Conference on the Intersections between Particle and Nuclear Physics, Steamboat Springs, Colorado, 1984.
51. J. Cugnon, Lectures presented at the Cargese Summer School, September 1984, to be published.

INDEX